The Emergence of Land Markets in Africa

Assessing the Impacts on Poverty, Equity, and Efficiency

The Emergence of Land Markets in Africa

Assessing the Impacts on Poverty, Equity, and Efficiency

EDITED BY

Stein T. Holden, Keijiro Otsuka, and Frank M. Place

RESOURCES FOR THE FUTURE
Washington, DC, USA

Printed in the United States of America.

An RFF Press book
Published by Resources for the Future
1616 P Street NW
Washington, DC 20036–1400
USA
www.rffpress.org

Library of Congress Cataloging-in-Publication Data

The emergence of land markets in Africa : assessing the impacts on poverty, equity and efficiency / edited by Stein T. Holden, Keijiro Otsuka, and Frank M. Place.
p. cm.
ISBN 978-1-933115-69-6
1. Land tenure—Africa. 2. Land use—Africa. 3. Farms—Prices—Africa. 4. Rural poor—Africa. I. Holden, Stein. II. Otsuka, Keijiro. III. Place, Frank, Dr.
HD963.E44 2008
333.33096--dc22 2008040548

The paper in this book meets the guidelines for permanence and durability of the Committee on Production Guidelines for Book Longevity of the Council on Library Resources. This book was copyedited and typeset by Publishers' Design and Production Services, Inc. The cover was designed by Ellen A. Davey. Cover photo credits, © iStock International, Inc. all rights reserved, and courtesy of Stein T. Holden.

ISBN 978-1-933115-69-6 (cloth)

Contents

Part III: Contract Choice, Poverty, and Efficiency of Land Use

Part IV: Land Markets, Land Tenure, and Land Management

Contributors

Editors

Stein T. Holden (stein.holden@umb.no) is a professor in development and resource economics at the Norwegian University of Life Sciences, Norway.

Keijiro Otsuka (otsuka@grips.ac.jp) is a professorial fellow at the Foundation for Advanced Studies on International Development (FASID) and a professor in economics at National Graduate Institute for Policy Studies (GRIPS) in Tokyo, Japan.

Frank M. Place (f.place@cgiar.org) is an economist in the World Agroforestry Centre (ICRAF), Nairobi, Kenya.

Contributors

Tekie Alemu (tekie_a@econ.aau.edu.et) is an associate professor in economics in Addis Ababa University and a researcher in Environmental Economics Policy Forum (EEPF) at the Ethiopian Development Research Institute (EDRI).

Daniel Ayalew Ali (dali1@worldbank.org) is an economist and researcher at the World Bank.

Samuel Benin (s.benin@cgiar.org) is a research fellow at the International Food Policy Research Institute (IFPRI), Washington, DC.

Mintewab Bezabih (mintw2000@yahoo.com) is a research fellow, Centre for the Economics and Management of Aquatic Resources, University of Portsmouth, UK.

Klaus Deininger (kdeininger@worldbank.org) is lead economist and land tenure advisor in the World Bank.

Hosaena Ghebru (hosaenag@yahoo.com) is a Ph.D. student in development and resource economics at the Norwegian University of Life Sciences and staff at Mekelle University, Ethiopia.

Menale Kassie (menalekassie@yahoo.com) has a Ph.D. in development and resource economics from Norwegian University of Life Sciences and is employed as a Post Doctoral Fellow at the International Crops Research Institute for the Semi-Arid Tropics (ICRISAT), Nairobi, Kenya.

Edward Kato (e.kato@cgiar.org) is a senior research assistant at the International Food Policy Research Institute (IFPRI), Washington, DC.

Rodney Lunduka (rodney.lunduka@umb.no) is a Ph.D.-student in development and resource economics at the Norwegian University of Life Sciences and a staff member at University of Malawi, Bunda College of Agriculture, Malawi.

Daniel Mpuga (pmpuga@worldbank.org) is an economist at the World Bank.

Ephraim Nkonya (e.nkonya@cgiar.org) is a research fellow at the International Food Policy Research Institute (IFPRI), Washington, DC.

Wilfred Nyangena (wilfred.nyangena@economics.gu.se) is senior lecturer at the School of Economics, at the University of Nairobi, Nairobi, Kenya.

Ragnar Øygard (ragnar.oygard@umb.no) is an associate professor in development and resource economics at the Norwegian University of Life Sciences, Norway.

John Pender (j.pender@cgiar.org) is a research fellow at the International Food Policy Research Institute (IFPRI), Washington, DC.

Million Tadesse (millionaytenfisu@umb.no) is a Ph.D. student in development and resource economics at the Norwegian University of Life Sciences and staff member at Southern Agricultural Research Institute (SARI), Awassa, Ethiopia.

Juliet Wanjiku (wanjikuju@yahoo.com) is a socioeconomist at the World Agroforestry Centre (ICRAF), Nairobi, Kenya.

Takashi Yamano (yamanota@grips.ac.jp) is a faculty fellow at the Foundation for Advanced Studies on International Development (FASID) and an associate professor in economics at National Graduate Institute for Policy Studies (GRIPS) in Tokyo, Japan.

Preface

Land issues in sub-Saharan Africa are again high on the policy agenda, as can be seen from the establishment of the Commission for Legal Empowerment of the Poor and the Global Land Tools Network by UNHabitat, as well as many countries implementing new land reforms with support from the World Bank and many donor countries. Yet our knowledge on these land policy-related issues are far from adequate in formulating sensible policies. Thus, the idea to write this book came about based on the observed gap in analysis of land markets in Africa as compared to the huge literature on the topic, and particularly on land rental markets in Asia. Despite this neglect of the importance of land markets in Africa in the literature, we found them to be more widespread than earlier thought and to be growing in importance. These facts, however, are not recognized in the land policies in many African countries.

This is the first book to address the issues of emerging land markets and their implications for poverty, equity, and efficiency across African countries, and it should form a basis for thinking about the future role of land markets in national land policies in Africa. All too often, land policies have been extreme, either prohibiting all land transactions or giving unrestricted freehold rights to a small elite at the expense of the poor. It is well known from the long experience in Asia that such policies are deterimental to both efficiency and equity of land use. We strongly argue that future policies in Africa should work with the markets, and that regulations should be imposed only with careful testing that they are having the intended effects.

The book will also be useful for teaching purposes in universities in both developed countries and developing countries in Africa, as it provides both comprehensive reviews of the literature and detailed case studies from several countries in Africa. We hope it will inspire students and researchers to go further, both in depth and width, to help uncover more policy-relevant findings that can help build a stronger basis for good policies. We hope also that it will help facilitate the dialogue between researchers and policymakers. While the presentations of the original research in the case studies include a substantial amount of technical and quantitative detail, we tried to ensure that our discussions of the objectives and the policy implications of our emprical work are more broadly accessible. We hope that readers across the policy, development, and environmental communities will find it useful to read our introductory and concluding chapters, as well as the discussions toward the end of each of the case studies.

Acknowledgments

This book is an outcome of international collaboration between researchers in Africa, Norway, Sweden, Japan, and the United States and from international organizations including the World Bank, International Food Policy Research Institute (IFPRI), and World Agroforestry Centre (ICRAF). Young researchers from Ethiopia, Kenya, Uganda, and Malawi were involved in the research in their own countries and provided valuable insights to the book.

The first coeditor of this book, Stein Holden, organized the project leading to the book publication. He requested the participation of the second coeditor, Keijiro Otsuka, who has been working on land tenancy issues particularly in Asia, and the third coeditor, Frank Place, who has been specialized on land tenure issues in Africa.

It has been a pleasure to work for this project where so many people have contributed so willingly as authors based on their interest in the topic. We were also fortunate in getting generous funding for two workshops that were instrumental to make the project come through. We are grateful to the Swedish International Development Cooperation Agency for this funding and to the University of Gothenberg for facilitating this through its Environment for Development initiative. Particular thanks go to Gunnar Köhlin who has supported and facilitated the project. Environmental Economics Policy Forum (EEPF), under the Ethiopian Development Research Institute (EDRI) hosted the first workshop in Addis Ababa in June 2005 and the World Agroforestry Centre (ICRAF) in Nairobi hosted the second workshop in September 2006. We would like to thank all the participants in the two workshops for all their valuable comments. This includes almost all the authors of book chapters as well as Marc Bellemare, Joan Kagwanja, Gunnar Köhlin, Ben Linkow, and Bekele Shiferaw.

The data used as the basis for the analyses in many of the book chapters have been collected through many different projects which have been funded in numerous ways. This project would not have been feasible without the access to all these data. Such a systematic analysis of data across many countries would have been very expensive if new data had been collected for our purpose. In short, this book is a value added product from many of these other projects.

We would like to thank for the good help from Qian Weng when preparing the book chapters into the RFF format and Mayuko Tanaka for the help with checking, organizing and compiling all the references for the book. Finally, we would like to thank Don Reisman for the facilitation and helpful comments when preparing the manuscript into the final book format

Stein T. Holden
Keijiro Otsuka
Frank M. Place

PART I

Conceptual Frameworks

Africa, showing case studies

CHAPTER 1

Land Markets and Development in Africa

STEIN T. HOLDEN, KEIJIRO OTSUKA, AND FRANK M. PLACE

The high rates of poverty and the need for sustainable growth and development in sub-Saharan Africa (SSA) remain most acute in rural areas. Rural areas are home to nearly 85% of Africa's poorest families and will continue to host the majority of poor for decades to come, in spite of expected urbanization trends. Identifying pathways for growth and investment inevitably point toward agriculture and land, most especially in the short to medium term (e.g., Commission for Africa 2005). There is increasing pressure to identify the strategies and mechanisms by which agriculture can be intensified, given the already small farm sizes (under 2 hectares for the majority of African farmers) and the high likelihood that farms will continue to shrink due to the current high rates of population growth.

Land Markets in the Context of Rural Poverty

Generating significant and sustained growth in African agriculture has been challenging, and nowhere can it be said to be completely successful. In fact, on the whole, the performance of the agricultural sector has been poor. The oft-quoted figures from the last part of the twentieth century clearly show a sector in stagnation. Per capita food production, although having increased significantly elsewhere in the world over the last 45 years, has only marginally improved in Africa, and this is almost completely due to expansion of cultivated area. On average, yields of cereal crops in SSA have increased by a meager 5.2 kg ha^{-1} y^{-1} over the past 33 years (FAOSTAT 2006). Much of this increase took place in the 1970s; however, after 1985, yields have remained flat, responding only to rainfall patterns. Thus, average cereal yield is still below 1 metric ton per hectare in sub-Saharan Africa. One of the key reasons for static yields is that fertilizer use leveled off in SSA as prices increased and subsidies were withdrawn (Otsuka and Kalirajan 2005, 2006). The growth in fertilizer use after 1990 has been less than one-quarter of the growth experienced between 1970 and 1990, and farmers are mining their soils of nutrients to produce food. Similarly, many studies have noted the lack of adoption by African farmers of other improved land-management practices. Thus, any increases in agricultural production have largely been met through opening up new land to cultivation and have been obtained at the cost of soil degradation as soils are mined for their nutrients.

Since agricultural productivity and growth in Africa has been low or even declining, in some countries there is a renewed interest in understanding the factors that promote or inhibit agricultural investment, including land tenure security and land markets. Furthermore, how to improve the poor's access to land is becoming a critical issue in the face of growing scarcity of land relative to rural population. The policy goals of improving land tenure security and land markets in Africa have been mentioned as important in the Commission for Africa Report (2005), the U.N. Millennium Project (2005), the FAO 2005 State of Food Insecurity Report, and the Comprehensive African Agricultural Development Programme (CAADP) (NEPAD 2003).

The formulation of beneficial and feasible land policies to improve the efficiency of markets and to increase agricultural investment and productivity is challenging, however. It requires a more critical understanding of the specific land-market imperfections, their effects on the access to land, and the way in which they interact with tenure security to create incentives for improved land management. This is further complicated by the different roles or functions of land and land-market transactions as well as broader market and policy context within which land markets operate.

The next section gives an overview of the different roles and functions of land in rural Africa. It is followed by a summary of the development of land markets in Africa, reviewing the broad evidence but also focusing on case studies from Ethiopia, Malawi, Uganda, and Kenya. Policy issues related to land markets, and the key research gaps that would inform these issues, are discussed later in this chapter, along with the general approach of how this book will address these research and policy issues.

Role of Land and Land Markets

There are three key functions of land in rural Africa and the demand or supply of land can reflect each of them. First, land is the basis for an agricultural livelihood and safety net. Households with few other livelihood options need land for crop, livestock, or tree-based farming. In sub-Saharan Africa, agriculture continues to provide the most important livelihood for the large majority of the rural population, and thus this function of land is the most important. The second function of land is as a valuable and tradable asset. It has long been observed that the elite of African society invested in land as a store of wealth and collateral for securing formal credit. The third major function of land is that it provides a residence and thus a cultural affinity for rural Africans. Rural areas are less costly to live in than urban areas, so even households that have diversified their livelihoods into urban sector jobs often wish to retain some land in the rural area. The same is true of soon-to-be retirees who will need to live in inexpensive settings.

The fundamental characteristic of land as a resource is that it is immobile in the physical sense due to the prohibitively high costs of moving land as compared to other resources. Nonland factors of production must therefore be brought to the land to initiate production, and then, after a necessary time delay from planting until harvest time, outputs have to be harvested and transported from the land to the places for consumption and sale. This spatial dispersion of production leads to high transaction costs, especially in areas where other factor markets also tend to be imperfect (Binswanger and Rosenzweig 1986). This spatial dispersion and immobility of land also ties up land users and typically makes them less able to migrate and access information, and renders land a less liquid asset for the purposes of mitigating income risks.

Importance of Land Markets

Scarcity is the basis for economic value and for the emergence of land markets. Africa has long been seen as a land-abundant continent. However, land scarcity is increasing in many parts of Africa, such as the East African highlands and parts of southern and central Africa, due to population increase and population concentration. Land scarcity is a necessary but not sufficient condition for the emergence of land markets. As pointed out by Zimmermann and Carter (1999), there must be sufficient heterogeneity across households in the economy such that a single price of land attracts willing sellers and buyers simultaneously. Imbalances in ratios between factors of production across producers, and the basic characteristic that these factors of production also serve as complements rather than substitutes in the production process, create opportunities for efficiency gains through trade in factors of production, as long as the transaction costs involved are not too high. Such imbalances in land holdings are evident in much of Africa, significantly in former colonial countries in southern Africa, but also in many other parts of Africa where imbalances grew from differences in power and influence within native groups.

Even though the conditions for land markets to be active may be present, it doesn't mean that efficiency of agricultural production requires well-functioning land markets. If markets for credit, purchased inputs, and labor were efficient, there is no need for a market for land, as one missing market does not lead to any loss of efficiency (Kevane 1996). But the reality is that markets for nonland factors of production are imperfect. Hayami and Otsuka (1993b) argue that because of the difficulty in monitoring hired workers in spatially dispersed and ecologically diverse agricultural environments, labor markets are imperfect.

Hired workers are employed only for simple tasks such as weeding and harvesting, but not for care-intensive tasks such as fertilizer application and water management. One of the more widely noted missing markets in rural Africa is credit in which very few smallholder farmers are able to access credit for agricultural investment. But other markets have also been observed to function poorly (Holden et al. 2001). The dominance of and seasonality in rain-fed agriculture due to limited investment in irrigation in Africa causes synchronized production with related patterns of credit, labor, and other input demand and output supply in which timing of operations strains the performance of the markets. These markets are also affected by covariate risk in production, risk-averse behavior of farmers, imperfect information, and transaction costs.

The fact that nonland factor markets are imperfect creates a potential for land markets to play a role to enhance production efficiency in order to equalize factor ratios across farms. The efficiency-enhancing effects of such markets depend on the degree of imbalances in the factor ratios before the land transaction, the substitutability and complementarities among factors of production, and the transaction costs involved in alternative factor markets.

Mechanization leading to economies of scale has played a modest role in tropical agriculture in certain types of export-oriented production, but also here the economies of scale arise mainly from the processing and marketing side and not so much from the production side itself (Binswanger et al. 1995). With increasing land fragmentation due to high-population pressures, economies of scale may also become important on very small farms when the use of oxen for plowing is an essential part of the production technology. A pair of oxen is required for plowing, and timing of the operation is crucial for land productivity, thus the markets for oxen plowing services tend to function poorly (Bliss and Stern 1982). Difficulties are also related to the vulnerability of animals to mismanagement and the critical timing of operations in which owners of oxen give priority to their own fields before they eventually are willing to provide plowing services to others.

Land Sales versus Land Rental Markets

It is very important to distinguish between land sales and land rental markets. Land sales markets may be nonexistent or thin for various reasons. In some countries, land sales markets are prohibited by law. One of the advantages of land rentals is that they can be used for temporary adjustment of factors of production as a response to shocks. Land sales have much more long-term consequences and require much larger amounts of money, which may depend on access to credit markets and consequently disfavor poor people on the demand side (Otsuka 2007). On the supply side, both sales and rental markets can be used by the poor to raise cash for emergencies or other investments, but in sales markets there is more risk for a poor family of receiving an unfavorable price and not regaining its former land holdings.

In addition, the nonproductive values of land—its asset value and long-term residence value—are capitalized into land sales prices but not land rental prices. As such, land sales prices can often be large multiples of land rental prices (Binswanger and Rosenzweig 1986). This makes land purchases all the more infeasible for the poor. Finally, in rural areas, full payment of land is normally required before taking possession of purchased land (because there are no alternative financing mechanisms). On the other hand, land rental markets have developed numerous mechanisms for compensation and cost sharing that facilitate the interests of a wide range of potential landlords and tenants (Hayami and Otsuka 1993b).

The Historical Context of Land Markets in Africa

Land markets were functioning in several African regions dating back as early as the first half of the 1900s. For example, in the Sukuma in Tanzania, sales were identified in the late 1800s (Malcolm 1953). Certainly by the 1960s, several studies found functioning land markets, notably in the highly populated areas of West Africa and East Africa. Among those reported by Feder and Noronha (1987) are Sokoto Nigeria, where market transactions accounted for about 30% of land holdings (Lunning 1965); the commonality of sales among the Nuba of Sudan (Roden 1971); and rental markets in cocoa areas of Ghana (Hill 1963). Further, Barrows and Roth (1990) cite earlier work by Mukwaya (1953) documenting that 58% of landholders in the Buganda region of central Uganda had purchased land.

Four cases from Ethiopia, Malawi, Uganda, and Kenya demonstrate the different ways in which land markets emerge, grow, and are affected by policy and other factors.

Ethiopia

Tenancy is far from new in modern-day Ethiopia. In part, tenancy was an important element of the feudal system, like in Ethiopia before the land reform in 1974. Particularly in southern and central Ethiopia land distribution was skewed and tenants operated a large share of the land. Various sources reported by Rahmato (1984) have estimated that more than 40% of the land was operated by tenants, who represented more than one-third of the total population in the country (even though very good historical data do not exist). Sharecropping was the dominant contract form. Contracts were oral and the rent was one-third or one-half of the output, depending on the land quality and the cost sharing.

Tenure insecurity was prevalent for tenants: "The threat of eviction, rather than the act in itself, was the potent weapon in the hands of the landlords, and the tenant over whom the danger of unemployment and destitution hung like the sword of Democles had no alternative but to accommodate all the demands of his landlord" (Rahmato 1984, *25–26*). The large landlords did not usually live in the rural areas and did not deal with their many tenants themselves but had agents who took care of this business for them.

It is understandable that the term *landlord* today is linked to this feudal history; Ethiopians use the same term for female-headed households that rent out their land because they lack oxen and the other necessary resources to farm their land themselves. The land reform in 1975 that led to a highly egalitarian land distribution was complemented with more or less frequent land redistributions to maintain the egalitarian distribution and to provide land to new households. The reform drastically changed the power balance between "landlords" and "tenants," although the basic contractual form in terms of share tenancy and use of oral contracts remained much the same. All land was made state property, and land rentals as well as hiring of labor were illegal under the Derg regime (1974–1991). After the change in government in 1991, short-term land renting and hiring of labor were legalized while land redistributions were mostly stopped with a few exceptions. Recent land reforms in the form of providing households with inheritable user certificates may also have contributed to tenure security and enhancement of land rental markets (Deininger et al. 2007; Holden et al. 2007a).

Malawi

Land tenancy had an important role in the colonial history of Malawi. Africans became tenants on European-controlled farms in the Shire highlands in a tenancy system called *thangata* (meaning "to assist" in Chichewa). The system implied an exchange of labor for land. The landlords provided land to the tenants and the tenants were expected to work for the landlords for a given period. The Africans perceived the system as a symbol of European oppression, whereas the settlers viewed it as reciprocal assistance. Abusive practices, such as withholding payment if a tenant failed to complete his or her daily or monthly task and payment of very low prices for tenants' crops, were common and contributed to the resistance against the system (Boeder 1984). These were compulsory work contracts with accompanying eviction threats based on the Native Rents (Private Estates) Ordinance of 1917.

Europeans started to settle in the central region in the 1920s, where land was more abundant. Unlike in the southern region, where the shortage of land created tenancy, it was a demand for cash income that attracted tenants. The tenancy system that was established, called "Visiting Tenancy System," may be seen as a form of contract farming and was fairly lucrative for the tenants compared to subsistence farming.

In the postcolonial period, tenants became an important component in the expansion of burley tobacco estates in the central region of Malawi. The number of tenants increased from 20,000 in 1980 to 105,000 in 1989 (Duncan 1990). Tenants rather than workers cultivated most of the land on the new estates. Several studies found that the average annual net return of tenants was better than those of estate workers but tenants also faced higher risk and their incomes were therefore more skewed, and many of them received no income (Vaughan and Chipande 1986; Nyanda and Shively 1989; Mkandawire et al. 1990; Nankumba 1990).

There is evidence of exploitative tenancy contracts contributing to high turnover rates (>30%) for tenants (Mkandawire et al. 1990). These included high interest rates on credit of 40 to 140% imposed on the tenants, whereas the estate landlords were favored with subsidized access to credit at negative real interest rates. Furthermore, the tenants were paid only a fraction of the tobacco auction price. This fraction was set as a maximum fraction and not a minimum fraction, and was determined by the Tobacco Association of Malawi (TAMA) (Mtawali 1989). The maximum fraction declined from 41% in 1983 to 19% in 1988. The regulation mechanism favored the estate owners at the expense of the tenants. It gave the owners the power to underpay the tenants, but this was also done through undergrading the tobacco purchased from tenants, delayed payment, and payment in installments (Nankumba 1990). One study found that tenants received cash payment of only 10% of the value of the crop after the credit had been paid on one estate (Nankumba 1990). Peasants were not allowed to cultivate tobacco, in order to reduce reservation utility of tobacco tenants. Tenancy was uncommon outside the tobacco sector in Malawi in the 1990s (Place and Otsuka 2001b).

The tobacco tenancy model as it was practiced in the 1980s seems not to be a good instrument for helping tenants out of poverty. The removal of the restrictions on smallholders' rights to grow tobacco in the 1990s may have made tenancy contracts less exploitative but recent studies on tenancy on tobacco estates are lacking. The New Land Policy in Malawi does not address tenancy issues, which is a serious defect in the policy. Land rental markets are also emerging in other parts of the country and it is of high policy relevance to investigate how these tenancy markets may help land-poor households to access land and how the land policy reforms should be shaped to make land rental markets more beneficial to the poor (Holden et al. 2006).

Uganda

Land markets have been active in Uganda for many decades. Although a new Land Act has been in effect since 1998, in order to understand the development of land markets it is necessary to reflect back on *de jure* and *de facto* land tenure systems that operated in the past. The tenure systems themselves were a major factor contributing to the way in which land markets developed.

The second major factor behind the development of land markets in Uganda were regional imbalances in population density and consequent migration patterns.

Among land tenure systems, a most unique one in Uganda is the "mailo" system. Under the Buganda Agreement of 1900, land across much of central Uganda was divided among the Protectorate Government, the Kabaka (the paramount chief of Buganda), chiefs, and notables (Kisamba-Mugerwa and Barrows 1989). Land was granted to these individuals in large blocks (square miles) and the resulting imbalance in control of land fostered subdivisions and soon led to the formation of a large tenant community on these lands. By 1928, laws were enacted to protect the tenants from eviction and from paying usurious rents. Later, land titles helped to spur sales of land by mailo owners. Tenants also acquired strong rights and were eventually able to transact those in land markets as well. In the 1980s, between 60 and 85% of mailo land was under a tenancy arrangement (Kisamba-Mugerwa and Barrows 1989).

In the rest of Uganda, more traditional customary systems of tenure were in operation in which land holdings were more egalitarian. As with other customary systems in Africa, land was acquired through allocation by chief or by clearing and then passed down to descendents through inheritance. In land-abundant regions, this practice continued to be the dominant form of land acquisition through the twentieth century. However, a number of regions saw a rapid growth in population and land shortage. This had two effects, one being the development of local land markets and the other being the development of national land markets for migrants. The case of southwest Uganda is interesting in terms of local land-market development. The population to arable land hectarage ratio increased to over 200 by the 1960s. Furthermore, the hilly terrain gave rise to a demand for parcels of land in different topographical locations. These forces led to the development of a strong land sales market. Bamwerinde et al. (2006) found that a higher percentage of land parcels were purchased than any other method of land acquisition. Baland et al. (2007) reported that migrants (many from the southwest) accounted for a large proportion of land purchases in the east-central area of Uganda.

Thus, land markets are not new to Uganda, but their character and strength vary from region to region and have been heavily influenced by previous tenure regimes. Interestingly, the high number of land sales in customary tenure areas has taken place although they were not legally recognized until the 1998 Land Act.

Kenya

Of all the case studies, Kenya's history is unique in that it has experienced a land registration and titling program, beginning in the early 1960s, and eventually covering most of the medium- to high-potential agricultural areas. This concept was conceived by the Swynnerton Plan of 1954, which lamented the poor functioning of land markets and the inability to channel landholdings to the more productive farmers. In addition to the registration program, a program of land redistribution was implemented in certain former white settler areas. The majority of these programs provided African farmers with between 5 and 20 hectares of land in areas that were likely to be neighboring on smallholder communities with smaller than average farm sizes. Although land-market activity in the pre-land registration period is not well documented, there have been a large number of studies of land tenure and land markets in the period that followed.

Few empirical studies focused on or found much evidence on land rental markets, and there is some evidence that land rental markets are relatively underdeveloped still today. Much of the studies concentrated on the functioning of land sales markets and in particular their implications on equity. Some studies indeed provided cases of distress land sales occurring, dispossessing the poor and women of their land rights (Wilson 1972; Collier and Lal 1986; Haugerud 1989). Some of these earlier findings were called into question by noting the continued practice of sanctioning of sales by elders (Migot-Adholla et al. 1994a) and in more recent larger sample studies (e.g., Place and Migot-Adholla 1998).

As for the vibrancy of land sales markets, evidence shows a mixed pattern across different regions. Pinckney and Kimuyu (1994) found land market activity to be relatively low in a highly populated area of central Kenya despite presence of land titles. Similarly, a study from a nearby region in central Kenya found that purchases accounted for between 6 and 13% of all land acquisitions (Place and Migot-Adholla 1998). In other areas, land purchases are more common (e.g., in the western province) but there, too, markets seem to have gone through active and thin periods. The land settlement program seems to have stimulated some land market activity, generally reallocating land from the land-rich to the land-poor.

Land Markets in Africa Today

Table 1-1 gives an overview of the existence of land markets in Africa, based on the sources listed below the table.

Land sales appear to be rare among smallholders in southern Africa (Zambia, Malawi, and Zimbabwe). For example, Nankumba (1994) and Place and Otsuka (2001b) found no land purchases among households in central and north-central Malawi, respectively. Place (1995) also found no purchases of land among 200 households in eastern and southern provinces of Zambia. Rentals are also rare, but greater than sales in several cases (Place 1995; Place et al 2001a; Zeller et al. 1997). Unruh (2001) reported significant land-market transactions (purchase, rent, pledge) in the cashew agroforestry system in prewar Mozambique. A general lack of market activity is not surprising in

TABLE 1-1 Overview of current land market status in various parts of Africa

Region	Population density	Land sales markets	Land rental markets
Southern Africa	**Low**	**No:** Zambia	**No:** Zambia
	High	**Low:** Southern Malawi	**Low:** Southern Malawi, Zambia
		Some: Central Malawi, Mozambique	**Some:** Central Malawi, Mozambique
Eastern Africa	**Low**		
	High	**Active:** Kenya, Uganda, Rwanda, Burundi, Southeast Somalia	**Active:** Ethiopia, Eritrea, Rwanda, Burundi
		Some: Tanzania	**Some:** Kenya, Uganda, Somalia, Madagascar
		No: Ethiopia, Eritrea	
Western Africa	**Low**		**Some:** Nigeria
	High	**Active:** Ghana	**Active:** Ghana
			Some: Benin, Cote d'Ivoire, Nigeria, Sierra Leone
Sahelian zone	**Low**		**Low:** Burkina Faso, Senegal, Niger
	High		

Sources: Benin: Burkina Faso: Matlon (1994), de Zeeuw (1997), Burundi: Place (1995), Cameroon: Lawry and Steinbarger (1991) and Manyong and Houndekon (2000); Cote d'Ivoire: Chaveau (2000), Eritrea: Tikabo et al. (2007), Ethiopia: Benin et al. (2005), Holden and Ghebru (2005), Pender and Fafchamps (2006), Teklu and Lemi (2004), Ghana: Delville et al. (2002), Migot-Adholla et al. (1994b), Quisumbing et al. (2001b), Kenya: Wangila (1999), Malawi: Holden et al. (2006), Place and Otsuka (2001a, 2001b), Mozambique: Unruh (2001), Niger: Gavian and Fafchamps (1996), Rwanda: André and Platteau (1997), Blarel (1994), Somalia: Roth et al. (1994), Tanzania: Pinkney and Kimuyu (1994), Uganda: Place and Otsuka (2001c), Bamwerknde et al. 2006, Baland et al. (2007), Zambia: Place (1995)

Zambia, which has one of the lowest population densities in Africa. In southern Malawi, population density is acute yet households do not claim to have bought or sold land, nor even to possess the right to do so (Place and Otsuka. 2001a). However, in 2006, Holden et al. (2006) found land rental markets to have emerged both in central and southern Malawi. Land-market activity is also reported in peri-urban areas of southern Africa (Adams et al. 1999).

Land purchases are generally much more common among smallholders in East Africa than elsewhere in Africa, but there is considerable variation. Almost 80% of smallholders have purchased land in southwestern Uganda (Place 1995). In the same place, the percentage of plots acquired through purchase equals that from inheritance. More recently, in a national sample in Uganda, Deininger and Mpuga (Chapter 7) found that 29% of land was purchased. High figures are also observed for areas in Kenya, Rwanda, Burundi, and Somalia. Place (1995) found that 16 to 25% of land parcels were acquired through purchase in three prefectures of Burundi. Similarly, purchases comprised 18 to 19% of the number of parcels and total farm area in two regions of Kenya (Migot-Adholla et al. 1994a). In the Shebelle Valley in Somalia, Roth and colleagues (1994) found that smallholders purchased 31% of their parcels and had rented in another 7%. Andre and Platteau (1998) found that 40% of households had both bought and sold land in Rwanda over the course of their years as farmers. Pinckney and Kimuyu (1994) also found significant sales among households in the northern Tanzania highlands. The importance of purchasing increased over time in Rwanda, but not necessarily in Kenya. Contrarily, because of enforced government policy in Ethiopia, land sales are nonexistent in rural areas, but such illegal sales are still common in some peri-urban areas and were in many cases disclosed during the land reforms (land registration and certification) in the Amhara and Oromiya regions in the period from 2003 to 2007 (Holden and Tefera 2008).

There is similar variation in the prevalence of land renting in east Africa, although in general the proportion of current land holdings under rent is lower than that found in Asia (e.g., Platteau 1996; Carter et al. 1994; Place 1995). For example, the percentage of 1,600 households from western Kenya hiring out labor or hiring in labor (32% and 45%, respectively) greatly exceeds the percentage that claim to have rented in land (6.5%) (Wangila 1999). Deininger and Mpuga (Chapter 7) found that 7% of agricultural land from a sample across Uganda was rented in. But there are pockets, especially in the highly fragmented areas of Rwanda and Burundi, where rented in parcels are significant (e.g., one-fourth of households had rented in parcels in central Burundi [Place 1995]). They are also very common in Ethiopia, where permanent land transfers are prohibited (Holden and Ghebru 2006; Pender and Fafchamps 2001, 2006; Teklu and Lemi 2004). Benin et al. (2005) found evidence of expansion in land rental market activity in the Amhara region of Ethiopia during the 1990s. Similarly, Tikabo et al. (2008) also found land rental markets to be very active in Eritrea, where land sales also are prohibited by law. Data from Rwanda suggest that temporary land transactions are complex in that a large percentage of market participants simultaneously rent in and rent out land (Andre and Platteau 1998; Blarel 1994). For example, Blarel (1994) found that among the lowest quintile of landowners in Rwanda, 64% rented land in and 23% rented land out. Thus, both the motivations of accessing land for production and using land to raise cash (i.e., distress rentals) appear to be in operation.

The frequency of outright purchases of land in West Africa appears to lie between the southern and eastern Africa figures. But in addition to these, more complex, indefinite, and fixed duration transfer of land (and tree) rights are commonplace, especially in the humid areas of Cameroon, Nigeria, Benin, Ghana, Cote d'Ivoire, and Sierra Leone (Adesina and Chianu 2002; Manyong and Houndekon 2000; Lawry and Stienbarger 1991; Kallon 1996; Chauveau 2000; Delville et al. 2002). The prevalence of market transactions is highest in areas characterized by in-migration. Delville et al. (2002) reported that almost two-thirds of households in the oil palm belt of Ghana had entered into share contracts. They also reported renting to be much more common than borrowing between relatives in a relatively sparsely populated area (<70 persons per square kilometer) in northeastern Nigeria. Migot-Adholla et al. (1994b) found 18% of parcels were purchased in a cocoa growing area, whereas 41% of parcels were rented or pledged in a very densely populated coastal area in eastern Ghana. The proportion of land acquired through purchase from individuals was between 4 and 5% in western Ghana (Quisumbing et al. 2001a). But this proportion is growing over time and the area acquired through rental markets is high, as much as 18.8% in migrant villages.

Studies from the 1980s suggest that land-market transactions in the drier Sahelian countries are very limited (e.g., Matlon 1994 for Burkina; Golan 1994 for Senegal). This appears to be affirmed by de Zeeuw (1997) for Burkina Faso, who noted that borrowing of plots continues without significant material compensation, but rather it is loyalty of the tenant that is desired by the landowner. However, Gavian and Fafchamps (1996) noted that land borrowing in Niger, accounting for about 19% of land plots, can be in the form of a market-like pledge in which cash is exchanged for land on a temporary basis.

Overall, it seems clear that land markets have emerged and gradually developed in many parts of sub-Saharan Africa, particularly in densely populated areas and in areas where population has been growing rapidly (Otsuka 2007). However, despite this set of data on prevalence of land markets, there is very little analysis of the roles and the implications of land markets. This book will fill that gap.

Key Policy Issues Concerning Land Markets in Africa

Land reforms are again high on the international policy agenda, as evidenced by the increasing number of land reform projects in the project portfolio of the World Bank and the recent establishment of the (High-Level) Commission for the Legal Empowerment of the Poor. Unlike the Asian land reforms, which aim to redistribute land from large owners to poor tenants and to regulate land rent at a low level, the major purposes of African land reforms are to identify means of increasing security of tenure—to reduce land conflict, to increase opportunities for land as collateral, and to promote agricultural productivity through trade and investment in land.[1] De Soto (2000) has contributed to this by attracting international donors' interests toward providing the poor with formal property rights to mobilize their "dead capital" as a basis for investment loans. However, the route toward well-functioning credit markets based on the formal property rights of the poor seems long and thorny in Africa. Also, de Soto's explanation (the subtitle of his book: *Why Capitalism Triumphs in the West and Fails Everywhere Else*) fails to explain how China has succeeded in having an impressive economic growth for many years without private property rights to land (Rozelle et al. 2005).

Important policy questions include the following:

1. To what extent do land sales markets alleviate or worsen inequality of land holdings? Are distress sales by the poor common?
2. To what extent do land rental markets alleviate or worsen inequality of land holdings? Do they provide other benefits to poor households (e.g., through the provision of credit)?
3. Are land rental and sales markets improving or worsening investment and agricultural productivity? What regulations are preferable to enhance investment and land management on rented land?
4. What institutional or policy reforms can enable rural land markets to better allocate land resources to the more productive farmers?
5. What regulations on land markets, if any, are required to protect the interests of the vulnerable, such as women, the illiterate, and migrants, with regard to land-market transactions? Are restrictions on size of traded land or remaining land beneficial or misguided?
6. What are the advantages and disadvantages of making rental markets more formal? Are there certain types of land rental contracts that should be promoted above others?
7. What have been the impacts of land registration and certification programs and how can these be made to be more effective?

The first three questions reflect the need to better understand the major concerns that policymakers have about land markets—namely, their equity and productivity impacts. Indeed, concern about how land markets will affect access to land by the poor is behind the policy of state ownership of land in Ethiopia and of land sales restrictions in Rwanda. Other countries, such as Nigeria and Zambia, mention the desire to address the functioning of land markets not from an equity concern but to increase agricultural productivity. One aim of this book is to demonstrate more clearly how different types of land markets may lead to different outcomes for equity and productivity. A major policy issue, then, is to determine if policies can be found that promote the positive aspects of markets while at the same time reduce the potential for undesirable outcomes to emerge. This issue is captured by questions (4) through (7) above.

Overview of Book and Key Research Questions

The goal of this book is to critically examine the functioning and impacts (in terms of efficiency, equity/poverty and sustainability consequences) of land markets across a number of densely populated countries in Africa to assess their importance and the extent to which their functioning is undermined by policy distortions or improved by conducive policies.

In order to generate a solid understanding of these issues, a set of case studies have been organized around three main themes: (1) allocative efficiency and participation in land markets, (2) contract choice and outcomes, and (3) land-market effects on land tenure security and land management. A common area of inquiry is on the production impacts of land markets and land tenancy contracts. Assessments related to poverty—Who benefits and loses from land markets? and How do land markets affect value of production?—are integrated into many of the case studies and is another common theme of the book.

Key Research Questions

Access to and participation in land markets

- Do all people have equal access to the land markets or are the poor rationed out or only participating under unfavorable conditions?
- Who participates by choice and who benefits?
- Who participates under distress?
- Do land sales markets generally provide means for the near-landless to acquire land or mainly for the wealthy to accumulate more land?
- Do land rental markets generally provide means for the near-landless to acquire land or mainly for the wealthy to accumulate more land?
- How do land markets affect the welfare of women and other vulnerable groups?
- To what extent do rental markets act as a buffer to prevent households from liquidating assets?

Allocative efficiency

- What is the extent of transactions costs in land markets and how efficient are land markets in terms of satisfying demand for land?
- Are land markets more or less efficient in equalizing land/nonland factor ratios than other factor markets?
- Is there an inverse farm size–land productivity relationship that can be potentially eliminated by land markets?
- What are the short-term efficiency consequences of land rental markets? Of sales markets?
- How do existing policy regulations affect the functioning of markets and can the efficiency of land markets be improved by policy means?

Contract choice

- What are the different types of rental contracts used?
- To what extent is contract choice a reflection of imperfect markets for credit and insurance and unequal bargaining power of the parties involved?

- How is contract choice related to the poverty of landlords and tenants?
- What are the productivity effects of different contracting arrangements?
- Do longer-term contracts encourage more investment in the land?
- What effects do different output-sharing or cost-sharing arrangements have on production?

Land tenure security and land management

- How do land markets alter tenure security and patterns of land rights?
- What are the implications of changes in tenure security and land rights on land investment?
- What are the longer-term investment and productivity growth consequences of land rental markets? Of sales markets?
- To what extent do land disputes discourage land transactions and long-term investments in land improvement?

Although each study is unable by itself to address all these questions, each question is addressed by more than one study. For many of the questions, through synthesis, we shall be able to understand how variation in land distribution, regulations, inheritance systems, and cultural norms (social capital) affect the answers.

NOTE

1. The distribution of land is also an important issue, especially in southern Africa. Here, there are concerns both with "excessive" and "underutilized" holdings of a few wealthy landowners as well as with "excessive" subdivision of landholdings among the smallholder farming community.

CHAPTER 2

Understanding Land Markets

Questions and Hypotheses

STEIN T. HOLDEN, KEIJIRO OTSUKA, AND FRANK M. PLACE

The understanding of land markets has multiple dimensions. Among these are the different thematic dimensions to land markets, such as the types of land-market transactions, the degree of access and participation, and the impacts of land markets on agricultural productivity, poverty, and natural resource management. Also is the dimension of spatial scale and the recognition that markets can be analyzed at mesoscales (e.g., at a village or even at a regional level) as well as at microscales (e.g., at a land-unit or even a household level). The individual papers rely mainly on microlevel data to draw conclusions about the performance of land markets at the household and community levels. A synthesis across the case studies in the final chapter yields more insights into the functioning of land markets at the larger scale.

In the first chapter, we briefly outlined the underlying fundamental characteristics that shape production relations in rural areas in Africa. These also are the basis for the conceptual model capturing the key factors contributing to land-market development and the related efficiency, equity, and sustainability implications. This chapter builds on those fundamentals. Following a more detailed presentation of key empirical issues and a conceptual framework, the chapter addresses the joint issues of access to and participation in land markets and how this affects asset- and land-poor households. The next section explores the allocative efficiency of land markets. The fourth section focuses more closely on the choice of contract in land rental markets and how this is related to equity of land holdings and efficiency of land use. Next, the impacts of land markets on land tenure and land management are examined. In each of these sections, there is a presentation of a conceptual or mathematical model, a review of the most important literature related to the topic, an identification of the major research hypotheses, and a discussion of methodological challenges and methods used in the book. Finally, the links to the different empirical chapters are discussed in the last section.

Key Issues in Land Markets

A major concern of policymakers is that by opening up all land markets, land concentration may increase and result in larger numbers of landless poor people. In particular, there is the potential for the poor losing land due to distress sales as a result of climatic, economic, or health-related adversities. The consequences of this could be increased migration to urban areas, for example. This is one of the reasons why land sales markets are prohibited in many countries. But how big is this problem? Under what conditions is it a problem and are prohibitions of land sales the best solution? Distress sales may take place because they are the best option of the poor whereas other alternatives such as insurance are infeasible, inferior, or nonexistent. A second reason for sales prohibitions may be for the interest of the state to keep control over the land and be able to access it easily without having to pay too much compensation to land users. There may also be development and growth benefits from such a property rights regime, like in China.

Although land rental markets are less likely to be regulated to the extreme that land sales markets have been, there have been many attempts to regulate and even prohibit such markets. One of the reasons for such regulation has been the historical experiences with exploitative contracts utilized by rich and powerful landlords to maximize their returns from poor tenants who lack or have very unfavorable exit options. Another reason has been that sharecropping has been prohibited because of the belief in Marshallian inefficiency as a necessary outcome of such contractual arrangements (Marshall 1890). However, one may also raise questions about the efficiency of often short-term rental contracts due to their inherent tenure insecurity. Similar to the questions surrounding land sales markets and their outcomes, the impacts of land rental markets on equity and efficiency in Africa are not yet well known. Descriptive evidence suggests that the distinctions in wealth and power between landlords and tenants in Africa are much less than in Asia and Latin America, but this has not been thoroughly scrutinized. In some African countries, like Eritrea and Ethiopia, the landlords are even poorer than their tenants, and this may also alter the "power balance" between landlords and tenants. This will also be one of the focuses of this book.

In terms of the efficiency debate with respect to land rentals, it is first important to note that the research in Asia does not generally support Marshallian inefficiency of share tenancies (Otsuka et al. 1992). In some of the most densely populated rural areas in Asia and Africa, farm size reduction has reached levels where further reduction will lead to inefficiencies. Already, many farms have become too small to provide a subsistence income for the owners. As households have increasingly looked for nonagricultural livelihood strategies, land renting becomes an important tool in their asset and investment strategy. Moreover, land renting offers some flexibility as a short- or longer-term option; it provides an exit option for some households and an entry opportunity for other households and may thus be a step in the ladder out of poverty.

But how good are land markets as a ladder for households to escape poverty or as a safety net for protecting the poor against disaster? Do all who want it have access to the market or are the poor rationed out because they are too poor to participate? How high is the "insurance premium" (and unattractive contract terms) the poor have to pay for using land markets as a safety net? Land markets may not always clear like ordinary markets that have one market clearing price. Imperfect information, transaction costs, and incomplete contracts may cause some to be rationed out and prices to differ for similar land quality and quantity. Sharecropping is often associated with rationing on the tenant side of the market. Too often, well-intended policies, such as prohibition of sharecropping, have failed to generate the intended outcome and often have results that are even counterproductive. So how can policies help reduce these problems and make land markets an important tool for poverty reduction?

And what are the short-term and longer-term efficiency and growth consequences of land rental markets? If, for the moment, we ignore the issue of contract choice, will a land rental market ensure that rented land is managed in the same way as owner-operated land? One expects that those who are willing and able to rent in land are also able to use the land more efficiently than the landlord who is willing to rent it out at the agreed price. But do the short-term productivity effects ensure that the rented land is conserved and invested such that its future productivity is sustained and/or enhanced? There are many studies on how tenure security may affect investment in land improvement and, hence, improve sustainability but there are few studies that study the management of rented versus owner-operated land. With short-term contracts there are reasons to question whether rented land is managed in a sustainable way. And this may have important policy implications in terms of having regulations that enhance more sustainable management without reducing efficiency of land use.

Conceptual Framework

Our general conceptual framework is illustrated in Figure 2-1. As can be seen, the emergence and development of land markets are related to rapidly changing driving forces such as population pressure and commercialization, to policy reforms and incentives, and to conditioning (less dynamic) factors such as land distribution, agroecology, customary tenure institutions, and cultural norms and traditions. These factors themselves cannot easily be isolated from one another as, for example, population pressure has been found to exert pressure on customary tenure institutions to change and undoubtedly has affected

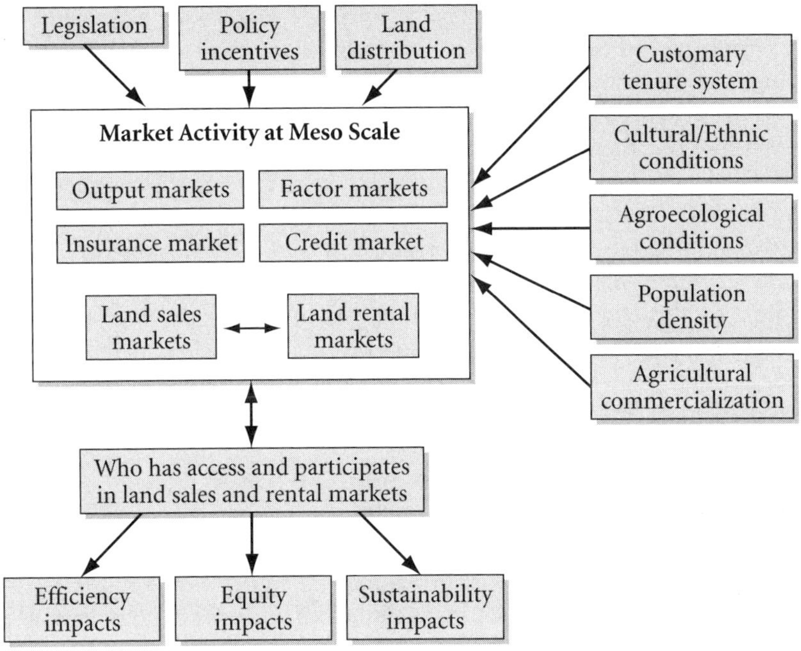

FIGURE 2-1 Conceptual model—land market development, participation, and impacts

the distribution of land (Migot-Adholla et al. 1991; Otsuka and Place 2001). The development of other markets also has a major influence on land markets, as they may be able to substitute, at least to a degree, for land markets.

We will start by looking at the major driving forces and how they shape land markets along with other markets. Despite clear trends of urbanization, absolute population growth rates in rural Africa remain high and population densities continue to increase. It is well known from many existing studies that the vast pool of unexploited land resources is vanishing throughout the African landscape, thus increasing the need for and the value of land. At the same time, there is mounting pressure on securing long-term tenure on land, as the resulting smaller farm sizes require more intensification of land through investment to generate adequate production and income for households in a sustainable manner. The commercialization of agriculture (through development of output and input markets) has had a similar effect on incentives for intensification and for increasing the value of land. Both commercialization and population density are highly correlated to agroecology, with positive correlations with the higher potential areas (e.g., highland areas and more humid zones). The combination of higher land scarcity value along with more private security of tenure has exerted pressure on the development of land markets.

Imbalances in factor ratios of land and other resources across households create inequalities in land productivities and, hence, provide strong motivation for land transactions. Households with relatively more labor to invest or relatively more capital (where there are profitable investments to make) will want to increase their land holdings to increase production. This is where other factor markets need to be considered. Households with abundant labor could as well hire out their labor to earn extra wages if labor markets functioned well. Credit markets can be seen as both a substitute and a complement to land markets. Credit markets could serve as a substitute for distress sales and distress land rentals in case of emergencies if they were working well. Credit markets could also enable households to purchase more inputs and increase production through existing production systems and patterns of land holdings. However, the existence of credit markets may also help land markets (and labor markets) function more efficiently. This is especially the case for land sales markets in which larger amounts of cash are required.

Land Market Access, Participation, and Equity

Based on our conceptual model in Figure 2-1, Figure 2-2 shows important household, market, and other institutional characteristics that may affect households' access to and participation in land rental or sales/purchase markets. The literature indicates that factors such as initial land and labor endowments, wealth/poverty levels, access to credit and other factor markets, extent of nonfarm livelihoods, land rights, transactions costs in relation to market participation and, of course, the cost of land are important in determining the nature of the participation in land markets. A broader perspective on land-market participation takes into account trust, reputation, and availability of potential partners (landlords and tenants) in assessing the functioning of the market. If the initial fixed costs for farming are high, this may also contribute to a land-market entry barrier. Likewise, transaction costs related to search for partners, negotiation, monitoring, and enforcement of contracts can hinder and reduce the degree of participation. Much of the literature in this area has focused on whether land markets reduce or exacerbate inequality of land holdings and wealth. In poor agrarian communities, the worsened inequality immediately leads to widespread and deepened poverty incidence. In this sense, the issue of equity and poverty are inseparable in the context of sub-Saharan Africa. The historical experience has implied that land rental markets have transferred land from land-rich households to land-poor and often landless households in situations in which the initial land distribution was inequitable due to historical power relations. There are several reasons why this pattern may not necessarily hold today. For example, mechanization in agricultural production has caused an increase in optimal farm sizes (e.g., in Norway, Japan, parts of India, and Thailand) and this has induced owners of small farms (landlords) to rent out their land to owners or tenants of larger farms who can take advantages of economies of scale.

Household Access and Participation

Factors in Selling/Renting out:
- Land abundance
- Migration/employment
- Lack of income/wealth
- Unfavorable access to other factor markets
- Adverse shocks
- Lack of credit access
- Rights of land transfer
- Tenure security
- Low transactions costs
- Price of land
- Availability of buyers/tenants
- Bargaining power
- Monitoring capacity

Factors in Buying/Renting in:
- Land shortage
- Lack of alternative livelihoods
- Income/wealth
- Capital or access to credit
- Favorable access to other factor markets
- Tenure security
- Low transactions costs
- Price of land/rationing
- Availability of sellers/landlords
- Reputation
- Farm skills and equipment
- Bargaining power
- Experience
- Social network

FIGURE 2-2 Factors affecting household participation in land markets

It has also frequently been stated that land sales markets lead to a concentration of land into the hands of the wealthy due to lack of credit and insurance markets, forcing the poor to sell their land under financial distress and preventing those without significant savings from buying land. But how are purchasers of land distributed among wealthy groups and landholders in Africa? How common are distress sales? We will look at some of the empirical evidence from Africa.

Review of Evidence from Africa

André and Platteau (1998) found that inequality of land ownership had increased over a period of five years (from 1988 to 1993) in western Rwanda, where two-thirds of the land sales were made by the land-poor to meet their subsistence and medical expenses. Conversely, in a study in Benin, Dubuisson (1998) found that land markets reduced inequity in the land distribution, as land was purchased by land-poor households from land-rich households. Similarly, Place and Migot-Adholla (1998) found that several landless households purchased land in Kenya. Baland et al. (2003) studied the distributive impact of land markets in 36 villages in eastern and central Uganda. They concluded that both the emerging land sales and land rental markets corrected initial inequality in land endowments, both across and within villages. It appeared that land rental markets played a compensatory role for insufficient sales markets in enabling households to adjust land sizes to their desired levels.

Concerning the effect of land purchase markets on equity between the rich and the poor (i.e., not simply the land-rich and land-poor), some studies find evidence that buyers of land are the wealthy and sellers are the poor (André and Platteau [1998] for Rwanda; Bruce [1988] on growing landlessness in Kenya; Zimmerman and Carter [1999]; and Dubuisson [1988] for Burkina Faso). It is further noted by Andre and Platteau (1998) that most sales could be described as distress sales. There is contrary evidence from other studies, indicating that land purchase markets are neutral or even favorable for the poor (Pinckney and Kimuyu [1994] for Tanzania and Kenya; Place and Migot-Adholla [1998] for Kenya; and Baland et al. [2007] for Uganda). In the Uganda and Kenya cases noted, a purchased parcel of land often represented the entire land holding of the farm and as such compensates for poor endowments of inherited land.

In the case of land rental markets, several studies find evidence that rentals contribute to more equitable operational land holdings (and factor ratios) between the rich and the poor. This was found in Rwanda (Blarel 1994; Andre and Platteau 1998), in Ghana (Migot-Adholla et al. 1994b), and in Ethiopia (Pender and Fafchamps 2001). Holden et al. (2006) also provided evidence of land-poor households accessing land through emerging land rental markets in Malawi, where the land distribution is inequitable. Moreover, if one assumes that migrants are generally poorer than natives in their destination village, tenancy arrangements in West Africa have contributed positively to equity (Quisumbing et al. 2003). However, as noted earlier, the poor are not only recipients of rented land—they are active lessors of land as well. As many as 25% of the most land-poor rent land in and out at the same time in Rwanda (Blarel 1994).

In Eritrea, Tikabo and Holden (2003) found that poor female-headed households tended to rent out their land to wealthier, usually male-headed households. A similar pattern of poor landlords and wealthy tenants has also been found in Ethiopia (Bezabih and Holden 2006; Holden and Ghebru 2006; Kassie and Holden 2006) and in Madagascar (Bellemare 2006). Also, the study by Laffont and Matoussi (1995) in Tunisia revealed that the tenants had higher levels of working capital than landlords. Although the 1975 land reform in Ethiopia and subsequent land redistributions created an equitable distribution of land within villages, the skewness in the distribution of land has again increased. Although land sales markets are prohibited, the land rental markets are very active, driven by inequities in nonland factors of production, like oxen and labor endowments. Oxen- and labor-poor, often female-headed households are induced or forced to rent out their land to households with more of these endowments. This typically leads to a more skewed distribution of operated holdings than owned holdings (Holden and Ghebru 2006; Kassie and Holden 2006). Similarly, in Eritrea, Tikabo and Holden (2003) found that land moved through the rental market to households with oxen and more labor endowments causing a more skewed distribution of operated land holdings.

In Eritrea and Ethiopia, it appears that the land rental market serves as an important opportunity for poor households who are too poor to farm their land efficiently themselves. By sharecropping out their land to households that are better endowed with nonland factors of production, they can get an income from their land in the form of food, which provides them with the food security that they need. However, laws have now been reintroduced in Ethiopia to restrict land renting to a maximum of 50% of the farm area. Similarly, the poor in western Kenya often rent out land for cash (Wangila et al. 1999). These can be described as "distress" rentals in many cases or "investment" rentals in others (where the landholder requires some capital to purchase labor or fertilizer, for example). In such cases, the equity of the distribution of operational land holdings worsens, but the equity of the distribution of working capital for investment improves. All in all, the ability to participate in land rental markets for shifting resources across factors of production would appear to be favorable for the poor.

Migot-Adholla et al. (1991) and Otsuka and Place (2001) found that population pressure and commercialization are generally associated with more individualized rights (comparing Ghana, Kenya, and Rwanda). Place (1995) also discovered a strong relationship between population pressure and land purchases, though not land rentals (comparing Burundi, Uganda, Malawi, and Zambia). Land markets and individualization of land rights seem to move in tandem. Land markets are most active where land transfer rights are individualized (much of east and central Africa fits this). For instance, the individual right of sale is hardly recognized by household respondents in Malawi or Zambia (Place 1995). This is not to say that the majority of people with rights to sell actually exercise the rights.

There are many instances in which high-value commercial enterprises have altered tenure and land market patterns. Cocoa in West Africa is well known for the establishment of share tenancies and ultimately purchases (Hill 1963; Berry 1975; Benneh 1970; Quisumbing et al. 2003). Tobacco appears to be affecting traditional matrilineal patterns in central Malawi (Place and Otsuka 2001a, 2001b). Land markets are well known to flourish in peri-urban areas where commercial opportunities are high. The link is not obvious or is nonexistent in other situations, however. For example, there was no evidence of increased market transactions in central Kenya where coffee production expanded. Perhaps because farms were already small, there was little scope for migrants to acquire land, and the demand for coffee land through inheritance was high among sons (Place and Migot-Adholla 1998).

Migration has seemingly spurred land market development. This is because migrants do not have access to the most common type of acquisition method—inheritance—in the new village. In Uganda, migrants are key purchasers of land (Baland et al. 2003) and the same can be said of migrants in commercial tree crop areas of West Africa. In southern Africa, migrants still claim to have been given land by chiefs, but in some cases the allocation may be a disguised purchase in which the migrant gives the chief a gift and the chief in turn requests land from one of his subjects. Ethnicity plays a role here because it is known that some areas are more receptive to migrants than others. Ugandan ethnic groups are generally known to be quite hospitable to migrants, land availability permitting, more so than many Kenyan ethnic groups.[1]

African states have been concerned about land ownership and markets since their independence, for social, economic, or political reasons. Laws prohibiting land sales were and are actually quite common in Ethiopia, Uganda (until just recently), and Rwanda. In other places, such as Kenya, sales were allowed but were required to have Land Control Board approval. These laws had varying levels of effect. In Ethiopia, the laws have had considerable impact, effectively squashing the land sales market, except in some peri-urban areas where illegal sales have become common. This was made effective by a policy of equity-oriented and periodic land redistribution. Ethiopia removed barriers against land renting in 1990 and such markets have expanded since (Pender and Fafchamps 2001). In Uganda, the 1975 Land Reform Act also abolished all rents in a move to protect the long-term *kibanja* tenants. However, in the case of Uganda and Rwanda, the state did not actively engage in enforcement, and certain land markets proceeded as if there were no laws stopping them.

The effect of land registration programs on land sales markets does not appear to be strong. Studies in Kenya found that sales were just as frequent before the registration programs as afterwards (Place and Migot-Adholla 1998; Green 1987). Roth et al. (1994) also found that land purchases were as common among registered as nonregistered farmers within and near a pilot land registration scheme in southwest Uganda. This may indicate that land registration programs also did not improve the access to credit.

Hypotheses

Based on this somewhat contradicting empirical evidence on the relationship between poverty, land distribution, and land markets, we have developed the following broad hypotheses:

HA1: Land rental markets contribute to enhancing allocative efficiency such that households rich in nonland resource endowments rent in land from households that are poor in nonland resource endowments relative to their land endowments.

HA2: Land rental markets reallocate land from land-rich to land-poor households.

HA3: Land rental markets serve as a safety net for poor tenant households.

HA4: Land rental markets serve as a safety net for poor landlord households.

HA5: Poor households may be rationed out of land rental markets.

HA6: Land markets can serve as a ladder out of poverty.

HA7: Land sales markets cause a concentration of land in the hands of the land-rich and wealthy due to distress sales by the poor and lack of credit markets to assist the poor in buying land.

HA8: Increasing population pressure cause emergence and expansion of both land purchase and rental markets.

Although HA2 can be a part of HA1, the latter considers the endowment of not only land and labor but also oxen and management ability. We also see that hypothesis HA5 may contradict hypotheses HA3 and HA4. Which hypotheses are rejected may vary across countries and locations and perhaps even across households within locations. We also aim to find out under which conditions the various hypotheses may or may not be true.

Methodological Approach

Land Markets and Vulnerable Groups. The impact of land markets on vulnerable groups can be assessed by examining the characteristics of households (e.g., resource endowments, gender, ill-health affected households, etc.) and individuals, and their extent of access and participation on either side of land markets. Analyses can be made on a variety of rights to land—access to land, and the ability to utilize land to generate income and as insurance—and how these may differ for vulnerable groups.

Land Distribution and Land Markets. This can be assessed through the use of Gini-coefficients that provide statistical measures of the equality of the distribution of land resources within and across communities before and after land-market participation. Inequality can also be assessed by econometric methods that regress land areas purchased or rented against land endowments, oxen holdings, or other wealth indicators. The same broad econometric approach can test for allocative efficiency, and this is presented in more detail in the following section.

Land Rental Markets and Allocative Efficiency

To test for the efficiency of land markets, we build on the basic models of Bliss and Stern (1982) and Skoufias (1995) that focused on the extent to which land rented in (or out) by a household adjusts to ensure that operational land holding is adjusted toward the desired land area.[2] This is econometrically estimated with the following linear model:

$$L_j = f\left(h(\overline{N}) - \overline{L}\right) = a_0 + f'h_N\overline{N} - f'\overline{L} = a_0 + a_1\overline{N} + a_2\overline{L} \qquad (2.1)$$

where Lj is net rented in area, desired cultivated area, $h(\overline{N})$, is a function of the endowment of nonland resources, $\overline{N}$, and the coefficient $a_2 = -f'$captures the degree of adjustment in operated land holding toward desired cultivated land area. The empirical literature has focused on whether this coefficient is equal to –1 or closer to zero. If $a_2 = -1$, the cultivated area (the sum of owned and rented in land areas) becomes independent of land endowment and dependent only on $\overline{N}$. On the other hand, if it is significantly different from –1, it is a sign of significant nonlinear transaction costs (see the next subsection for further details). In the following paragraphs, we will expand on this model by developing a more general model with transaction costs and introduce nonlinearities to show how they may affect the size of the adjustment coefficient on land.

Before that, it is worth looking at some of the key references on the analysis of allocative efficiency of land markets (Bell and Sussangkarn 1988; Bliss and Stern 1982; Skoufias 1995). Bliss and Stern studied the land rental market in Palanpur village in India using a theoretical model to assess whether the land adjustment in the rental market was complete ("desired") or less than desired due to transaction costs in the market. They found evidence of incomplete adjustment

Bell and Sussangkarn (1988) suggested that the market for tenancies may not clear in Walrasian fashion because of adverse selection and moral hazard problems, as some households that want to rent land may be rationed. They suggested that this is due to high transaction costs, particularly for landlords. Their study was in an Indian setting with absentee landlords, a well-functioning labor market, and many landless households. Similarly, Bliss and Stern (1982, *54–65*) also argued that sharecropping involved moral hazard problems and a strong likelihood of rationing. This is because sharecropping does not involve a market-clearing price in the land rental market, leading to an asymmetry between landlords and tenants in the form of rationing on the tenant side of the market.

Skoufias (1995) developed a model with transaction costs that he tested on a dataset from India with censored tobit models on each side of the market. He found signs of significant asymmetries across the two sides of the market and the presence of transaction costs in the market.

A Household Model with Transaction Costs

A farm household has initial endowments of (uniform quality) land $\overline{L}$ and nonland $\overline{N}$ resources. L_j and N_j are land and nonland resources transacted in the land and nonland factor markets. The land rent is given by R_0 and the fixed and variable transactions costs are F^L and V^L, respectively. We assume that the transaction costs are nondecreasing in land transacted, $L_j = L - \overline{L}$. The nonland factor can be transacted at price w_j, which depends on market price w, and fixed (F^N) and variable (V^N) transaction costs in the nonland factor market. The household's income y is, therefore, given by

$$
\begin{aligned}
y = {} & pq(\bar{L}+L_j,\bar{N}+N_j) - R_0 L_j - F^L\left[L_j\right] - \\
& V^L(L_j) - wN_j - F^N\left[N_j\right] - V^N\left(N_j\right) \\
F^L\left[L_j\right] & = \begin{Bmatrix} F^L > 0 \text{ if } L_j \neq 0 \\ F^L = 0 \text{ if } L_j = 0 \end{Bmatrix}, \\
F^L_{Lj} & = 0, \text{ fixed transaction costs} \\
V^L\left(L_j\right) & = \begin{Bmatrix} V^L > 0 \text{ if } L_j \neq 0 \\ V^L = 0 \text{ if } L_j = 0 \end{Bmatrix}, \\
V^L_{Lj} & = \begin{Bmatrix} V^L{}_{Lj} > 0 \text{ if } L_j = L_b > 0 \\ V^L{}_{Lj} < 0 \text{ if } L_j = -L_s < 0 \end{Bmatrix}, \text{variable transaction costs}
\end{aligned}
\tag{2.2}
$$

where F^L_{Lj} and V^L_{Lj} are the derivatives of the fixed and variable transaction costs with respect to land transacted, and similarly for fixed and variable transaction costs for the nonland factor.

The household maximizes utility from income:

$$
\max_{Lj,Nj} U = \begin{Bmatrix} U\Big(pq(\bar{L}+L_j,\bar{N}+N_j) - R_0 L_j - F^L\left[L_j\right] \\ -V^L(L_j) - wN_j - F^N\left[N_j\right] - V^N\left(N_j\right)\Big) \end{Bmatrix} \tag{2.3}
$$

where U is a twice-differentiable quasi concave utility function with positive and negative first- and second-order derivatives, respectively. Four land strategies are possible for households owning some land. The optimality conditions for these land strategies are summarized in Table 2-1.[3]

The four land strategies in Table 2-1 are also illustrated in Figure 2-3 in the case of linear variable transaction costs, no fixed transaction costs, and

TABLE 2-1 Optimality conditions for alternative land strategies

	Land strategy	*N optimal*	*L optimal*	*Optimality conditions*
1	Pure landlord	$N^* < N^1$	$L_j{}^* = -L_s{}^* = -\bar{L}$	$pq_L(0,N^*) < R_s = R_0 + V^L_{Lj}$
2	Landlord-cultivator	$N^1 < N^* < N^2$	$-\bar{L} < -L_s{}^* < 0$	$pq_L(\bar{L} - L_s{}^*, N^*) = R_s = R_0 + V^L_{Lj}$
3	Pure owner-cultivator	$N^2 < N^* < N^3$	$L_j^* = 0$	$pq_L(L,N^*) = R_0,\ R_s < R_0 < R_b$
4	Tenant-cultivator	$N^* > N^3$	$L_b{}^* > 0$	$pq_L(L + L_b{}^*, N^*) = R_b = R + V^L_{Lj}$

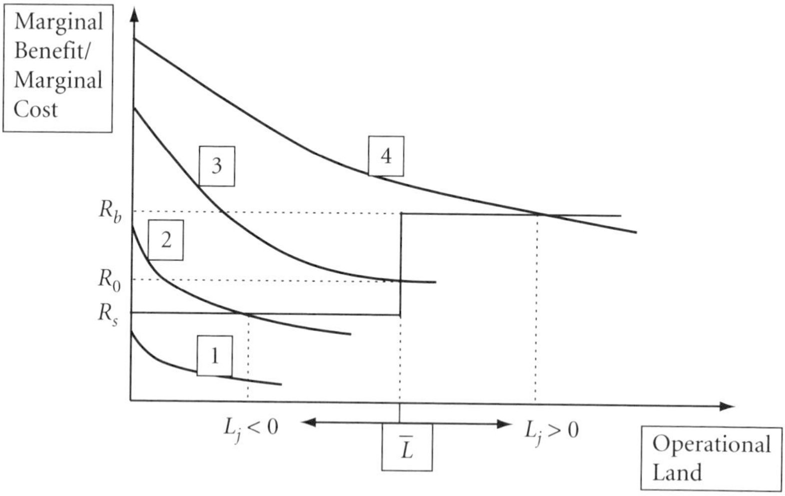

FIGURE 2-3 Alternative land strategies

optimal level of use of the nonland factor. Because of the variable transaction cost, total cost of renting in is R_b, which is the sum of market rent (R_0) and variable transaction cost. Thus, renting in is chosen only if the marginal product of land is sufficiently high (see case 4 in Figure 2-3). On the other hand, net rent revenue per unit of land is R_s, which is lower than R_0 by the transaction cost. Renting out part of the land will occur if the marginal product is sufficiently low (see case 2) and renting out all land will be chosen if the marginal product is very low (see case 1). Autarchy is the best choice if the marginal product curve cuts the vertical portion of the "effective" renting cost curve (see case 3).

Figure 2-3 illustrates the following points: (1) a linear variable transaction cost is sufficient to bring about nonparticipation in the land rental market, (2) transaction costs lead to higher land productivity on land operated by tenants than on land operated by landlord households, and (3) a reduction in the transaction costs should lead to a reduction in the share of households not participating in the land rental market and in the land productivity difference between landlords' owner-operated land and tenants' owner-operated and rented in land. This would imply that the distance between R_b and R_s would become shorter and this would reduce the probability of a household choosing land strategy 3 (pure owner-operator). Thus, in a heterogeneous sample of households, a smaller share of them would fall in this category.

We can also show that in the case of linear variable transaction costs the adjustment coefficient will be equal to –1 for households that operate in the market. A significant deviation from –1 is therefore not necessarily a sign of transaction costs *per se* but indicates a deviation from linear functional form

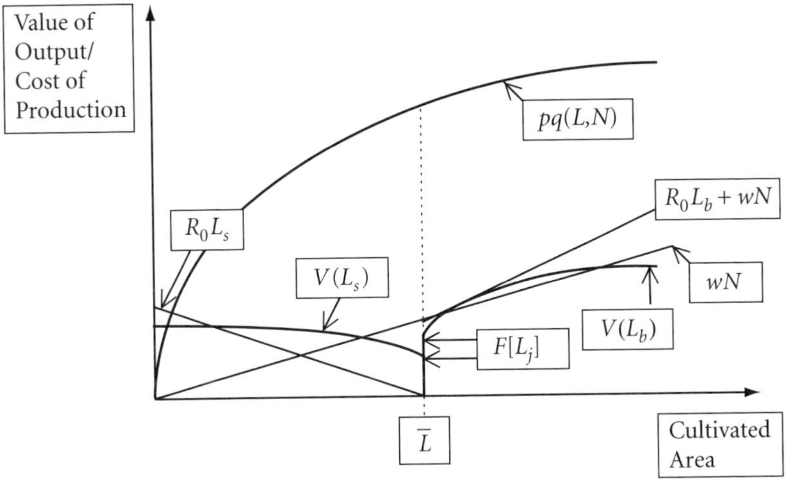

FIGURE 2-4 Fixed and concave variable transaction costs in the land rental market

(see Tikabo et al. [2008] and the appendix at the end of this chapter for a proof). If transaction costs are convex (marginal transaction costs are increasing), it can be demonstrated that the absolute value of the adjustment coefficient will be less than 1 and will be zero or larger than 1 (switch or "jump" adjustment) when transaction costs are nondecreasing but concave (marginal transaction costs are decreasing). This can be confirmed from the fact that the total cultivated area is insensitive to small changes in the land endowment.

Now is a good opportunity to resume our discussion on the nature of transaction costs. These costs involved in the conducting of land-market transactions include search costs in relation to finding willing and suitable contract partners and suitable land to rent in or out in the case of landlords and tenants. Transactions costs also include negotiation costs related to the formulation of contracts and eventual disputes that may occur, and enforcement costs related to payment and land management. The spatial dispersion of land and potential contract partners and the fact that farms themselves may be fragmented in several plots implies that the transaction costs tend to depend on the sizes of and distances to plots. Absentee landlords therefore face much higher fixed costs related to monitoring the land management of their tenants, but these would also depend on the frequency of such monitoring and whether the landlords can enlist a local person to do this for them. Sharecropping contracts have their own challenges, as the tenant may have an incentive to shirk. In order to avoid this, the landlord may have to monitor more closely and more frequently, or the landlord may contract only with tenants that he or she trusts, or perhaps the landlord will use some punishment mechanisms like threat of eviction. In Figure 2-4 we illustrate the case with fixed and concave

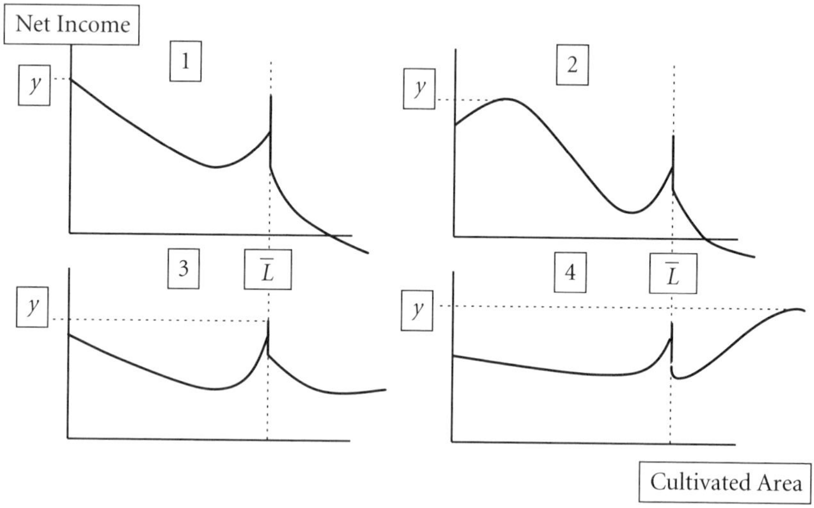

FIGURE 2-5 Choice of land strategies when there are fixed and variable transaction costs

nondecreasing transaction costs in the land rental market. For simplicity we have assumed a linear function for the nonland factor costs (wN) that are increasing in land because factors of production are complements. We refer to equation (2.2) and the related text for an explanation of the different curves in the graph. Aggregation of these functions according to equation (2.2) gives the net income after deduction of transaction costs and other costs. Households that want to rent in or rent out land (move from $\overline{L}$ in Figure 2-4) face initial fixed transaction costs, $F^L[L_j]$, that cause an entry barrier and a jump in the cost function both if they want to rent in and rent out land.

Such nonlinear transaction costs may cause the second-order conditions for optimality not to be satisfied and the optimal outcome may be a corner solution. The four alternative outcomes illustrated in Table 2-1 and Figure 2-3 for the case of linear transaction costs may also be illustrated in the case of nonlinear transaction costs like in Figure 2-5, where the curves illustrate the net income (y) after deduction of transaction costs. Households will choose the land strategy that maximizes their net income, and this may imply that they (land strategy 1) rent out all their land, (land strategy 2) rent out part of their land, (land strategy 3) farm all the land themselves and do not participate in the land rental market, or (land strategy 4) farm their own land and rent in additional land. This graph also illustrates the case with nonconvexities that one may observe "jumps" in and out of the market (between extreme points) based on relatively small changes in prices, costs, or production functions, for example.

Since crop production involves multiple carefully timed operations, particularly in rain-fed agriculture, and possibly requiring transportation of

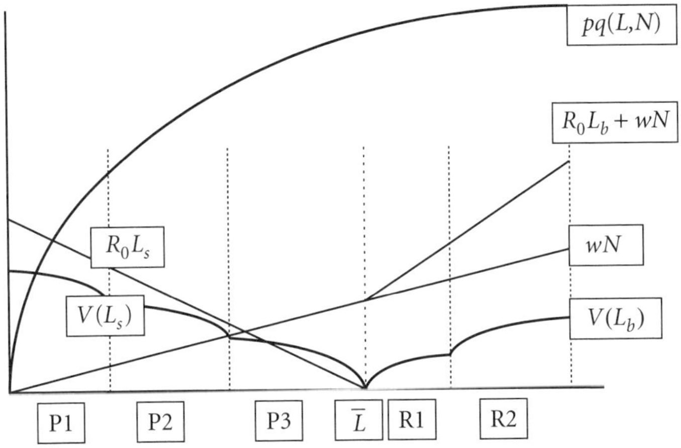

FIGURE 2-6 Land fragmentation and plot-specific transaction costs in the land rental market

animals or equipment (like oxen and plows), the distance and size of plots are likely to affect total production and transaction costs. Thus, in areas with fragmented and dispersed plots, land transactions involving whole plots will be the more common because transaction costs for renting in or renting out one extra plot are likely to be concave. We illustrate this in Figure 2-6 for a household owning three plots (P1, P2, and P3) and considering renting in two additional plots (R1 and R2). Again, we refer to equation (2.1) and the related text for explanation of the curves. The "bumpy" curves ($V(L)$) are the accumulated nonlinear transaction costs for rented in and rented out plots. After subtraction of these transaction costs to arrive at net income (following equation 2.1), it is likely that maximum net income is at a "break point" for the transaction cost curve, and such break points occur at every whole plot point for areas cultivated or rented out or in.

A basic underlying assumption of the analysis so far is that there are constant returns to scale in production. This has been a generally accepted assumption in tropical agriculture but it does not hold when mechanization is important. Economies of scale may also become relevant in tropical agriculture where oxen are used for plowing and farm sizes are very small, like in Ethiopia. This influences the degree to which it is advisable to subdivide farms and farm plots. Spatial dispersion of small plots also adds to transportation costs of going to and from plots with farm equipment and animals when repeated and multiple operations are required and timing is crucial for productivity. This would also provide an economic rationale for regulations introduced in the form of minimum farm sizes and promotion of consolidation of holdings through voluntary exchange of plots.

Review of Evidence from Africa

If factors of production are reallocated to more efficient producers through land-market transactions, this should lead to a less skewed distribution of relative factor ratios across farms after the land transactions than before the transactions. The initial distribution of land and nonland factors of production and the level of transaction costs would determine the degree of factor adjustment in the factor markets. We review some of the empirical evidence from Africa regarding the efficiency of land markets and resulting factor allocations.

Kevane was the first to test the allocative efficiency model of Bliss and Stern in Africa, in western Sudan, where he found the adjustment in the land rental market to be efficient (Kevane, 1997). In a study in Eritrea, Tikabo et al. (2008) found significant transaction costs in the land rental market causing only partial adjustment. They also found significant asymmetries in the market and selection bias on the tenant side of the market, where different factors appear to affect the decision of whether to participate or not than the degree of participation. Teklu and Lemi (2004) analyzed rental market behavior on the landlord side in southern Ethiopia using a Heckman selection model. They also found an adjustment coefficient that indicated significant transaction costs (significantly less than 1) in relation to land adjustment and also significant selection bias. Households with more labor force and with more healthy heads of households were found to rent out less land.

Pender and Fafchamps (2006) also tested for transaction costs and the allocative efficiency of the land rental market using a dataset from 1994 in Arsi in eastern Ethiopia. Using a Censored Least Absolute Deviations (CLAD) model for the test, they did not find significant inefficiency in the land rental market. However, this could be due to the relatively small sample size (only 72 uncensored observations) and the fact that the CLAD model is demanding with respect to sample size. They found that oxen endowment significantly affected land renting activity, showing that the oxen factor market was imperfect and that the land rental market helped to equalize factor ratios across households.

Holden and Ghebru (2006) analyzed the adjustment in the land rental market in Tigray in northern Ethiopia with a focus on the role of kinship and trust in relation to access and participation in the market. Building on Fafchamps (2004), they developed a theoretical model for entry and participation in the land rental market as a landlord or as a tenant. The basic idea is that trust among contracting partners builds up gradually over time and may depend on kinship relationships (Holden and Ghebru 2006). They found a high share of kinship contracts in Tigray and found that tenants had easier access to the land market in communities with a high share of kinship contracts, but adjustment appeared to be no more efficient for contracts among kin than among non-kin. Overall, they found significant transaction costs in the market and indications of nonconvex transaction costs creating entry barriers while those participating in the market appeared to face lower transaction costs.

Hypotheses

HB1: There are significant transaction costs in the land rental markets causing allocative inefficiency.

HB2: High fixed costs related to farming due to indivisibility of factors of production (such as oxen for plowing) in combination with small farm sizes increase the importance of the land rental market to achieve efficiency in production.

HB3: High transaction costs in the land rental market (and labor market) explain the inverse farm size–land productivity relationship because land-rich large farmers do not rent out land to land-poor small farmers up to the point where the ratios of land to nonland factors are equalized.

HB4: Land fragmentation and nonconvex transaction costs in relation to land cause land transfers to be lumpy (primarily transactions of whole plots).

HB5: Kinship contracts enhance allocative efficiency in the land rental market.

HB6: There is asymmetric adjustment in the rental market related to sharecropping contracts and higher transaction costs (entry barrier) on the tenant side than on the landlord side.

Methodological Issues

Efficiency of Land Markets. Determinants of market participation using households as observation units in (land sales and) land rental markets should be analyzed with probit models separately on the two sides of the market. Censored tobit models should be tested against double-hurdle models to assess whether access and degree of participation should be disentangled. Test for selection bias in relation to participation should be made if a two-stage model is found to be appropriate. Rationing may cause other factors to be at work in relation to whether to participate or not and how much to participate in the market. For example, trust, reputation, and endowment characteristics may matter for whether and which households to rent out land to or rent in land from, whereas the amount of land rented may depend on the farm characteristics, such as farm size, fragmentation (number of plots and their sizes), location of each of the plots related to the home of the tenant and the landlord (transportation costs and monitoring costs), and the technology. Appropriate censored regression models should be applied to assess the degree

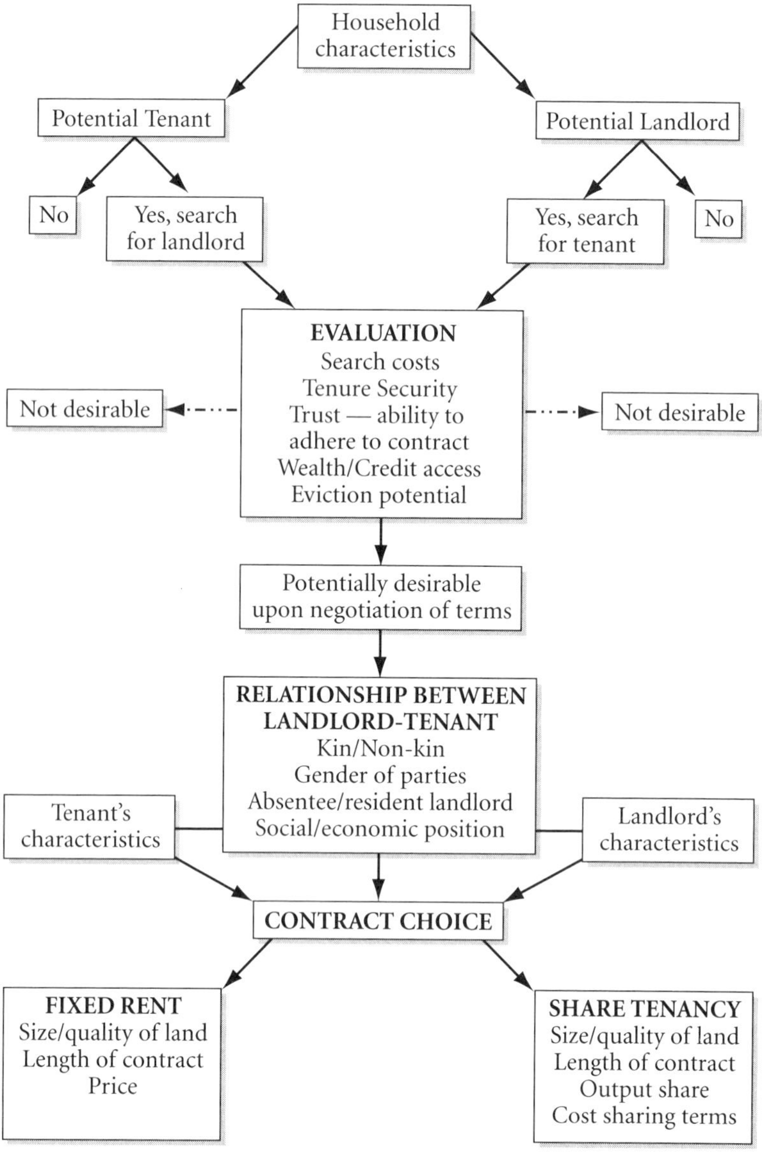

FIGURE 2-7 Conceptual model of rental contract evaluation and choice

of participation in land rental markets if there is no selection bias. In case of selection bias, appropriate selection models should be used. Assessment of transaction costs in the land market should be made by observation of the size of the parameters for owned farm size (or land endowment). Symmetry tests across the two sides of the market should be performed (see Bliss and Stern 1982; Skoufias 1995; Tikabo et al. 2008).

Choice of Land Rental Contract, Poverty, and Production Efficiency

The decisions of households regarding whether to participate in the land rental market as tenant or landlord, and if they decide to participate, what type of contract to choose, are illustrated in Figure 2-7. The figure shows the different types of transaction costs that potential tenants and landlords face when considering to operate in the market. Their ability to find a good partner may depend on their own characteristics, availability of information, trust, reputation, social networks, and security of tenure, which again may depend on past and present land policies, cultural norms, and power structures.

Contract choice may depend on imperfections in credit markets, risk and insurance markets, the relative bargaining power and preferences of landlords and tenants (e.g., degree of risk aversion), their mutual trust and monitoring ability, and their relative poverty. The transaction costs analysis in relation to land rental markets that we developed in the previous section needs to be integrated with the analysis of contract choice and production efficiency. Pender and Fafchamps (2006) developed a model that includes land, labor, and capital as factors of production. They assume the capital market to be missing and that transaction costs are prevalent in the markets for land and labor. The model also includes risk and sharecropping or fixed rent contracts allowing the output share of the tenant to be flexible. The landlord faces a transaction cost in relation to enforcing the tenant's effort and the amount of capital input the tenant applies. Transaction costs are also nondecreasing in the amount of land that is rented out and nonincreasing in the share of the output that the tenant receives.

Some of the basic implications of the Pender and Fafchamps (2006) model are (1) fixed-rent contracts are preferred over share tenancy by landlords when transaction costs of monitoring tenancies are sensitive to the share of the output going to the tenant (the implication is that sharecropping may be more common among relatives and where the level of trust is high); (2) the usual Marshallian inefficiency results unless the tenant's effort can be fully enforced by the landlord; and (3) a positive marginal cost of monitoring the tenant's effort will cause the landlord to restrict area rented out such that at equilibrium the tenant will have higher land productivity than the landlord (i.e., due to allocative inefficiency of the land market). This result is similar to what we found with variable transaction costs in a market with fixed-rent contracts (see Figure 2-3). This seems, therefore, to be a general result that can be tested. However, sharecropping has two contrasting effects on the efficiency of tenant production as compared to that of landlords. The presence of sharecropping may further restrict the amount of land rented out as compared to fixed renting leading to a decrease in allocative efficiency (and higher land productivity on tenants' land (Pender and Fafchamps 2006)), but this may be partly offset by Marshallian inefficiency on rented in land of share tenants. Irrespective of which effect dominates, with Marshallian inefficiency, the owner-operated plots of tenants tend to have higher productivity than the sharecropped plots.

We take this analysis further by assessing how kinship relationship (trust), threat of eviction, gender, bargaining power, and coercion may affect land productivity and productivity differentials of owner-operated and rented in and rented-out plots. Explanations for productivity differentials may be related to selection of partners in the rental market (assortative matching[4]); another may be selection of plots to rent out (which also affects the quality of plots that can be rented in by tenants); a third is related to whether contracts are voluntary or whether moral obligations (trust) and coercion play a role; and a fourth is related to the duration of contracts and the conditions for contract renewal that may discourage or enhance tenants' efforts and input use. Finally, we assess how contracts may contribute to or substitute for imperfections in credit and insurance markets that may be particularly severe for poor households living in risky environments.

The Economics of Share Tenancy

The inefficiency view of share tenancy asserts that since the marginal return to the work effort for the tenant is proportionally less than the marginal product of his effort (conventionally called the marginal product of labor) under the share contract, the work incentives to the tenant are thwarted. In contrast, it is commonly accepted that the fixed-rent leasehold contracts—in which the tenant pays a fixed sum to the landlord and claims the residual—and owner-operated farming do not distort work incentives.

In order to clarify this argument rigorously, let us assume that the production function is characterized by

$$q = \theta f\left(L, N, I\right) \tag{2.4}$$

where q is output; θ represents the state of nature, which is influenced typically by the vagaries of the weather; and L, N, and I are land, work effort (or labor), and purchased input, respectively. Following Stiglitz (1974), θ is assumed to take multiplicative form for simplicity. Note that if uncertainty represented by θ is absent, work effort can be inferred by the landlord from the information on q, L, and I, so that there is no information asymmetry about the tenant's work effort. Also for simplicity, the amount of land is assumed to be determined by the landlord. The tenant's income (Y) is expressed by the following linear form:

$$Y = \alpha q + \beta - \gamma P_I I \tag{2.5}$$

where output price is normalized to be unity; α is output-sharing parameter, which is $(0 \leq \alpha \leq 1)$; β is a parameter representing the fixed payment that corresponds to a fixed-wage component if $\beta > 0$ and a fixed-rent component if $\beta < 0$; and γ is the cost sharing parameter ($0 \leq \gamma \leq 1$) and P_I is the price of purchased inputs.

An implicit assumption of the inefficiency view of share tenancy is that it is prohibitively costly for the landlord to observe and enforce the tenant's work effort, so that the tenant determines her or his effort to maximize her or his own utility without fear of punishment for shirking labor. Assuming that the tenant, who is a risk averter, maximizes the expected utility, which is a quasi concave function of income and labor (N = labor) (i.e., $EU = EU[Y, N]$), with positive and negative marginal utilities of income and labor, respectively, the first order-order optimum conditions can be shown as

$$\partial EU / \partial N = EU_Y \theta \alpha f_N + EU_N \leq 0 \tag{2.6}$$

$$\partial EU / \partial I = EU_Y \theta \alpha f_I - EU_Y \gamma P_I \leq 0 \tag{2.7}$$

where subscripts refer to the partial derivatives. Note that the specification of equation (2.7) implicitly assumes that the landlord can observe the amount of purchased input; otherwise, the landlord cannot accurately share $(1 - \gamma)$ of the cost. Indeed, if purchased inputs are unobservable for the landlord, cheating (e.g., resale) can take place (Hayami and Otsuka 1993a).

The landlord is supposed to maximize his or her utility with respect to α, β, γ, and L subject to equations (2.6) and (2.7) and the following reservation utility constraint:

$$EU \geq V \tag{2.8}$$

which assures that the expected utility to be obtained from the contract is at least as large as the given reservation utility (V) obtainable elsewhere. The optimum share parameter is likely to be greater than zero but smaller than one, because if α is zero there is no work incentive for the tenant, whereas if α is unity the risk-averse tenant must shoulder all the production risk (Otsuka et al. 1992). If the tenant is risk neutral and, hence, the risk sharing does not matter, α is chosen to be unity, so that the work incentive is undistorted.

In practice, like owner operators, the tenant engages not only in simple tasks, such as harvesting, but also in care-intensive activities, such as water control. In other words, the tenant's effort includes his or her conscientious efforts, which are not easily observable. Given a spatially dispersed and ecologically diverse production environment in agriculture, it will not be an easy task for landowners to supervise a tenant's effort, unless the field is located near to the landlord's residence or his or her own-cultivated fields. It is no wonder, then, that the inefficiency view of share tenancy is widely accepted.

This argument can be illustrated by Figure 2-8. For simplicity of exposition, we assume that labor supply, represented by the marginal rate of substitution (MRS) between income and labor (i.e., $-[EU_N/EU_Y\theta]$), is upward-sloping

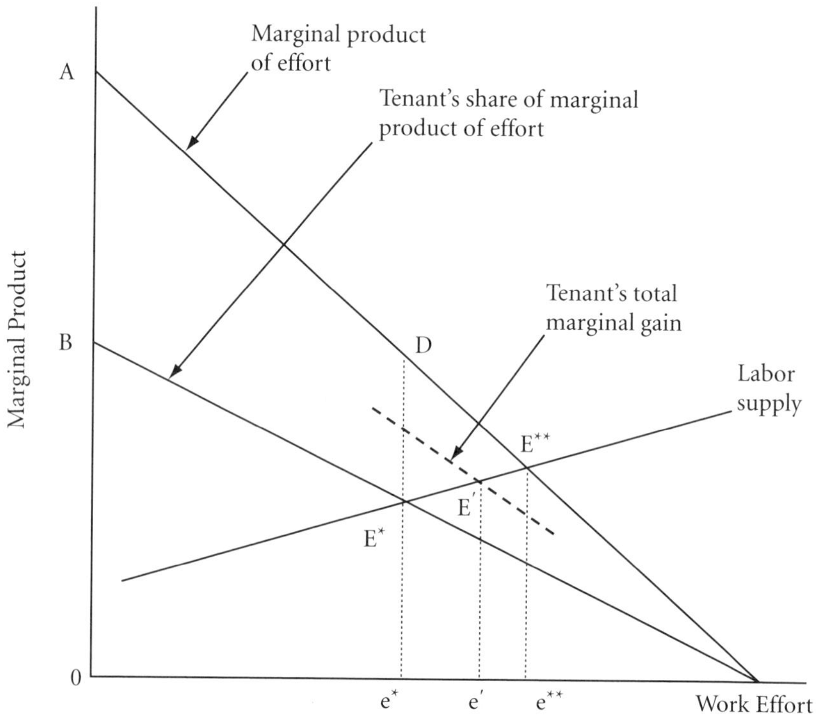

FIGURE 2-8 Determination of tenant's work effort under share tenancy

(see equation [2.6]). For a given amount of land and purchased input and a given output sharing rate, the share tenant applies his or her work effort up to e*, which corresponds to point E*, where his share of marginal product of effort is equated with MRS. Work effort under the share contract is lower than Pareto-optimum work effort, e**. Such optimum effort can be realized under fixed-rent tenancy ($\alpha = 1$) and owner operation, which ensure the equality between the marginal product of effort and the MRS at point E**. Thus, the inefficiency thesis predicts that work effort and, hence, yield per unit of land are lower for land under share tenancy than under fixed-rent tenancy and owner-operator farming.

Note that if a fixed-rent tenant is constrained by capital due to credit market inefficiency, the location of the marginal product curve may be lower due to the inadequate application of purchased input. On the other hand, if the cost of purchased input is shared between the tenant and the landlord, the application of purchased input may be higher under share tenancy than under fixed-rent tenancy. In this case, yield of a share tenant is not necessarily lower than that of a fixed-rent tenant.

The source of the incentive problem lies in the difficulty the landlord has in observing the tenant's work effort. If the work effort is observable and enforceable without cost, as assumed by Cheung's (1969) model of share tenancy, the landlord can specify and enforce the tenant's work effort at the desired level. In equilibrium, no inefficiency is shown to arise under share tenancy in this situation.

The assumption of costless enforcement is unrealistic, yet the assumption of prohibitively costly enforcement seems too restrictive. A model of Hayami and Otsuka (1992) introduces the monitoring function of a tenant's work effort, which posits that the probability of detecting a tenant's shirking is an increasing function of the degree of his or her shirking (i.e., the difference between the actual level of effort and the level specified by the landlord) and the landlord's supervision time. The efficiency of monitoring will depend on the landlord's ability (e.g., associated with his or her farming experience, size of landholding, and residential proximity to his or her tenant farms) and the technological characteristics of production (e.g., traditional versus new technology and food crops versus commercial crops). As long as the monitoring of a tenant's work effort is feasible but costly, as it certainly is, the equilibrium effort level will lie between e^* and e^{**} in Figure 2-8.

Whether the equilibrium effort level is closer to e^* or to e^{**} depends not only on the probability of detecting a tenant's shirking but also on the level of punishment. If there is no punishment, the share tenant will supply his or her effort up to e^*, even if the landlord specifies the effort level e^{**} in the contract. Thus, the threat of penalty must exist to enforce the terms of the contract effectively. A synthesis of the agrarian contract literature by Binswanger and Rosenzweig (1986) argues that the loss of reputation may play a role of punishment in agrarian communities, where people are long settled and know one another quite well. If a tenant's dishonest behavior is detected in such a community, not only will the tenant's current contract be terminated but the tenant will also lose his or her reputation as someone to contract with in the future. Thus, in equilibrium the landlord sets the terms and conditions of the contract so as to assure the tenant utility higher than his or her opportunity utility, so that the fear of losing lucrative future contractual opportunities will represent a heavy penalty on a tenant's shirking.

Johnson (1950) also argued that the threat of a landlord's refusal to renew the share contract could be an effective enforcer of share contracts in U.S. agriculture. If so, an increase in the tenant's work effort brings about not only marginal returns from increased production in the current period but also an additional gain associated with a lower probability of losing future contractual opportunities (see e' in Figure 2-8). A threat of eviction will be particularly effective if the reservation utility is very low, corresponding to households reliant on subsistence farming. In such a case, it may be possible that the tenant allocates more work effort per unit of land than the owner operator.

Review of the Literature

Determinants of Contract Choice. Ackerberg and Botticini (2000), in their study from Medieval Europe, suggest that imperfect capital market, risk, and fear of asset abuse favored spread of sharecropping. In their study in Tunisia, Laffont and Matoussi (1995) found that the working capital of tenants and landlords was important for contract choice. Boadu (1992) in his study of cocoa farms in Ghana found increasing frequency of sharecropping to coincide with higher risk in cocoa production. In Eritrea, Tikabo and Holden (2003) found that credit market imperfections and differences in level of wealth among landlords and tenants affected contract choice.

Contract Choice and Efficiency. Sharecropping caught the interest of the classical economists as an inefficient institutional arrangement (Marshall 1890). However, seminal contributions by Cheung (1969) and Stiglitz (1974) provided new light and renewed interest in analyzing the efficiency implications of sharecropping. A large theoretical literature has developed since then and many empirical studies have been carried out, especially in Asia. Many of the studies in Asia have been reviewed by Otsuka and Hayami (1988), Otsuka et al. (1992), and Otsuka (2007). They do not find any general evidence of Marshallian inefficiency and they explain this by the endogeneity of contract choice: Sharecropping may occur mainly in cases in which it works efficiently, but alternative contracts will be preferred in which sharecropping would be inefficient. Another problem is that many of the studies are flawed due to endogeneity bias and a failure to control for unobservable household and plot characteristics.

Still, there are some studies that identified significant Marshallian inefficiency, while using credible methodologies. The most famous among these is the study done in India by Shaban (1987), who found a 16% difference in yields on sharecropped and owner-operated land of the same tenants. He compared productivity on tenants' owner-operated and sharecropped-in plots while controlling for plot and household characteristics. Otsuka (2007), however, suggests that the Marshallian inefficiency found in India is due to policy distortions ("land-to-the-tiller" policies) causing sharecropping contracts to be short term and therefore inefficient.

Relatively few studies have been done on the efficiency of sharecropping in Africa. However, we will give a brief overview of the existing studies.Laffont and Matoussi (1995) found significant evidence of Marshallian inefficiency in a study in Tunisia. Due to financial constraints some tenants preferred sharecropping to fixed-rent contracts and such contracts were associated with lower land productivity.[5] One may therefore attribute the lower efficiency on sharecropped plots to the financial constraint rather than to the sharecropping contract *per se*. Instead, sharecropping may improve the allocative efficiency in the land rental market by relaxing the financial constraint.

There are three studies with the same dataset from Arsi in Ethiopia but with different theoretical and methodological approaches. Pender and Fafchamps (2006) in their empirical analysis failed to find evidence of higher yields on rented out plots of landlords than on their owner-operated plots, possibly due to the small sample size that would have required the yield difference to be as high as 40%. Similarly, they also failed to find evidence of Marshallian inefficiency when comparing input levels and output values of owner-operated and rented in plots of tenant households, using household fixed effects to control for unobservable household characteristics. Here also, the small sample size of only 64 fixed-rent plots, 31 sharecropped plots, and 149 owner-operated plots may be the main reason for lack of significant differences.

Earlier tests of the same dataset by Gavian and Ehui (1999) found total factor productivity to be lower on contracted land (fixed rent, sharecropped, borrowed or gifted land) than on owner-operated land, but they found no significant difference in input use between the different contract types (Gavian and Ehui 1999). Using a stochastic frontier approach on the same dataset, Ahmed et al. (2002) found significant technical inefficiency on sharecropped land but not so on land under fixed-rent contracts. This demonstrates that the use of different methods applied to the same dataset can lead to different conclusions. However, the household fixed effects approach used by Pender and Fafchamps (2006) must be considered to be the preferred approach, as it controls for unobservable heterogeneity of households. The requirement of the stochastic frontier models that only strictly exogenous variables should be included in the first stage, while input use and crop choice were included in this stage in the analysis by Ahmed et al., variables that are clearly endogenous, may give reason to be cautious when interpreting the results. It is also not clear why technical inefficiency should be higher on rented land than on owner-operated land, since Marshallian inefficiency from theory should lead to a reduction in input use (allocative inefficiency rather than technical inefficiency).

Among other studies in Ethiopia, Holden et al. (2001) found land productivity to be significantly higher on rented land than on owner-operated land. However, this study combined rented in and rented out plots and therefore did not correct for unobservable household characteristics. The finding could, however, possibly be explained by significant transaction costs, as suggested by the model of Pender and Fafchamps (2006).

In Eritrea, Tikabo and Holden (2004) found land productivity to be significantly higher for owner-tenants than for pure owner-operators and owner-landlords, showing that land is transferred from less efficient to more efficient land users (Tikabo and Holden 2004). The fact that the productivity differential persisted after participation in the land rental market indicates that there are significant transaction costs in the land rental market. Using household fixed effects, they also found significantly higher land productivity on plots with cost-sharing contracts than on owner-operated and pure sharecropped plots of tenants, whereas there was no significant difference between the owner-operated and sharecropped plots.

In a study in the Amhara region in Ethiopia, using community-level data Benin et al. (2005) found a decline in the use of fixed-rent contracts to be associated with increasing landlessness and preference for sharecropping among the young landless households.

Contract Choice and Equity. The study by Laffont and Matoussi (1995) in Tunisia revealed that tenants had higher levels of working capital than landlords. They also found that the amount of working capital of landlords and tenants affected contract choice. Tenants with more working capital and landlords with less working capital were more likely to have fixed-rent contracts rather than sharecropping contracts. This may indicate that sharecropping is a response to credit constraints on the tenant side, whereas fixed-rent contracts may be a sign of distress rentals due to credit constraints on the landlord side. Asset wealth of tenants appeared to play a less significant role in contract choice than the financial liquidity constraints.

Tikabo and Holden (2003) found that wealthier tenants and poorer landlords were more likely to have fixed-rent contracts, and less wealthy tenants and more wealthy landlords were more likely to have cost-sharing contracts, in line with the credit constraint hypothesis. However, if poverty also is correlated with risk aversion such that the wealthier are less risk averse, this could also explain the same pattern of contract choice and matching in the tenancy market. The study was unable to distinguish between these possible alternative explanations.

Bellemare (2006) studied and tried to explain the presence of reverse share tenancy in Lac Alaotra in Madagascar, where it was common that poorer landlords rented out land to wealthier tenants. He tested three hypotheses to explain why poor landlords preferred sharecropping: (1) tenure insecurity (asset risk); (2) price risk (risk sharing); and (3) limited liability. All three may explain the occurrence of sharecropping when landlords are risk aversive and tenants risk neutral. However, he also shows that sharecropping would be optimal if the more plausible assumption that both landlord and tenant are risk aversive is introduced. The theoretical analysis also ignores the role of liquidity constraints of landlords and tenants as an alternative explanation of the dominance of sharecropping.

Hypotheses

HC1: Contract choice is determined by the bargaining power of the contract parties such that those who are more powerful get a larger share of the benefits.

HC2: Contract choice is associated with resource pooling such that it balances factor ratios for mutual benefit.

HC3: Sharecropping and cost-sharing contracts are a response to missing or imperfect credit markets.

HC3a: Fixed-rent contracts are chosen by landlords who need urgent cash (distress rentals) when credit markets do not work. (Note that fixed rents are paid in the beginning of crop season.)

HC3b: Fixed-rent contracts are preferred by tenants with access to credit.

HC3c: Sharecropping contracts are chosen by tenants who want to delay payment of land rent until after harvest.

HC4: Fixed-rent contracts dominate in areas with better market access and where cash cropping is common because of better access to cash and credit in these economies.

HC5: Sharecropping dominates in food-insecure and drought-prone areas (response to risk).

HC6: Sharecropping tends to be chosen when the landlords are able to enforce efficient land management by tenants.

HC7: Sharecropping is associated with Marshallian inefficiency.

HC8: Sharecropping among kin is inefficient because such contracts may be driven by other concerns than production efficiency.

HC9: Poor female landlord households face higher transaction costs in the land rental market and are associated with inefficient land use.

HC10: Opportunities for contract renewal enhance input use on sharecropped land and this counteracts Marshallian inefficiency.

Methodological Issues

Determinants of Contract Choice. Ordered or multinomial logit and probit models may be used to identify the factors associated with contract choice in which these factors are identified from theory.

Contract Choice and Land Use Efficiency. The effect of contract choice on land use efficiency is best tested with plot-level data of households with rented in and owner-operated plots with control for land quality and unobservable household effects (Shaban 1987).

Contract Choice and Equity. The correlation between household endowment and market access characteristics and contract choice can be tested using probit, ordered probit, or multinomial logit models while attempting to control for endogenous matching.

Impacts of Land Markets on Land Tenure and Land Management

A conceptual model of the link between land markets, investment in land improvements, and productivity is given in Figure 2-9. In some cases, the effects

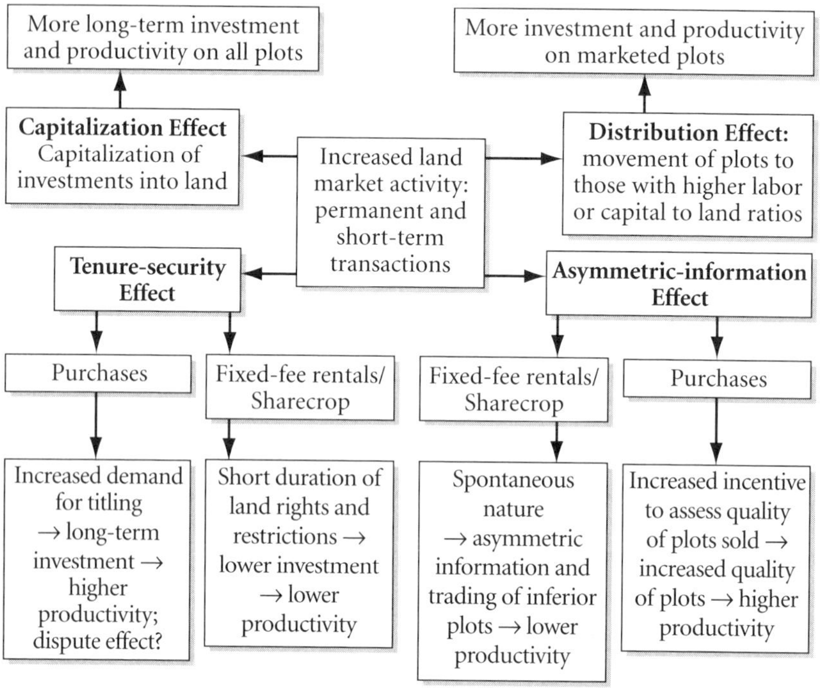

FIGURE 2-9 Conceptual model—land markets, land tenure, and land management

of markets are similar, whether the markets are for more permanent purchases of land or shorter-term rentals. In other cases, the effects of land purchase and rental markets are expected to differ. Four main pathways are identified. The first, *capitalization effect* (in the upper left quadrant), is that land-market activity will lead to greater capitalization of the value of investment into land values and will thus provide increased incentives for longer-term investment, notably visible and lumpy investments. The incentive effect would apply to all land—assuming that all land would potentially be traded—in a community. The second, *distribution effect* or allocative efficiency effect (in the upper right quadrant), is that land markets may facilitate the movement of land resources to households with higher labor/land or capital/land ratios and thus increase investment and efficiency (and reduce the variance of yields across farms). The movement of leased land may also be coupled with input credit from the landlord, in the case of sharecropping. This effect was discussed earlier in the context of allocative efficiency. The third effect, the *tenure-security effect* (lower left quadrant), posits that farmers who purchase land will have more incentive to title this land and privatize it away from local customary tenure arrangements that may impinge on individual decision making. On the contrary, rented plots are typically of limited duration and carry with them significant limitations in terms of investments that can be made by the tenant. The fourth effect, the *asymmetric-information effect*, hypothesizes that where investments are not capitalized into land prices and where land quality is difficult to assess, the types of land traded may be of inferior quality. This is likely to be much more the case for land rentals than land purchases because the spontaneous nature of rental markets, along with the low price, reduces information gathering efforts. If this is the case, then one would expect lower investment on traded plots, especially rented plots.

So it is obvious that the overall effects of land markets on land investment are complicated. The effects of improved markets for land purchases are mainly positive (although we have not discussed noneconomic reasons for acquiring land, some of which may lead to overaccumulation and inefficient use of land). The case of land rental markets is much less clear, with some positive and some negative linkages.

Review of the Literature

Land Markets and Tenure Security. Land rights are strongly linked to the mode of acquisition of land. Land that is purchased is often found to have more individualized rights than land that is acquired through other permanent means (e.g., gift, inheritance, etc.). But this is not always the case, as households with inherited land may often claim the right to sale—for example in Kenya (Migot-Adholla et al. 1994a) and in Uganda (Place 1995). Purchased land is less likely to be subjected to competing claims than land inherited (Quisumbing et al. 2003). Just as in the case of inherited land, purchased land is hardly ever expropriated from rural households by traditional authorities (an exception being conflicts arising between indigenous groups and land-acquiring migrants, notably in the West African tropics). However, purchased land is more secure than inherited land from claims by extended family members (Place and Otsuka 2000, 2002 for Uganda). Land acquired from within the family that is seen as underutilized can be subjected to heavy pressures for use rights, especially in land-scarce areas. But all land, even that which is purchased, may be subject to secondary rights of family or community members, unless investment in fencing is undertaken. Such rights include those of grazing during the dry season, the collection of grasses, or the picking of fruits from naturally growing trees (Baland et al. 2007).

The breadth and duration of rights are much inferior on land rented or borrowed compared to those on land permanently acquired (case studies in Bruce and Migot-Adholla [1994] and Platteau [1995]). Otsuka and Place (2001) observed that tenants and other farmers with weak property rights may attempt to increase their security of tenure over such parcels through lengthy cultivation or investments. Such behavior may be encouraged by customary institutions that confer security of tenure based on investment (Otsuka and

Place 2001). In eastern and southern Africa, tenancy arrangements are made strictly as temporary. However, in West Africa, landholders allowed tenancies to evolve into permanent land transfers in areas where perennial tree crops predominate (Quisumbing et al. 2003). This was undertaken in order to attract migrants to provide much needed labor and to increase land productivity and income for both the landlord and tenant.

Although households and individuals may claim to possess a range of secure rights to land, that may not be sufficient to forestall unwanted and costly land disputes from arising. Studies of land disputes in rural Africa are surprisingly rare. The data on the presence of land conflicts have not been well summarized and are mixed in terms of the frequency and severity of disputes. In a study of four countries (Rwanda, Kenya, Ghana, and Uganda), disputes were found to vary in degree of frequency—from fewer than 10% of parcels in Rwanda and Ghana to 25% to 30% in Uganda and Kenya (Bruce and Migot-Adholla 1994). Place (1995) found that land disputes were common throughout much of Uganda, Zambia, and Burundi, with the percentage of households ever having been involved in a dispute ranging from 21% in Zambia to 43% in Burundi. Nearly 20% of Burundian households reported that they had lost land in a dispute. Similarly, Deininger and Castagnini (2006) found that the extent of land disputes among households in five different regions of Uganda was high and the adverse impact was large, affecting about one-third of households.

Land Markets and Land Management. Considerable evidence shows that tenants do not make long-term natural resource management investments (e.g., Place and Hazell 1993). An exception to this is the longer-term tenants who invest in tree crops in West Africa (Hill 1963; Berry 1975; Quisumbing et al. 2003) or the multigenerational kibanja tenants in central Uganda (Kisamba-Mugerwa and Barrows 1989). These tenants differed from the more common types in Africa in that the expected duration of occupation is long, even matching that of landowners in some cases.

With respect to purchased land, there have not been many empirical studies of the direct relationship with investment in land improvements. Research has instead focused on the effects of land title, land rights, or land size on investments, all of which can vary between purchased land and inherited land. Quite a number of studies have found some relationships between land rights and land investments (e.g., Place and Hazell 1993 for Kenya, Rwanda, and Ghana; Deininger and Jin 2006 for Ethiopia), but these did not hold for all types of investments and other studies do not find such similar effects. There is even less effect between land rights and short-term input use, such as fertilizer (e.g., Holden and Yohannes 2002 for Ethiopia; Pender et al. 2004 for Uganda). Studies of the impact of land title on investment in land have generally not found strong direct linkages (e.g., Carter et al. 1994). Exceptions are

Deininger et al. (2007) and Holden et al. (2008), who find significant investment impacts of land certification in Ethiopia. In terms of the size of parcel, in which case purchased land is often larger than land that is inherited or rented, several studies have found a positive relationship between parcel or farm size and investments, canal maintenance in Somalia (Roth et al. 1994), tree crops in Uganda (Roth et al. 1994), and incidence of tree planting in Zambia (Place 1995).

Hypotheses

HD1: Longer-term land investments are lower on rented land than land acquired through purchase.
HD2: Short-term investments are preferred on rented land.
HD3: Rented land is associated with less sustainable land management practices, including conservation structures.
HD4: Tenants invest more in rented land in order to strengthen their property rights and probability of contract renewal.
HD5: HD1 will be offset by the allocative efficiency effect of land markets so that production efficiency on rented land is not lower than on permanently held land.
HD6: Land with land disputes is associated with lower use of farm inputs.
HD7: Poor female-headed households are more likely to suffer from land disputes.

Methodological Issues

Land Markets, Input Use, Conservation, and Investment. The impact of land markets on land investments and input use is appropriately tested using household and plot-level data to assess differences on rented, purchased, and inherited land (e.g., use of fertilizer, manure, conservation technologies, planting of perennials, etc.). Care needs to be taken to account for the endogeneity of land acquisition—for example, the purchase of land may be highly related to wealth levels of households (which in turn is likely to strongly influence investment). In addition, consideration should be paid to the relationships between different types of land investments and hence the formulation of an appropriate structural model.

Overview of Our Research

With our empirical studies in Ethiopia, Kenya, Malawi, and Uganda, we would like to make comparisons across these countries by drawing on the similarities and differences between these in their histories and basic characteristics.

First, all four countries can be characterized as having high population density and thus facing land scarcity, which is one of the basic requirements for land markets to play a role. Second, land distribution is much more skewed in Kenya, Malawi, and Uganda than in Ethiopia, which may have crucial consequences for the relationship between land markets and poverty. Third, land sales markets are prohibited in Ethiopia and in customary land in Malawi, whereas recent land reforms have opened land sales in Uganda. This permits looking at the interactions between land sales and land rental markets in Kenya and Uganda. Fourth, Malawi has interesting internal contrasts between the patrilineal areas in the north and the matrilineal areas in the south, which may have important implications for the functioning and role of land markets under the different inheritance systems. Fifth, Ethiopia does not have a colonial history, like the other countries have, but a history with feudalism with rich landlords and poor tenants in the south of the country until 1975, whereas in the north of the country there was a more communal (rist) land tenure system. The north of the country is also more semi-arid and drought prone, which may affect the nature of the land rental markets. Sixth, most farming systems in Ethiopia use oxen for plowing while hand hoe–based systems dominate in the other countries, and this may affect the elasticity of substitution among factors of production and the need for a land market to enhance allocative efficiency.

Drawing on these differences, we will discuss a number of broader research questions and hypotheses that we may be able to answer and test to some extent. These questions include

1. Do land markets have an important role to redistribute land from land-rich to land-poor households in countries with an in-egalitarian land distribution, but this is not the case in countries with more egalitarian land distribution?
2. Do land rental markets play a more important role in countries where land sales markets are prohibited than in countries where they are not?
3. Does formal legalization enhance, and traditional (customary) land rights inhibit development of land markets, and can land markets undermine the rights of people with only customary rights?
4. Is sharecropping more common in food-insecure areas and in high-trust societies?
5. Are land rental markets more active in areas with oxen-based cropping systems than in hand hoe–based cropping systems?
6. Are land markets good for poor women and female-headed households?
7. What policies are most relevant to think of to enhance the role of land markets related to improving production efficiency, investment, and poverty reduction under the observed variations in the countries of study?

We will discuss these research questions at the end of Chapter 14.

NOTES

1. Migot-Adholla et al. (1990) note that it is difficult for a Kikuyu from a particular lineage to acquire land in the area of another lineage.
2. Although their models pertain only to land rental markets, land sales markets could similarly be examined using the same methodology.
3. For simplicity we have assumed that fixed transaction costs are zero here.
4. Positive assortative matching implies that partners with similar characteristics prefer each other and negative assortative matching implies that partners with different characteristics prefer each other.
5. The study did not control for possible systematic differences in land quality between rented plots under different contracts and owner-operated plots.

APPENDIX CHAPTER 2

The Kuhn-Tucker conditions for the utility maximisation problem (equation (1) become (A1) wrt land rented in or out, and (A2) for the hiring in or out of the non-land factor when four regimes are possible for each factor: 1) all is rented out, 2) a part is rented out, 3) non-participation in the factor market, and 4) the factor is rented in

$$\left\{\begin{array}{l} U_y\left(pq_L(\bar{L}+L_j,N)-R_0-V^L{}_{Lj}\right)\leq 0,\left\{\begin{array}{c} L_j>0 \textit{ if } j=b \\ -\bar{L}\leq L_j<0 \textit{ if } j=s \end{array}\right\} \\ \text{and} \\ \left[U_y\left(pq_L(\bar{L}+L_j,N)-R_0-V^L{}_{Lj}\right)\right]=0 \text{ for } L_j>0 \text{ and } -\bar{L}<L_j<0 \\ (\bar{L}+L_j)\left[U_y\left(pq_L(\bar{L}+L_j,N)-R_0-V^L{}_{Lj}\right)\right]=0 \text{ for } -\bar{L}\leq L_j<0 \\ U_y\left(pq_L(0,N)-R_0-V^L{}_{Lj}\right)<0 \text{ for } L_j=-\bar{L} \\ \text{and } U_y\left(pq_L(\bar{L},N)-R^*\right)=0 \text{ for } L_j=0, \text{ where} \\ R_s=\left(R_0+V^L{}_{Lj}\right)|_{Lj<0}<R^*<\left(R_0+V^L{}_{Lj}\right)|_{Lj>0}=R_b \end{array}\right\} \quad \text{(A1)}$$

$$\left\{\begin{array}{l} U_y\left(pq_N(L,\bar{N}+N_j)-w-V^N{}_{Nj}\right)\leq 0,\left\{\begin{array}{c} N_j>0 \text{ if} j=b \\ -\bar{N}\leq N_j<0 \text{ if } j=s \end{array}\right\} \\ \text{and} \\ \left[U_y\left(pq_N(L,\bar{N}+N_j)-w-V^N{}_{Nj}\right)\right]=0 \text{ for } N_j>0 \text{ and } -\bar{N}<N_j<0 \\ (\bar{N}+N_j)\left[U_y\left(pq_N(L,\bar{N}+N_j)-w-V^N{}_{Nj}\right)\right]=0 \text{ for } -\bar{N}\leq N_j<0 \\ U_y\left(pq_N(L,0)-w-V^N{}_{Nj}\right)< \text{ for } N_j=-\bar{N} \\ \text{and } U_y\left(pq_N(L,\bar{N})-w_0\right)=0 \text{ for } N_j=0, \text{ where} \\ w_s=\left(w+V^N{}_{Nj}\right)|_{Nj<0}<w_0<\left(w+V^N{}_{Nj}\right)|_{Nj>0}=w_b \end{array}\right\} \quad \text{(A2)}$$

A simplified illustration of the comparative statics analysis follows. To avoid excessive mathematics we look at the case of a tenant and ignore transaction costs in the non-land factor market.[1] We then get the following model:

$$\underset{Lj,N}{\text{Max}}\, y=pq\left(\bar{L}+L_j,N\right)\text{-}F\left[L_j\right]\text{-}R_0L_j\text{-}V\left(L_j\right)\text{-}wN$$

A closer look at the alternative assumptions for the functional form of the transaction costs has the following implication:

$$V_{LjLj} = \left\{ \begin{array}{l} = 0, \text{ linear variable transaction cost} \\ > 0, \text{ increasing marginal variable transaction cost} \\ < 0, \text{ decreasing marginal variable transaction cost} \end{array} \right\} \text{ if } L_j > 0$$

The Kuhn-Tucker conditions for the decision to participate as a tenant and how much to participate in the land rental market become (assuming away the fixed transaction costs that otherwise would require a third equation to be included to ensure non-negative overall returns to the participation in the rental market for land):

$$\frac{\partial y}{\partial L_j} = pq_L - R_0 - V_{Lj} \le 0, L_j \ge 0, \left(pq_L - R_0 - V_{Lj} \right) L_j = 0$$

$$\frac{\partial y}{\partial N} = pq_N - w$$

Total differentiation yields;

$$\begin{pmatrix} pq_{LL} - V_{LjLj} & pq_{LN} \\ pq_{NL} & pq_{NN} \end{pmatrix} \begin{pmatrix} dL_j \\ dN \end{pmatrix} = \begin{pmatrix} -q_L & 0 & -pq_{LL} & 1 \\ -q_N & 1 & -pq_{NL} & 0 \end{pmatrix} \begin{pmatrix} dp \\ dw \\ d\bar{L} \\ dR_0 \end{pmatrix}$$

The comparative statics results are as follows;

$$\frac{dL_j}{dp} = |H|^{-1} \left(-pq_L q_{NN} + pq_N q_{LN} \right) > 0$$

$$\frac{dL_j}{dw} = |H|^{-1} \left(-\left(pq_{NL} \right)^2 \right) < 0$$

$$\frac{dL_j}{d\bar{L}} = |H|^{-1} \left(-pq_{LL} q_{NN} + pq_{NL}{}^2 \right) < 0$$

$$\frac{dL_j}{dR_0} = |H|^{-1} \left(pq_{NN} \right) < 0$$

It is easy to see that the functional form of the variable transaction costs in the land rental market does not affect the sign of the responses in area rented as long as the Hessian remains positive. However, a closer inspection of $\frac{dL_j}{d\bar{L}}$ yields

$$\frac{dL_j}{d\bar{L}} = \frac{-pq_{LL}q_{NN} + pq_{NL}^2}{\left(pq_{LL} - V_{LjLj}\right)q_{NN} - pq_{NL}^2}$$

Whether it is larger or smaller than –1 hinges on whether $pq_{LL} - V_{LjLj} \lesseqgtr pq_{LL}$. The response will be equal to –1 if the variable transaction costs are linear ($V_{LjLj} = 0$).

$\frac{dL_j}{d\bar{L}} > -1$ if $V_{LjLj} > 0$, that is when marginal transaction costs are increasing.

$\frac{dL_j}{d\bar{L}} < -1$ if $V_{LjLj} < 0$, that is when marginal transaction costs are decreasing

The same result holds if we look at the landlord side, just remember that L_j is negative for landlords. The same result also holds if we introduce transaction costs in the market for the non-land factor. We only get the following expansion:

$$\frac{dL_j}{d\bar{L}} = \frac{-pq_{LL}\left(pq_{NN} - V_{NjNj}\right) + p^2q_{NL}^2}{\left(pq_{LL} - V_{LjLj}\right)\left(pq_{NN} - V_{NjNj}\right) - p^2q_{NL}^2}$$

If there are nonconvexities in both markets (decreasing marginal transaction costs) it is more likely that the requirement of the Hessian to be positive will not hold. With production functions that are near linear this is also more likely to be the case.

NOTES

1. In reality there must then exist another non-land factor of production that we for simplicity of notation have suppressed, for which there is no market. This is necessary to create a rationale for adjustment in the land rental market with transaction costs

PART II

Land Markets, Allocative Efficiency, and Poverty

Who Has Access to Land?

CHAPTER 3

Land Rental Markets

Transaction Costs and Tenure Insecurity in Rural Ethiopia

KLAUS DEININGER, DANIEL AYALEW ALI, AND TEKIE ALEMU

In situations in which the distribution of land ownership is different from the operationally optimal level, reallocation mechanisms—such as rental markets—that transfer this factor to its most productive use play a key role in increasing total production and enhancing economic efficiency. Historically, when most of the land was cultivated under rudimentary traditional technology and nonagricultural labor markets were virtually nonexistent, the majority of such land transactions occurred between large landlords, often absentees, and small tenants with few alternative opportunities, essentially compensating for a relatively skewed distribution of land ownership. Increasing complexity of the agricultural production processes implies that imperfections in other markets, which are frequent in rural areas of developing countries, will affect the nature and direction of rental contracts. Moreover, with economic development, a number of other disequilibrating factors such as changes in the demographic structure, nonagricultural demand for land, and an array of other reasons for land transactions—from producers' desires to take advantage of new opportunities from off-farm employment to temporary migration caused by differential agricultural ability—have started to complement, and in many cases substitute for, the "traditional" model of land leasing from large to small landlords (Sadoulet et al. 2001). As a result, land rental markets have become quite active even in countries such as China and Vietnam where, in view of the prevailing highly egalitarian ownership distribution of land, the traditional model of land leasing would lead one to expect limited levels of land rental activity.

As is the case for many developing countries, land in Ethiopia remains to be a key productive asset and the main source of income for a large segment of the population. Consequently it is not only the level of land rental activity but also the productive use of the land thus transferred that is of relevance in such circumstances. In fact, economists have long been concerned about the efficiency implications of rental market operations with a view toward identifying policy options that have the potential to make everybody better off by helping improve the efficiency of outcomes. A key motivation for doing so has been that, in addition to the limited investment incentives conveyed by short-term rental contracts, share tenants will, in any given season, receive only part of their marginal product and thus have limited incentives to supply effort. A large number of theoretical and empirical studies have aimed to identify conditions—mainly in terms of other market imperfections—under which sharecropping may be a rational strategy and to quantify the productivity implications, in a given production period or in terms of longer-term investment, of such an arrangement as compared to owner-cultivation.

Although such analysis can provide useful insights and has led to a number of policy interventions, it takes the renting decision as given and makes few predictions about the extent to which rental markets' responses to exogenous changes are optimal and fully utilize available opportunities. Nonetheless, exploring the renting decision explicitly is of policy relevance because the incidence of exogenous changes to which land markets need to respond has sharply increased and is likely to become of even greater importance with economic development. Moreover, the transaction costs that need to be incurred in renting out land may well affect land owners' decisions (e.g., adopting new technology and engaging in off-farm employment, etc.), and could thus have implications for growth in the nonfarm economy.

In this chapter, we use a large panel dataset from Ethiopia that is significantly more detailed than what had been available earlier to explore the extent to which land markets in this country help farmers capitalize on existing and emerging opportunities. Ethiopia is of interest because, given the scant endowment of land that is under significant threat of degradation and high levels of climatic risk, making the most productive use of its available land resources is important to escape the threat of starvation and dependence on food aid. At the same time, it is difficult to imagine sustained growth of rural incomes in the country without a significant transfer of labor out of the agricultural sector—something that will be contingent on well-functioning land rental markets.

We use the friction model developed by Rosett (1959) and Skoufias (1995) to assess whether land rental markets allow households to attain their desired level of cultivated land irrespective of their endowment. Our results, consistent with what is reported in Chapters 4, 5, 6, and 7 of this book, point toward the existence of large amounts of friction that prevents landowners from using rental markets to adjust to their optimum size.

In the remainder of the chapter we discuss features of the rural economy in Ethiopia to motivate the study and relate it to the conceptual framework in Chapter 2 by providing the background for some of the analytical hypotheses. Next, we describe the sample and use descriptive statistics from our data to generate the basic information used at household level. We then present results from the econometric analysis and discuss potential implications. Finally, we conclude by relating our results to the literature and the other chapters in this book and highlight areas for future research.

Background and Conceptual Framework

To motivate our analysis, we focus on key features of land tenure in Ethiopia, in particular the relative scarcity of land, the high risk associated with agricultural production, and the fact that land is state owned and relatively equally distributed. Despite recent efforts to increase tenure security by demarcating individual land holdings in a very participatory manner, land policy remains ambiguous, and we show how this may limit individuals' and households' desires and abilities to participate in nonagricultural activities. This is followed by a discussion of the conceptual framework.

Distinguishing Features of Ethiopian Land Tenure

Prior to 1975, land in Ethiopia was largely concentrated in the hands of absentee landlords, which was particularly true in the southern parts of the country. Tenure was highly insecure, arbitrary evictions posed a serious threat, and many lands were severely underutilized. The land tenure system of the time was characterized by great inequality, which, through its negative impact on production and investment, not only affected productivity but was also considered to have been the most important cause of political grievances that eventually led to the overthrow of the imperial regime. The Marxist government, soon after ousting the imperial regime, transferred ownership of all rural land to the state for distribution to cultivators with only user rights through local peasant associations (PAs). This was soon followed by a massive collectivization drive of peasant agriculture. However, contrary to the case of China (Dong 1996), the collectivization program was not associated with high levels of infrastructure investment. As a result, most of the agricultural land remained rain-fed and subject to degradation and soil erosion (Kebede 2002). Transferability of land through lease, sale, exchange, or mortgage, among others, was prohibited, and inheritance was possible only within immediate family members.[1] In spite of all these restrictions, however, farm households were involved in informal rental arrangements—in the form of largely share but also fixed rent tenancy—at the very likely risk of losing land allocated to them by the PAs. The ability to use land was contingent on proof of permanent physical residence, thereby preventing migration from rural areas. More important, tenure security was undermined by the ability of peasant associations and other authorities to redistribute land, an ability that was in some cases used for political ends (Ege 1997).

Although it is committed to a free-market philosophy and enacted a number of new pieces of legislation, the government that took power in 1991 did not fundamentally alter the ambiguity surrounding many aspects of land policy in the country. This is illustrated by the fact that the right of every Ethiopian who wants to engage in agriculture to receive a piece of land for free is anchored in the 1995 Constitution. Although it is recognized that this provision may conflict with other goals such as the desire for greater tenure security and well-functioning land rental markets, a 1997 federal proclamation, formulated to set out the overall legal framework, essentially transferred responsibility for enacting laws regarding the nature of land rights, their transferability, and matters of land taxation to regional governments (FDRE 1997).[2] Given their relevance for the topic at hand, we briefly review the regional policies with respect to administrative land redistribution and land rental markets.

Despite repeated public announcements against land redistribution, regional positions are not at all clear-cut even beyond the use of eminent domain for establishment of irrigation with due compensation for land and improvements, which is clearly enunciated in the southern Nation, Nationalities and Peoples region (SNNPR), Amhara, and Oromia (ANRS 2000; ONRS 2002; SNNPR 2003). Proclamations in Amhara and the SNNPR explicitly provide for other types of redistribution but make them the responsibility of local communities and require that they be supported by research that it will not lead to land fragmentation and does not adversely affect land productivity. No clear statement on redistribution is included in the Tigray proclamation (TNRS 1997). Proclamations in all regions highlight that the rights of persons who obtain a nonagricultural job or are absent from the village longer than a certain period will fall back to the village.

Households' use rights now include the rights to lease land to a third party (i.e., rental is officially allowed) (Pender and Fafchamps 2006), but regions impose various restrictions on the amount of land that can be leased or the duration of contacts (Nega et al. 2003, Beyene 2004). Oromia, the SNNPR, and Tigray allow farmers to rent out up to 50% of their holding and stipulate maximum contract terms of 2 to 5 years for traditional and 10 to 20 years for modern farming renters (ONRS 2002; SNNPR 2003; TNRS 1997). Current policy in Amhara, the region where our data were collected, is more progressive, as a recently passed proclamation allows leasing of land for up to 25 years irrespective of the technology used (ANRS 2006). Since all our data are from before 2005, the changes brought about by this new development will not be reflected, but suffice it to say that even after its passage, a large amount of uncertainty remains. For example, a clause that would have allowed mortgaging was removed from the draft proclamation before its enactment. More important, regulations are silent on whether long-term rental contracts entered into by a person who subsequently finds a nonfarm job will have to be honored or not.

The hypothesis that there is lack of clarity on legal provisions is supported by our data and other surveys. On average, 47% of the households in our data expect to gain or lose land due to administrative land redistribution or reallocation in the next five years. This could limit investment incentives (Deininger and Jin 2006) and, especially if authorities are believed to interpret rental as a signal that land is no longer needed, the propensity to engage in land rental markets will be reduced. With weak judicial institutions to interpret regulations or adjudicate in case of dispute, it is not surprising to find land conflict to be frequent and bureaucratic discretion relating to land matters pervasive (Rahmato 2003).

To reduce such conflicts and respond to widespread tenure insecurity, the government has recently launched an ambitious program of land certification.[3] This program has made rapid progress, covering more than 6 million households by late 2005 in a highly participatory and low-cost process that includes delineation of borders and issuance of land use certificates in the name of husband and spouse. Evidence on farmers' willingness to pay for land registration (e.g., 84% in a sample from Amhara) (Solomon 2004) suggests that this is indeed a step in the right direction. However, unless land ownership and the ways in which it can be exercised (including land transfers) are defined more clearly and the scope for interference is limited, the impact of such initiatives may well be significantly below the potential.

Optimality of Adjustment

It is important to explore whether land rentals impose significant transaction costs (e.g., for search of partners and contract enforcement) and, with the pervasive imperfections in labor and draught animal markets, to explore if land rental will allow households to attain their "optimum" land size (Bliss and Stern 1982; Bell and Sussangkarn 1988). Studies of this issue in Ethiopia yield contradictory results. Some argue that land markets will, at best, facilitate partial adjustment to the desired level of cultivated land (Teklu and Lemi 2004; Holden and Ghebru 2006; Chapter 4 by Ghebru and Holden); others are unable to reject the hypothesis of frictionless adjustment in rental markets, albeit based on a very small sample that is limited to households renting out land (Pender and Fafchamps 2006).[4] To allow for existence of a regime of nonparticipation over which the amount of net land rented in will not respond to small changes in land endowments, we use a friction model that explicitly makes allowance for a regime of nonparticipation in land rental markets.

Empirical Specification. Based on the basic methodology developed in Chapter 2, the reduced form of the net land rented in equation is a function of the endowment of land and nonland resources.[5] Depending on households' participation status, net rented in area will take positive, zero, and negative values. An appropriate econometric model that also allows for asymmetry is given by

$$L_i = \begin{cases} -\alpha_n + \beta_n' Z_i + \varepsilon_i & \text{if } \varepsilon_i < \alpha_n - \beta_n' Z_i \\ 0 & \text{if } \alpha_n - \beta_n' Z_i \le \varepsilon_i \le \alpha_p - \beta_p' Z_i \\ -\alpha_p + \beta_p' Z_i + \varepsilon_i & \text{if } \varepsilon_i > \alpha_p - \beta_p' Z_i \end{cases} \tag{3.1}$$

where L_i is net rented in area; subscripts n and p denote the rental market participation status of household i with negative net rented in area and positive net rented in area, respectively; the parameters α_n and α_p include the constant term, a_0, of the basic (latent) model in Chapter 2 and the unobserved transaction costs that relate the latent to the observed variable; Z_i is a vector of household-level endowments of land and nonland resources; β is a vector of coefficients that combine the slope of the adjustment function and the marginal responses of desired cultivated area to the underlying input endowment; and ε_i is the unobserved household specific error term. In the empirical application, Z_i consists of household-level variables that are expected to determine a household's participation in the tenancy market and its desired amount of net rented in area. The list includes the demographic structure of the household (number of male adults, number of female adults, number of children, and the age and sex of the household head); key assets (number of bulls and oxen, value of other livestock owned, and number of rooms in dwellings); the area of land and its quality; and the literacy level of the head of the household as a proxy for human capital endowments. This formulation implicitly assumes that households' endowments with land, labor, and oxen are exogenously determined. This is a reasonable assumption in our context, because endowments of land use rights have been assigned by the village some time ago and land cannot be sold, oxen markets are highly imperfect, and instantaneous adjustments in family size are impossible. District dummies and a time trend are included to capture differences in social and infrastructure endowments that affect agricultural productivity, prices, access to markets, and time-varying factors that affect rental market functioning.

The parameters of the model are estimated using maximum likelihood (ML) under the assumption of an identically, independently, and normally distributed error term with mean zero and constant variance σ^2 (Skoufias 1995). The ML estimator is consistent and asymptotically efficient. However, it is not possible to distinguish the threshold values related to transaction costs from the intercept terms, implying that the magnitude of the estimated intercept terms does not have a straightforward economic interpretation.

The econometric model in equation (3.1) can be used to test a range of hypotheses regarding the role of transaction costs and market imperfections in explaining the performance of the land rental market. First, ascertaining that the coefficients of owned land area equals to –1 provides evidence for efficient and complete adjustment toward desired cultivated area through the land rental market, whereas its rejection suggests that there is significant friction preventing full adjustment in this market (Skoufias 1995). Second, we can test the symmetry of the adjustment function for households who rent in land and rent out land through a straightforward test of the equality of the coefficients of land area owned across the two sides of the rental market as well as the symmetry of all or subsets of the estimated coefficients of the two sides of the rental market. Finally, we can make inferences regarding nontradability of other inputs by testing whether the estimated coefficients of these inputs are statistically different from zero (Pender and Fafchamps 2006).

Who Participates in Land Rental Markets?

The panel data point toward a relatively high level of land rental activity, together with an almost exclusive reliance on sharecropping whereby the vast majority of transactions occur between relatives. Moreover, the supply of land to the rental market is concentrated among female-headed households who lack sufficient oxen for draught power, suggesting that, contrary to what is observed in other countries, land rental actually contributes toward concentration of operational holdings as compared to land ownership.

Data and Household-Level Evidence

We use data from three rounds of a longitudinal survey of rural households in the Amhara region of the Ethiopian highlands that comprise production information during the main agricultural season (*meher*, i.e., September to February) of the 1999, 2001, and 2004 agricultural years. In each of these periods, the survey, which was implemented by the Department of Economics of Addis Ababa University in collaboration with the Department of Economics of Göteborg University, covers about 1,520 randomly selected households from 12 villages (*kebeles*) in six districts (*woredas*) in South Wollo and East Gojam zones of the Amhara region. Among others, data for every production year include detailed information on household resource endowments, crop production inputs and outputs at the plot level, and demographic characteristics of households and their participation in land, labor, and credit markets. In the last round, additional modules, including detailed information on household consumption and rental contracts, were added.

Table 3-1 provides household-level information for the whole sample, and separately for those leasing out, leasing in, and remaining in autarky. A first observation of interest is that land markets are very active, with almost half the households (46%) participating in rental markets and only 54% remaining in autarky. The fact that our sample is relatively balanced (20% leasing out land and 26% leasing in) also highlights that the vast majority

TABLE 3-1 Household level descriptive statistics

Variable	*Total*	*By Land Market Participation Status*				
		Rented out		*Autarky*	*Rented in*	
Household characteristics						
Female head dummy (%)	15.37	39.91	***	11.82	3.29	***
Head literate (%)	41.31	29.13	***	41.24	51.14	***
Age of head in years	48.38	51.61	***	48.71	45.11	***
No. of rooms in dwellings	1.95	1.74	***	1.95	2.13	***
Owned area in ha	1.08	1.04		1.11	1.06	
Number of adult males	1.49	1.04	***	1.57	1.70	***
Number of adult females	1.46	1.33	***	1.51	1.46	*
No. of dependents (< 15 or > 60)	2.72	2.26	***	2.74	3.05	***
Agricultural production						
Net land rented in	0.01	–0.68	***	0.00	0.60	***
Cultivates any land (%)		63.3				
Owned cultivated area in ha	0.94	0.36	***	1.10	1.05	
Total cultivated area in ha	1.09	0.36	***	1.11	1.65	***
Share of good quality land (%)	38.43	36.67	*	39.44	37.72	
Owns bulls or oxen (%)	67.3	23.7	***	72.4	91.2	***
No. of bulls/oxen (for owners)	1.19	0.39	***	1.21	1.78	***
Value of other livestock in ('000) Birr	1.20	0.57	***	1.20	1.69	***
Distribution over districts (*Woredas*)						
Woreda1 (Machakel), East Gojam (%)	15.44	13.07		12.56	23.38	
Woreda2 (Gozamin), East Gojam (%)	15.91	16.86		16.64	13.61	
Woreda3 (Enemay), East Gojam (%)	15.98	21.33		12.26	19.54	
Woreda4 (Tehuledere), South Wollo (%)	20.57	19.04		23.16	16.35	
Woreda5 (Tenta), South Wollo (%)	15.96	13.53		20.03	9.32	
Woreda6 (Habu), South Wollo (%)	16.14	16.17		15.34	17.81	
Number of observations	4268	872		2301	1095	
Percentage of sample	100	20.4		53.9	25.6	

Source: Own computation from AAU/UG Amhara Panel Survey.

Note: Unit of observation is a household in a given year. Except for negligible attrition (less than 3% between rounds), this implies that the same household contributes three observations.

Stars indicate that means for the rent out and rent in group, respectively, are significantly different from those for the group remaining in autarky. * significant at 10%; ** significant at 5%; *** significant at 1%.

of transactions are between households in the same village, contrary to what is often observed in other samples.[6] A second observation of interest is that, 40% among those renting out (compared to 12% among those in autarky and 3% among those renting in) are households headed by females. The fact that more than 70% of all female-headed households rent out land implies that one important function of land rental is to transfer land from resource-poor female-headed households to resource-rich male-headed ones.[7] Third, with no significant differences in area owned between those renting in and renting out land and those in autarky (1.04, 1.06, and 1.11 hectares, respectively), the operation of land markets leads to a concentration of operational land holdings compared to the land ownership structure—different from what is observed in most other countries. In fact, the Gini coefficient is only 0.40 for land ownership but 0.47 for operational holdings. This is low compared to the high levels of inequality found in Latin American countries, but it is higher than that in China, where rental markets equalize the operational land distribution.[8] Finally, although one would expect to find those renting in land to have better access to family labor, it is rather surprising that renters also have significantly higher endowments of human and physical capitals (literacy and education levels), contrary to most other countries, which suggests that there may be barriers to entry into nonfarm activities.

Plot-Level Characteristics

Information on output and input as well as source of traction and, for rented ones, the type of transaction partner for the 23,215 plots (out of which slightly more than one-third [8,562] belong to owner-cum-sharecropper households who both own and rent in land) is provided in Table 3-2.[9] The first 4 columns are based on the whole sample, columns 5 through 7 refer to the subsample of owner-cum-sharecroppers, and column 8 indicates statistical significance of the difference in means between owned and sharecropped plots. The table illustrates that sharecropping is the predominant contractual arrangement (more than 90% of rented plots transferred under sharecropping rather than fixed rent). The table also indicates that most transfers are limited to close kin: 60% of plots were exchanged between relatives or in-laws, 30% between neighbors or members of the same village association,[10] and only 10% between unrelated individuals. About 21% of plots are transacted in rental markets, a share that is significantly lower than that of households involved in rental transactions due to the fact that rental transactions normally involve only part of a household's endowment. With close to 50%, the share of output accruing to landlords is comparatively high. Although detailed data were not collected on the nature and formality of contracts, the mean rented in plot in the sample had been in the possession of the current tenant for about four years, consistent with the average duration of contracts of three years as reported in the third-round data.

TABLE 3-2 Plot characteristics for different samples

	Total Sample				*Owner-Cum-Sharecroppers*			
	All	*Own cult.*	*Shared in*	*Shared out*	*All*	*Own cult.*	*Shared in*	
Relation w. partner (TP)								
TP relative (%)	48.42		49.10	49.84	48.90		48.90	
TP in-law (%)	10.65		11.02	11.19	10.64		10.64	
TP neighbor (%)	17.41		18.47	18.43	15.48		15.48	
TP member of VA (%)	12.55		10.97	10.64	14.12		14.12	
TP unrelated (%)	10.97		10.45	9.90	10.86		10.86	
Harvest share received (%)	51.11		53.06	49.18	53.29		53.29	
Plot characteristics								
Plot size in ha	0.24	0.24	0.28	0.27	0.26	0.25	0.28	***
Years of possession	16.34	17.72	4.01	17.91	14.06	16.81	3.89	***
Good soil quality (%)	38.03	38.91	35.40	33.86	36.24	36.66	34.71	
Medium soil quality (%)	40.85	40.66	40.51	42.98	40.96	41.19	40.11	
Poor soil quality (%)	20.98	20.28	24.09	23.02	22.67	21.99	25.18	***
Flat land (%)	69.28	68.56	73.44	69.92	68.75	67.52	73.27	***
Gently sloped (%)	26.74	27.47	23.31	25.35	27.33	28.27	23.87	***

Steeply sloped (%)	3.86	3.84	3.18	4.64	3.91	4.20	2.85	***
Irrigated (%)	3.58	3.93	2.34	2.23	3.34	3.74	1.87	***
Source of traction								
Own pair of oxen (%)	58.18	56.30	69.88		67.34	66.76	69.48	**
Oxen exchange for labor (%)	15.52	16.73	7.19		10.93	12.05	6.78	***
Oxen sharing (%)	19.24	19.29	20.13		18.85	18.26	21.01	***
Gift/support (oxen party) (%)	5.26	5.72	2.17		2.00	1.95	2.17	
Oxen rental (%)[a]	1.29	1.22	0.53		0.70	0.76	0.50	
Hoe (%)	0.80	0.92	0.13		0.39	0.49	0.07	**
Pair of oxen days per ha	52.27	51.71	56.20		56.47	56.82	55.21	
Number of observations	23215	18292	2234	2307	8562	6740	1822	

Source: Own computation from AAU/UG Amhara Panel Survey

Significance levels reported for t-tests of the equality of the means between owned and shared-in plots for the sub-sample of owner-cum-sharecroppers. ** significant at 5%; *** significant at 1%

[a]Rental includes provision of animals by the sharecropping partner or other arrangements.

Although traditional technology requires a pair of oxen, only about 60% of plots are cultivated with own oxen, and the considerably higher share of own oxen on shared in plots suggests that, as noted earlier, shortage of oxen could indeed be a key reason for households to engage in land rental. The lack of rental markets for oxen can be attributed to synchronic timing of activities and moral hazard that could lead renters to overuse and underfeed animals. Thus, the main source of traction power for about 30% of cultivated plots was either oxen-sharing (*mekenajo*) or exchange of labor for oxen services.

High Transaction Costs in the Land Rental Market

In this section, we will quantify the extent of rationing in the land rental markets (i.e., whether high transaction costs prevent producers from adjusting to their optimum holding size). Maximum likelihood estimates of the parameters for intensity of participation in tenancy market as presented in Table 3-3 facilitate a number of insights.[11] Most important, noting that the coefficient of owned land area corresponds to the slope of the adjustment function in equation (2.1), the positive (negative) and significant coefficients of owned area in the leasing out (in) equations suggest that, once households' endowments with fixed factors are accounted for, land markets facilitate an adjustment toward households' desired level of cultivated area.

The numbers in Table 3-4 illustrate that both the hypotheses of the symmetry of the coefficients on both sides of the market and, more importantly, full adjustment (i.e., the equality of these coefficients to 1 or –1) are strongly rejected. Although suppliers appear to be slightly less constrained than those demanding land through the market, estimated coefficients suggest that, on average, farmers are able to realize only 22% of the desired amount of land leased, which points toward large amounts of friction in the land rental market. This is in marked contrast to the study by Pender and Fafchamps (2006) that was based on a much smaller sample (161 observations) in which only leasing in was observed and that was unable to reject the hypothesis of perfect adjustment. However, the finding is in line with findings in Chapter 4 (Tigray, Ethiopia), Chapter 5 (Kenya), Chapter 6 (Malawi), and Chapter 7 (Uganda). To put this figure into perspective, note that in India, where a large number of policies impose *de jure* restrictions on land rental (Deininger et al. 2008), the corresponding coefficient was 78% (Skoufias 1995). This suggests not only that these policies were often honored in the breach, but it also highlights the surprisingly large magnitude of friction in the case of Ethiopia and the other countries that have been studied in this book. We interpret this as an indication that the transaction costs of land rental markets, many of which may be related to the policy environment as discussed earlier, prevent participation. Even for those able to overcome the obstacles to entry, the level of adjustment afforded by such markets is far from the desired one.

TABLE 3-3 Determinants of net land rented in: Maximum likelihood estimates

Variable	Rented Out	Rented In
Owned area (ha)	0.266***	–0.182***
	(4.66)	(3.46)
Proportion of good soil quality	–0.060	–0.008
	(0.96)	(0.14)
Number of dependents	–0.053***	0.065***
	(2.64)	(3.46)
Number of adult male	–0.081**	0.070**
	(2.56)	(2.40)
Number of adult female	–0.001	–0.043
	(0.03)	(1.25)
Number of bulls and oxen	–0.622***	0.284***
	(9.15)	(4.25)
Value of other livestock owned × 10^{-3} (Birr)	–0.091***	0.035**
	(2.94)	(1.99)
Number of rooms of the household	–0.026	0.073***
	(0.90)	(2.65)
Age of household head (years)	0.011***	–0.011***
	(6.00)	(5.20)
Female headed household	0.615***	–0.652***
	(7.42)	(5.86)
Household head can read and write	0.027	0.001
	(0.47)	(0.03)
Woreda2	–0.269**	0.298***
	(2.52)	(3.25)
Woreda3	–0.272***	–0.190*
	(2.62)	(1.94)
Woreda4	–0.784***	0.321***
	(6.07)	(2.68)
Woreda5	–0.982***	0.238**
	(7.34)	(2.08)
Woreda6	–0.462***	–0.074
	(4.00)	(0.69)
Time trend	0.098***	0.134***
	(3.43)	(4.49)
Constant	–0.608***	0.380***
	(3.53)	(2.67)
σ	1.011***	
	(9.50)	
Log likelihood	–4659.95	
Wald chi2(23)	146.51***	
Observations	4268	

Note: Coefficients for leasing-out (except constant and dummies) are multiplied by –1 for ease of interpretation.

Robust z statistics adjusted for clustering in parentheses. * significant at 10%; ** significant at 5%; *** significant at 1%

TABLE 3-4 Wald tests of Equality of coefficients on opposite sides of the land rental market

Hypothesis Tested	*Wald statistic chi²(r)*
Symmetry of all coefficients (except woreda dummies), r = 11	33.87***
Symmetry of subset coefficients (resource endowment: labor, land, oxen, assets), r = 8	33.42***
Owned area	4.05**
Owned area = −1, r = 2	246.55***
Good soil quality	0.77
Dependants	0.24
Male adult	0.08
Female adult	1.08
Oxen	22.25***
Other livestock	2.79*
Rooms	1.91
Age	0.11
Female headship	0.11
Literacy	0.17
Equality of intercept terms	
Woreda1[a]	0.91
Woreda2	0.81
Woreda3	9.07***
Woreda4	6.53***
Woreda5	16.94***
Woreda6	9.97***

* significant at 10%; ** significant at 5%; *** significant at 1%

[a]Woreda1 is the district included in the constant term (the reference category). The intercept term for the other districts is, therefore, the sum of the coefficients of the respective district dummy and the constant term.

A second finding of interest relates to the signs and significance levels of the coefficients on other factors, all of which should equal zero under the assumption of perfect markets.[12] The coefficient on ownership of draft animals is highly significant and positive (negative) on the demand (supply) side of the market, suggesting that households with more (less) animals are likely to rent in (out) more (less) land in line with the hypothesis that land leasing decisions help improve utilization of households' imperfectly tradable endowment with oxen. As might be expected, the extent of land transactions is significantly affected by female headship and the number of male adult members and dependents. This is due in part to the fact that (even though females participate in other activities such as weeding and harvesting) ox plowing is a male task and that it would be difficult for females to obtain the necessary male labor (plus oxen) input through the market.

The value of other livestock and the number of rooms in the household can be interpreted as a proxy for wealth and access to credit and working capital. Their coefficients are positive (negative) in the rent in (out) equation pointing toward credit market imperfections as another factor that may prevent households from leasing in land.

While the regressions suggest that land markets transfer land from older to younger households and thus substitute for administrative redistribution of land, households' literacy level is not significant. This suggests that, in contrast to other countries where off-farm labor markets are often found to be a driving factor underlying land market activity, higher levels of education do not increase the propensity of renting out land in our sample. Finally, intercept terms and the district dummies are significantly different from zero (Table 3-3), pointing toward differences in the extent of land rental market functioning across villages.

Conclusions and Implications

Study of land rental markets in Ethiopia is of interest not only because of the relevance of such markets for nonagricultural development but also because the country is characterized by an equal allocation of land but high levels of production risk lead to sharecropping as the most prevalent contractual form. At the same time, transaction costs imply that many households are rationed out from rental market participation or, even if they are able to participate, are not able to use rental markets to attain their optimum operational land holding. This is of relevance because, in rural areas all over the world, land markets that can adjust flexibly to the changes brought about by new technology, greater integration with marketing and processing, and better off-farm opportunities will affect the extent to which rural producers will be able to take advantage of these new opportunities. For example, lack of clarity in the policy as to whether land will be lost if the owner takes up nonagricultural employment may constitute a strong impediment to out-migration from rural areas (Rahmato 2003). Although they do not allow us to unambiguously identify the source of such friction, our results are not inconsistent with this interpretation.

The finding of large friction being associated with land rental market participation implies that a more in-depth discussion of factors preventing the use of land rental markets would be of great policy relevance. In conducting such analysis it would be useful to assess factors that prevent households from entering into land rental market transactions at all or supplying or demanding certain amounts of land (including renting out all of their land); contracting with certain types of partners (e.g., close kin); or restricting the duration of contracts.

If a large part of the observed friction in Ethiopia's land rental markets is indeed policy induced, more detailed analysis is needed—with the goal of identifying factors that prevent adjustment to households' desired level of operated land through rental markets. If a national sample were available, the fact that policies are formulated at the regional level and, at least on paper, have undergone considerable change over time could help separate the impacts of policy from those of other relevant factors. Indeed, our data suggest that, during the period of concern, households' perceptions of the likelihood of receiving or losing land through administrative redistribution have undergone significant changes, possibly as a result of the revision of proclamations (laws) or land right certification programs. Combining "objective" legislative provisions with households' subjective perceptions of their land rights could thus provide insights on how awareness of new legislation spreads through the population and the way in which it affects economic outcomes. Such analysis—which could, for example, help determine whether better dissemination of existing legislation or improvements in the legal framework would be most effective—is left for future research.

NOTES

1. The inhibitions imposed by the reform of the military regime were so massive that they even included the prohibition of using hired labor on own land.
2. A new federal rural land administration and use proclamation was enacted in 2005, partly to establish some consistency. Although its relevance for the analysis reported here is limited, since our data were all collected before its enactment, it does not appear to clarify many of the provisions that had remained ambiguous in earlier legislation.
3. "In order to protect the user rights of farmers, their land holdings should be registered and provided with certificate of user rights. In this regard, a guarantee *may* be given to the effect that land will not be re-divided for a period ranging from 20–30 years. Some regional states have already started this aspect of the land use policy and it is a step in the right direction" (FDRE 2002, *53*; italics added).
4. Related empirical studies show that imperfection on land rental markets leads to gendered productivity differentials; more specifically, it results in lower productivity on own-operated land of female-headed households than of male-headed households (Holden et al. 2001), and that female-headed households rent out their land to less productive tenants than do male-headed households (Bezabih and Holden 2006, Chapter 9 by Holden and Bezabih). See Chapter 2 of this book for a detailed review of empirical evidences from other African countries.
5. Using the basic model, Bliss and Stern (1982) applied ordinary least-squares to obtain regression estimators by pooling participants and nonparticipants together. This approach has been criticized because it ignores the nonlinear nature of the participation decision in the tenancy market due to the presence of transaction costs and the possibility of asymmetries across the two sides of the market (Bell and Sussangkarn 1988, Skoufias 1995). Such asymmetries may arise from

differences in the slope of the adjustment function and/or from differences in the marginal responses of desired cultivated area to the household's endowment of nonland resources.

6. Almost all the households involved in the land rental market exclusively participated in either renting out or in land with less then 1% of households reporting to simultaneously rent out and rent in land.
7. Excluding female-headed households from the analysis does not change the result that land is predominantly transferred from resource-poor to relatively resource-rich households. The descriptive statistics on the restricted sample can be obtained from the authors upon request.
8. The Gini coefficient for the ownership distribution of land is 0.92 for Venezuela and 0.87 for Brazil (Deininger and Squire 1998) as compared to a Gini of 0.36 for the distribution of use rights, which was further reduced through transfers, in three provinces of China (Deininger and Jin 2005).
9. Given the small number of plots transferred under fixed rent arrangements and the fact that these are not used in our subsequent plot-level analysis, descriptive statistics for these are not reported separately but can be obtained from the authors upon request.
10. Village associations consist of self-help groups and networks such as agricultural mutual aid groups, rotating savings and credit associations, rotating feast groups, burial associations, and so on, that are formed among persons of common kin, neighborhood, or faith.
11. The dependent variable is the amount of net land rented in (i.e., rented in minus rented out) in hectares. To facilitate interpretation, estimated coefficients in the rented out equation (except for the district dummies, the time trend, and the intercept term) are multiplied by –1. A positive (negative) coefficient in the leasing-out column can thus be interpreted as an increase (decrease) in the amount of land rented out in response to a positive change in a given explanatory variable.
12. Note that it is difficult to disentangle the effect of the adjustment function from the marginal responsiveness of the desired cultivated area to the key inputs.

CHAPTER 4

Factor Market Imperfections and Rural Land Rental Markets in Northern Ethiopian Highlands

HOSAENA GHEBRU AND STEIN T. HOLDEN

When markets are perfect, renting and buying land are closely related. Tenants pay rent to landlords and buyers pay equivalent amounts to banks as interest to loans. In a world, however, with asymmetric information (costly or unavailable) and imperfections in capital and various agricultural input markets, there are good reasons for farm households to prefer transferring land through land rental markets rather than through the formal land sales markets (Sadoulet et al 1998; Chapters 1 and 2).

Besides the acute market imperfections for nonland factors of production due to high transaction costs, the Ethiopian constitution of 1995 gives the land rental markets an exceptional role. The constitution which bans any land sales make the land rental market the core venue of land transfer among farmers alongside the alternative land transfer mechanisms of inheritance, land takings for investment purposes, and redistribution of communal lands

Rural land tenancy markets have received ample attention, both theoretically and empirically but much of the attention has focused on the efficiency aspects of alternative land tenure contracts (Otsuka et al. 1992; Singh 1989; Otsuka and Hayami 1988; Pender and Fafchamps 2006; Stiglitz 1974; Sadoulet et al. 1998; Gavian and Ehui 1999 and Ahmed et al. 2002). The majority of studies, however, fail to address the primary reasons behind the firsthand decision of households' participation in the tenancy market in an imperfect world.

In such a world, due to adverse selection and moral hazard problems, the tenancy market may not clear with a market clearing price. This may be due to high transaction costs for the landlords and may cause households who want to rent in land to be rationed out (Bell and Sussangkarn 1988). Bliss and Stern (1982) suggested that sharecropping involves moral hazard problems and a strong likelihood of rationing. Sharecropping implies absence of a market clearing price in the land rental market and this leads to an asymmetry between the two sides of the market (landlords and tenants) and to rationing on the tenant side of the market. Empirical studies by Skoufias (1995), Tikabo and Holden (2004), and Tikabo et al. (2008), and other chapters in this book essentially assess whether factor market imperfections create a rationale for the land rental market, and whether transaction costs in the land rental market prevent participation and complete adjustment in this market. In our study we provide additional evidence on the extent of rationing of participants and nonparticipants in the market.

The empirical literature on the land-leasing behavior of households is dominated by studics conductcd in South and Southeast Asia, leaving sub-Saharan Africa (SSA) to be still underexplored. With particular reference to Ethiopia, despite the highly debated issue of the land tenure system that has lasted for several decades, few studies have been conducted that address the allocative efficiency of land tenancy markets. Exceptions include Ahmed et al. (2002), Pender and Fafchamps (2006), Deininger (2003), and Teklu and Lemi (2004), but the findings from these earlier studies are inconclusive because of data limitations and methodological weaknesses. This chapter adds to these studies by assessing the role of factor market imperfections (transaction costs) in dictating the likelihood and intensity of participation on both sides (demand and supply sides) of the tenancy market. Specifically, the chapter empirically explores the extent to which rationing problems may affect adjustment and allocative efficiency in the land rental market.

The following three broad objectives will be discussed: (1) to investigate factors that affect farm households' likelihood of participation (market entry); (2) to assess the efficiency of factor ratio adjustment through the land rental market (intensity of participation): and (3) to examine the extent of transaction costs, rationing, and asymmetry in the land rental market. We address these issues by analyzing household survey data from a random sample of 400 households from 16 villages in the Tigray region in the northern highlands of Ethiopia collected in 2003.[1] Conceptually we build on the theoretical models in Chapter 2.

Hypotheses

Our basic hypothesis is that, in an attempt to adjust farm input combinations, the persistence of relatively high imperfections in nonland factor markets as compared to the informal land markets enhances farm households' participation and level of transaction in the land rental market. Therefore, the distribution of nontradable or semi-tradable inputs may play a dominant role in rural household decisions to participate in land rental markets as well as their level of transaction. Our first hypothesis (H1) becomes: Landlords with low endowments of semi-tradable or nontradable inputs relative to their land holdings will look for potential tenants with high endowments of such farm inputs.

The semi-tradable/nontradability and skewed distribution of the nonland farm inputs among farm households cause farmland (physically immobile and fairly distributed factor) to be more mobile than nonland farm factors in an economic sense (contrary to their ease of physical mobility and highly skewed distribution) (hypothesis H2).

Significant transaction costs in the land rental market lead to incomplete adjustments of operational land related to owned land and other nonland semi-tradable and nontradable resources (hypothesis H3).

Kinship networks are important when property rights are weak and tenure insecurity is prevalent, as kinship relations provide mutual insurance and reduce market frictions (Sadoulet et al. 1998). Kinship ties among farm households, therefore, trigger more participation and enhance the degree of participation in the informal land rental market (hypothesis H4).

The asymmetry in the land rental market favors the landlord side of the market (hypothesis H5) such that tenants and potential tenants tend to be rationed and they thus achieve only partial adjustment toward their desired cultivated area.

The fragmented land holdings of farm households due to the rugged topography and land distribution policies of the past in Ethiopia, in combination with economies of scale on very small plot sizes in oxen-based cropping systems, cause nonconvex transaction costs in the land rental market. This leads to a preference for transacting whole plots rather than fine-tuning the areas rented in and out through the subdivision of plots for renting (hypothesis H6).

Estimation Methods and Data

Estimation Methods

In order to test our hypotheses we used four dependent variables to identify the determinants of participation in the informal land rental market as well as farm households' intensity of land transactions. Two dummy dependent variables TENT and LLORD (identifiers for participants as tenants and landlords in the land rental market, respectively) and two variables, land rented in, LLI, and land rented out, LLO, have been used.

Our hypothesis was that oxen are important for tenants' access to the land rental market, whereas the area rented in is unaffected by the number of oxen owned by tenants. This implies that it is relevant to assess whether participation or access is determined by different variables than those determining the degree of participation. If trust and reputation are important determinants of transaction costs in the market, earlier market participation may also be crucial for access but it may be less important for the degree of participation.

Consequently, we tested alternative econometric models to find out which specifications were most appropriate. First we tested censored tobit versus double-hurdle models, and since the tobit models were rejected in favor of the double-hurdle models, we tested for selection bias on both sides of the market separately. To test the robustness of results, alternative specifications were used, including models with and without a dummy variable for earlier market participation in the trade selection equation as well as in the second stage. Similarly, we ran the models with and without the oxen dummy variables in the second stage. Tables 4-5 and 4-6, seen later in this chapter, present the results without oxen dummy variables in the second stage. The estimated models with oxen dummy variables in the second stage are available upon request. Oxen are essential for land cultivation and are used in pairs. Oxen are therefore a lumpy essential input. Households without oxen may fail to cultivate their land because they cannot afford to buy oxen, and because the rental market for oxen plowing is missing, making it impossible to hire oxen. We also found a positive correlation between oxen and farm size. The lumpiness of oxen, then, makes it relevant to run alternative models with dummies for households with different numbers of oxen.

In all cases the censored tobit models were rejected in favor of double-hurdle models. On the tenant side, we found significant selection bias in all specifications. On the landlord side, we found no significant selection bias except in one of the specifications (not included here). Consequently, in order to control for selection bias related to unobservable characteristics, we used Heckman two-stage selection models.

Data and Descriptive Analyses

The data used in the analysis come from a stratified random sample of 16 villages (with a random sample of 25 farm households from each) covering the five administrative zones of the Tigray region of Ethiopia. The stratified sampling of villages was based on agricultural potential, population pressure, access to irrigation, and market access. The survey was conducted from May to July 2003 and data were collected for the 2002–2003 crop season.

Table 4-1 shows that the average farm size (*Lando*) is only 1.106 hectares and, on average, each household is endowed with 1.502 units of adult male labor (*Totmal*) and 1.701 tropical livestock units excluding oxen (*Tluox*) (see Table 4-2 for an explanation of variable names). The mean oxen endowment (*Ox*) of the surveyed households is as small as 0.89 (far less than the minimum requirement of a pair of oxen to fully engage in self-sufficient crop cultivation activity).

TABLE 4-1 Descriptive statistics of key variables

Variable Name	Total Sample (n = 372)		Landlords (n = 89)		Nonparticipant (n = 174)		Tenants (n = 109)		P-tenants (n=58)	
	Mean	Std. Dev.	Mean	Std. Dev.	Mean	Std. Dev.	Mean	Std. Dev.	Mean	Std. Dev.
Household demographic variables										
HHsex	0.723	*0.45*	0.449	*0.50*	0.724	*0.45*	0.945	*0.23*	0.810	*0.40*
HHage	54.011	*14.55*	56.494	*16.82*	54.506	*14.32*	51.193	*12.46*	53.017	*12.87*
HHedu	0.333	*0.47*	0.225	*0.42*	0.310	*0.46*	0.459	*0.50*	0.431	*0.50*
Adulteq	3.542	*1.50*	2.788	*1.43*	3.590	*1.54*	4.080	*1.23*	3.728	*1.55*
Totmale	1.263	*1.01*	0.753	*0.91*	1.270	*1.00*	1.670	*0.92*	1.379	*0.99*
Totmalp	1.502	*1.61*	0.689	*0.89*	1.678	*1.85*	1.884	*1.45*	2.107	*2.19*
Totfemale	1.255	*0.80*	1.393	*0.76*	1.328	*0.87*	1.028	*0.64*	1.310	*1.05*
Totfemp	1.684	*1.69*	1.788	*1.59*	1.867	*1.84*	1.306	*1.45*	1.992	*2.27*
Kins3	0.183	*0.39*	0.438	0.50	0.000	*0.00*	0.266	*0.44*	0.000	*0.00*
Asset endowment & other physical variables										
Ox	0.890	*0.97*	0.225	*0.52*	0.776	*0.75*	1.615	*1.08*	0.690	*0.73*
Oxp	0.984	*1.27*	0.209	*0.52*	0.954	*1.12*	1.665	*1.55*	1.014	*1.24*
Tluox	1.701	*2.07*	0.735	*1.32*	1.490	*1.75*	2.827	*2.51*	1.362	*1.41*
Tluoxp	1.789	*2.25*	0.661	*1.16*	1.734	*2.22*	2.798	*2.52*	1.934	*2.25*

Lando	1.106	*0.75*	1.085	*0.60*	1.014	*0.67*	1.269	*0.94*	0.978	*0.72*
Landop	1.152	*1.10*	0.370	*0.54*	1.014	*0.67*	2.010	*1.39*	0.978	*0.72*
Irrgp	0.022	*0.09*	0.024	*0.10*	0.026	*0.10*	0.016	*0.06*	0.021	*0.08*
Nli	0.046	*0.68*	-0.715	*0.54*	0.000	*0.00*	0.741	*0.62*	0.000	*0.00*
Vinput	120.179	*166.00*	70.438	*218.23*	117.095	*123.55*	165.715	*165.12*	137.255	*160.66*
Vasset	770.898	*2046.8*	760.202	*2374.8*	463.126	*1182.7*	1270.936	*2674.1*	843.069	*1654.6*
Mktd	155.323	*104.02*	129.438	*88.97*	175.747	*112.60*	143.853	*95.15*	149.828	*98.84*
ROAD	34.753	*46.92*	31.517	*45.41*	36.989	*47.45*	33.826	*47.50*	33.603	*43.79*
OFDIST	23.695	*26.15*	34.401	*32.51*	18.298	*20.48*	23.568	*25.98*	17.587	*23.15*
Offma	0.728	*0.45*	0.820	*0.39*	0.713	*0.45*	0.679	*0.47*	0.741	*0.44*
Lrmpb	0.672	*0.47*	0.944	*0.23*	0.362	*0.48*	0.945	*0.23*	0.328	*0.47*
Geographical (village) level variables										
Agrieco	0.298	*0.46*	0.472	*0.50*	0.195	*0.40*	0.321	*0.47*	0.190	*0.40*
Vgini	0.235	*0.07*	0.236	*0.08*	0.232	*0.07*	0.238	*0.07*	0.233	*0.06*
Irrig	0.231	*0.42*	0.303	*0.46*	0.178	*0.38*	0.257	*0.44*	0.190	*0.40*
Mkt1	0.508	*0.50*	0.618	*0.49*	0.431	*0.50*	0.541	*0.50*	0.362	*0.48*
Popd	0.360	*0.48*	0.461	*0.50*	0.299	*0.46*	0.376	*0.49*	0.310	*0.47*

Note: See Table 4-2 for definitions and descriptions of variables.

TABLE 4-2 Definition and description of variables used

Variable Name	*Variable Description*	*Variable Type*[a]
Household level variables		
Hhhsex	Gender of the household head (1 = female, 0 = male)	d
Hhhage	Number of years age	c
Hhheduc	(1 = household head able to read and write, 0 = otherwise	d
Adulteq2	Adult equivalent family size	c
Kins3	Farm households who transact with kin-related partners only	d
Totmal	Number of adult males	c
Totmalp	Adult males per hectare of owned land	c
Totfemale	Number of adult females	c
Totfemp	Adult females per hectare of owned land	c
Kinrel	(1 = transact with kin-related partners only, 0 = otherwise	d
Ox1	Farm households with one ox	d
Ox2	Farm households with two oxen	d
Ox3	Farm households with three oxen	d
Ox4	Farm households with four oxen	d
Oxp	Oxen per hectare of owned land	c
Tluox	Tropical livestock unit other than oxen	c
Tluoxp	Tluox per hectare of owned land	c
Fs	Owned farm size	c
Fs-kins	Interaction variable of farm size with transactions among kin-related partners only	c
Irrlp1	Portion of irrigated land	c
Mktd	Household distance to major market (in minutes)	c
Road	Household distance to nearest road service (in minutes)	c
Ofdist	Distance to owned farm plots	c
Offma	Lagged off-farm employment activities	d
Lrmpb	Previous participation in the land rental market; 1 = yes, 0 = no	d
Lrmpbfs	Interaction variable of previous participation with farm size	c
Village level variables		
Ecol	1 = if village is located above 2,000m above sea level, 0 = otherwise	d
Vgini	Gini coefficient showing land distribution within villages	c
Vilkin	Share of contracts that are among kin partners in the village	c
Irrg	Village access to irrigation projects	d
Mkt1	1 = if a village is located > 10km from major market, 0 = otherwise	d
Pop1	population density (1 if it is > 200 persons/km^2, 0 = otherwise)	d

[a]d = discrete, c = continuous

As shown in Table 4-1, various farm household characteristics in our dataset provide some indications on the factors that affect the functioning of the land rental market, with the transacting households (tenants or landlords) having some distinct variations from the nontransacting households. We see that landlord households are more likely to be headed by females (55.1% against 27.8% for all households), but only 5.5% of the tenants were female-headed households. Landlord households also tended to have fewer oxen and other livestock than average, whereas tenants had more oxen and other livestock than average, pointing toward a situation with poor landlords and relatively wealthier tenants. Landlord households were also found to have less than average endowments of male labor, but tenants had an above-average amount of male labor and a below-average amount of female labor. The incidence of illiteracy was also higher among landlord households (77.5%) as compared to the 69% and 54% of the autarky and tenant household groups, respectively.

When it comes to land distribution, landlords did not have significantly more or less land than average, but tenants had a slightly above-average amount of land. It is therefore not typically the landless households that access land through the land rental market in Tigray. This may indicate that it is the inequitable distribution of nonland resources in form of oxen, other livestock, and male labor that is driving participation in the land rental market. A question then becomes: How will this situation, with relatively less wealthy and often female-headed landlord households and more wealthy and usually male-headed tenant households, affect their relative bargaining position and the extent of rationing in the tenancy market? Furthermore, we see that landlord households rented out, on average, about two-thirds of their land (*NLI* = net land rented in = –0.715 hectare), causing their average operational land holding (*Landop*) to be about 0.37 hectare while the average tenant increased his or her operational holding to 2.01 hectares.

The informal land tenancy market involved slightly more than half of the total respondents as landlords and tenants (24% and 29% of the total sample, respectively). Moreover, a large share of the respondents were reported as rationed households (about one-third of the total sample) that operated less than their optimal level (rationed tenants and potential tenants).

As shown in Table 4-3, nearly 53% of the sample respondents participated in the informal rural land tenancy market, which is active in all the sampled communities although with considerable degree of variations across villages, ranging from 16% (*Hagere-selam* village) to 83% (*Samre* village) among the sample respondents in each village. Using the data from the 2003 household survey, it is shown that most of the cultivated farm land area (81%) is under owner cultivation.

An overview of the extent of rationing on the two sides of the market is presented in Table 4-4. We see that rationing is much more prevalent on the tenant side, as 56% of the tenants against only 5% of the landlords considered themselves to be constrained in the market. In addition, as many as 33% of

TABLE 4-3 Description of sampled districts (Tabia's)

No.	Tabia	Woreda	Population Density (Persons/km^2)	Distance to Market (in Km)	Mean Rainfall	Irrigation Project	Land Distribution (Gini coeff.)	Tenancy Market Participation
1	Hintalo	Hintalo wajirat	80.2	14	03.7	1	0.222	66%
2	Samre	Seharti samre	248.9	1.25	557.5	0	0.363	84
3	Mahbere-Genet	Enderta	441.5	8	552.1	1	0.118	60
4	Mai-Alem	Enderta	429.6	6	552.1	0	0.192	48
5	Seret	D. Tembien	707	12.5	420.4	0	0.156	56
6	Kihen	Wukro	160.6	23	420.4	0	0.169	65
7	Genfel	Wukro	166.5	4	596.8	1	0.191	67
8	Emba-Asmena	Tsaeda-emba	631.1	7	419.05	0	0.345	56
9	Hagere-selam	Gulo Mekeda	749.4	39	761.4	0	0.276	16
10	Debdebo	Ahferom	161	6	668.52	0	0.199	65
11	Mai-Keyahti	Ahferom	636.6	16	736.6	0	0.253	31
12	Adi-selam	Mreb-leke	206.8	29	579.32	0	0.160	20
13	Hadegti	L. Adiabo	130.8	9	832	0	0.292	40
14	Tsaeda-Ambora	L. Adiabo	41.8	20	596.55	0	0.252	62
15	Mai-adrasha	T. Koraro	440	5.2	893.55	1	0.249	62
16	Adi-Menabir	T. Koraro	236	21	783.4	0	0.281	59

Source: Hagos and Holden (2003).

TABLE 4-4 Rationing on the two sides of the market

Supply Side of the Market		*Demand Side of the Market*	
Number of constrained landlords	4 (5%)	Number of constrained tenants[a]	61 (56%)
Amount of land rented out	63.6 hectares	Amount of land rented in	80.75 hectares
Desired land to rented out	65.1 hectares	Desired land to rented in	126.5 hectares
Unfulfilled gap	1.5 hectares	Unfulfilled gap	45.75 hectares
% of the Actual—desired[b]	98%	% of the Actual—desired LLI	64%
Number of potential landlords	8 (5%)	Number of potential tenants	58 (33%)
Total sample respondents	89	Total sample respondents	109

[a]Farmers were explicitly asked, given their farming capacity, if they want to rent in more amount of land.

[b]How much of their desired adjustment did farm households manage to satisfy through participation in the tenancy market.

the nonparticipating households wanted, but had failed, to enter the market as tenants, whereas this was the case for only 5% of the nonparticipants who wanted to participate in the market as landlords. The gap between the stated desired cultivated area and the actual rented area was also much larger on the tenant side than on the landlord side. This information appears to support our hypothesis H5 about asymmetry and rationing in the tenancy market on the tenant side.

We also found that land transactions tended to be lumpy. Households preferred to rent out the entire plot. This indicates that the decision for each plot is whether to owner-cultivate or rent out each plot rather than deciding directly on area to rent out and then subdivide plots accordingly afterwards. There is typically a substantial distance from the homes of households and a number of their plots, although one plot tends to be located near or around the house. Typically, owners would first rent out the more distant plots, keeping the homestead plot as the last one to rent out. Since cultivation of a distant plot will require multiple visits and bringing along oxen and plows for repeated plowings and transportation of other inputs, and the output the related costs would be of a noncovex nature, the incentive would be not to further subdivide plots in line with our hypothesis H6.

Results and Discussions

We will now turn to the econometric analysis that allows a more rigorous test of the allocative efficiency of the land rental market and the factors affecting it, while controlling for other variables as well as potential selection bias.

Likelihood of Participation

Supporting our basic hypothesis (H1), the relative endowment of nonland farm inputs were found to significantly influence the household decisions to participate in the land rental market. As shown in Tables 4-5 and 4-6, the oxen dummies showing different numbers of oxen ownership (*ox1*, *ox2*, *ox3*, and *ox4*) were found to have strongly significant (at 5, 1, or 0.1% levels) effects on the participation in the market, whereas the effect of other livestock was less significant (significant at 5% level only at most). This may be because oxen are crucial for land cultivation, and landlords may not be willing to rent out their land to potential tenants who have no oxen. However, having one ox may be sufficient to go into collaboration with another farmer who also has one ox to form a pair of oxen needed for plowing. This may explain why the one ox dummy (*ox1*) had a significant positive effect on leasing in as well as a significant negative effect on leasing out decisions.

TABLE 4-5 Heckman two-step selection models for rented in plots: With bootstrapped standard errors

	ModH0	*ModH1*	*ModH2*
	b(se)	*b(se)*	*b(se)*
Lli (**Land rented in equation**)			
Fs	0.331(0.10)****	0.317(0.09)****	0.302(0.39)
Road	0.001(0.00)	0.001(0.00)	0.001(0.00)
Adulteq2	–0.127(0.10)	–0.103(0.09)	–0.103(0.09)
Hhhsex	–0.179(0.33)	–0.122(0.34)	–0.123(0.32)
Hhhage	–0.002(0.01)	–0.003(0.01)	–0.003(0.01)
Hhheduc	–0.040(0.12)	–0.017(0.11)	–0.016(0.13)
Totmale	0.076(0.11)	0.062(0.11)	0.062(0.11)
Totfemale	0.273(0.16)*	0.241(0.16)	0.240(0.15)
Tluox	0.030(0.03)	0.032(0.04)	0.032(0.03)
Ofdist	0.002(0.00)	0.002(0.00)	0.002(0.00)
Irrlp1	0.037(1.18)	0.091(1.12)	0.095(1.10)
Offma	–0.001(0.12)	0.019(0.11)	0.020(0.11)
Irrg	0.126(0.14)	0.138(0.14)	0.139(0.13)
Mkt1	–0.202(0.12)*	–0.196(0.12)*	–0.197(0.12)
Vgini	0.242(0.80)	0.513(0.78)	0.511(0.84)
Pop1	0.070(0.13)	0.097(0.13)	0.097(0.13)
lrmpbfs			0.015(0.36)
Mills lambda	–0.319(0.13)**	–0.322(0.11)***	–0.319(0.15)**
Constant	0.677(0.46)	0.511(0.44)	0.509(0.47)
Trade selection equation			
Fs	–0.049(0.16)	0.014(0.20)	0.014(0.19)
Road	0.001(0.00)	0.001(0.00)	0.001(0.00)

TABLE 4-5 *(Cont.)*

	ModH0 b(se)	*ModH1* b(se)	*ModH2* b(se)
Adulteq2	0.030(0.09)	0.040(0.09)	0.040(0.10)
Hhhs	–0.929(0.33)***	–0.952(0.37)**	–0.952(0.38)**
Hhha	–0.020(0.01)***	–0.014(0.01)	–0.014(0.01)*
Hhhed	0.063(0.20)	0.117(0.24)	0.117(0.25)
Totmalp	0.094(0.09)	0.121(0.10)	0.121(0.12)
Totfemp	–0.002(0.08)	–0.014(0.10)	–0.014(0.09)
Ox1	0.469(0.22)**	0.637(0.28)**	0.637(0.27)**
Ox2	1.060(0.28)****	1.080(0.33)***	1.080(0.32)****
Ox3	2.441(2.70)	2.702(2.73)	2.702(2.87)
Ox4	7.851(0.62)****	7.298(0.52)****	7.298(0.56)****
Tluoxp	0.091(0.05)*	0.068(0.07)	0.068(0.07)
Ofdist	–0.002(0.00)	–0.004(0.01)	–0.004(0.01)
Irrlp1	–1.502(2.31)	–0.873(3.20)	–0.873(2.60)
Offma	–0.034(0.22)	0.012(0.27)	0.012(0.26)
Irrg	–0.414(0.29)	–0.331(0.34)	–0.331(0.32)
Mkt1	0.237(0.22)	0.035(0.24)	0.035(0.26)
Vilkin	3.080(0.84)****	1.895(0.81)**	1.895(0.90)**
Vgini	–1.921(1.53)	–1.715(1.73)	–1.715(1.79)
Pop1	0.213(0.23)	0.078(0.26)	0.078(0.25)
Lrmpb		1.634(0.35)****	1.634(0.35)****
Constant	–0.976(0.72)	–2.249(0.85)***	–2.249(0.81)***
Prob > chi2	0.000	0.000	0.005
Number of obs.	372	372	372

TABLE 4-6 Heckman two-step selection models for rented-out plots: With bootstrapped standard errors

	ModH0 b(se)	*ModH1* b(se)	*ModH2* b(se)
Llo (**Land rented out equation**)			
Fs	0.633(0.19)***	0.644(0.18)****	0.491(0.43)
Road	0.001(0.00)	0.000(0.00)	0.000(0.00)
Adulteq2	–0.020(0.07)	–0.023(0.07)	–0.023(0.07)
Hhhsex	0.262(0.20)	0.303(0.17)*	0.315(0.18)*
Hhhage	–0.001(0.00)	0.000(0.00)	0.001(0.00)
Hhheduc	0.043(0.12)	0.058(0.11)	0.063(0.11)
Totmale	0.137(0.11)	0.123(0.11)	0.122(0.11)
Totfemale	–0.039(0.09)	–0.028(0.08)	–0.026(0.09)
Tluox	–0.156(0.06)***	–0.180(0.05)****	–0.190(0.06)***

(continued)

TABLE 4-6 *(Cont.)*

	ModH0 *b(se)*	*ModH1* *b(se)*	*ModH2* *b(se)*
Ofdist	0.001(0.00)	0.002(0.00)	0.002(0.00)
Irrlp1	–0.200(0.60)	–0.173(0.62)	–0.180(0.55)
Offma	0.086(0.15)	0.122(0.12)	0.138(0.13)
Irrg	0.197(0.14)	0.232(0.13)*	0.237(0.14)*
Mkt1	0.035(0.12)	0.051(0.11)	0.063(0.11)
Vgini	1.251(1.03)	1.350(0.89)	1.331(0.96)
Pop1	0.043(0.10)	0.069(0.09)	0.077(0.09)
Lrmpbfs			0.152(0.43)
Mills lambda	–0.116(0.21)	0.022(0.13)	0.075(0.22)
Constant	–0.406(0.76)	–0.692(0.59)	–0.773(0.67)
Trade selection equation			
Fs	–0.049(0.22)	–0.015(0.21)	–0.015(0.22)
Road	0.001(0.00)	–0.001(0.00)	–0.001(0.00)
Adulteq2	–0.048(0.10)	–0.031(0.09)	–0.031(0.10)
Hhhsex	0.284(0.31)	0.339(0.33)	0.339(0.37)
Hhhage	0.012(0.01)*	0.020(0.01)**	0.020(0.01)**
Hhheduc	0.009(0.27)	0.098(0.27)	0.098(0.27)
Totmalp	–0.232(0.13)*	–0.238(0.14)*	–0.238(0.15)
Totfemp	0.043(0.09)	0.054(0.10)	0.054(0.10)
Ox1	–0.800(0.23)****	–0.718(0.27)***	–0.718(0.29)**
Ox2	–1.240(0.99)	–1.268(0.99)	–1.268(0.75)*
Ox3	–5.965(0.33)****	–6.514(0.41)****	–6.514(0.39)****
Ox4	–5.708(0.47)****	–5.964(0.44)****	–5.964(0.51)****
Tluoxp	–0.125(0.09)	–0.190(0.10)*	–0.190(0.10)**
Ofdist	0.006(0.01)	0.006(0.00)	0.006(0.00)
Irrlp1	–0.154(1.96)	–0.322(2.08)	–0.322(2.00)
Offma	0.402(0.26)	0.426(0.26)	0.426(0.24)*
Irrg	0.315(0.32)	0.264(0.35)	0.264(0.37)
Mkt1	0.445(0.25)*	0.383(0.27)	0.383(0.28)
Vilkin	1.706(0.93)*	0.034(0.96)	0.034(0.95)
Vgini	–0.310(1.84)	–0.746(2.05)	–0.746(1.78)
Pop1	0.330(0.21)	0.151(0.28)	0.151(0.27)
Lrmpb		1.595(0.60)***	1.595(0.39)****
Constant	–1.983(0.74)***	–2.820(1.04)***	–2.820(0.84)****
Prob > chi2	0.000	0.000	0.000
Number of obs.	372	372	372

* significant at 10%; ** significant at 5%; *** significant at 1%; **** significant at 0.1%.

This result supports our basic hypothesis that the nontradability/semi-tradability of nonland farm inputs triggers farm household participation in the tenancy market. The result may indicate that there are higher transaction costs in the oxen rental market than in the land tenancy market. This finding is supported by the fact that 43% of the sample respondents were reported to have no oxen and yet only 11.5% of them engaged in the oxen rental market. The significant (5 or 10% level) and negative coefficient of ownership of other livestock per unit of own land (*tluoxp*) in the supply side of the market indicates that households with more endowment of animals other than oxen are less likely to rent out their land. Households with relatively higher endowment of such animals may be forced to operate their own land to produce enough fodder to feed their animals, since, more often than not, fodder (crop residues) is taken by plot tillers (tenants) unless it is agreed to be shared with the landlord during the tenancy contract.

The coefficients showing household relative endowments of labor (*Totmalp* and *Totfemp*) were found to be statistically insignificant in the leasing in models, whereas male labor was significant (10% level only) and with a negative sign in two of the leasing-out models. This may imply that transaction costs are relatively lower in the labor market, but the fact that the sex of household head (*hhhs*) and age of head of household (*hhha*) variables were significant and with negative signs in the tenant models (Table 4-5) are also signs of imperfections and significant transaction costs in the labor market. This may be particularly important for female-headed households that lack male labor, since females are not allowed to plow with oxen. Perhaps this is because landlords have less trust in female-headed households as land managers. Potential female-headed tenant households may therefore have been totally rationed out of the market even if they had the necessary oxen endowment for farming. The insignificance of the labor variables may also partly be explained by the relative abundance of labor in most households.

The age of the household head (*Hhha*) variable could also have been interpreted as an indicator variable for farm experience. However, it was found to have a significant effect on both sides of the market such that the older the age of the household head, the higher the probability of renting out land, and the younger the household head, the higher is the likelihood of leasing in land. Older age, then, seems to capture less working capacity than more and better farm skills.

The regression results also show previous participation in the market for tenancies to have a significant positive effect on current probability of participation in both sides of the market. This attests our *a priori* expectation that households well acquainted with the existing tenancy market (having insiders' market information) are more likely to face a lower entry barrier in the tenancy market than others. Perhaps this is due to the high initial cost (fixed transaction cost) an agent may face when first attempting to enter the market. We refer the reader to Holden and Ghebru (2006) for a more comprehensive theoretical treatment of the relationship between kinship, transaction costs, and market participation and to Holden et al. (2007a) for a more rigorous analysis of the change in market participation over time.

In line with our hypothesis H4, kinship (social network) was found to have a positive impact in reducing the entry barriers for tenant households. We hypothesized that trust may be higher among kin and therefore factor adjustment through land market would be smoother in kin contracts. This could imply that the land rental market functions better in villages with a high share of kinship contracts. The highly significant and positive coefficient of the *vilkin* variable in Table 4-5 shows that the land rental market functions significantly better in villages with high share of kinship contracts by improving the access to the market for tenants. The *vilkin* variable was significant in only one of the models and at 10% only on the landlord side. This shows that kinship contracts may primarily help potential tenants to enter the market because rationing is more severe on this side of the market.

The regression results on the supply side (Table 4-6) provide some weak evidence that households that engaged more in off-farm income-generating schemes were more likely to rent out their land (significant at 10% level only in some model specifications). This may also indicate that there were imperfections in the labor market since extra ("high-pay") off-farm income was not used to hire ("low-pay") labor to substitute for lost household labor in farming. There are additional transaction costs related to hiring and monitoring hired labor, and such costs may be even higher for households engaged in off-farm work than for other households.

Intensity of Market Participation

Area Rented Out. We found (Table 4-6) that the tropical livestock units other than oxen (*tluoxp*) variable had a strongly significant negative effect on land area rented out by landlord households. Households with more livestock may need the land themselves for fodder production and may also be more wealthy and depend less on renting out land as a source of income and food.

The Heckman selection models in Table 4-6 also indicate that the amount of land rented out was positively correlated with farm size. This result is robust across the various model specifications except when the earlier market participation (*lrmpbfs*) variable is included in the rent out equation. To examine whether landlord households adjust their factor endowment smoothly through the land rental market or not, we assess the size of the coefficient on the farm size variable and whether it is significantly different from 1.0 in the rent out models, which is expected with a perfect land tenancy market with zero transaction costs, a linear response to nonland factors of production, and constant returns to scale. In the first two models in Table 4-6, in which the earlier market participation (*lrmpbfs*) variable was left out, the coefficient was significantly lower than 1 (0.63–0.64) and significantly different from 1, but when we controlled for earlier market participation, the coefficient was reduced to 0.49 while the standard error became inflated, making the coefficient insignificantly different from 0 and 1. The earlier market participation variable was highly significant in the trade selection equation but was insignificant in the rent out equation, probably due to its multicollinearity with farm size, making it appropriate to leave it out in the second stage. In the models that included oxen dummy variables in the second stage (not shown here), the earlier market participation variable was significant in the trade selection as well as in the rent out equation, and its inclusion in the second stage also made the Mills lambda variable highly significant. Again, the farm size variable became insignificantly different from 0 and even changed signs to become negative. We therefore rely on the other model specifications in our analysis. With the inclusion of oxen dummy variables, the coefficients on farm size increased to 0.70 but still significantly less than 1 (at 10% level only). This may be interpreted as there being significant transaction costs that cause only partial adjustment on the landlord side. Such transaction costs may be related to searching for and finding trustworthy tenants.

Area Rented In. The farm size variable was the only variable that was highly significant in the land rented out models in Table 4-5. Based on our theoretical model in Chapter 2, we expected that the area rented in would be lower the larger the own farm size is. Perhaps surprisingly, we found that the farm size variable was highly significant and with a positive sign in the first two model specifications. It remained positive, with about the same size in the third model specification in which the earlier market participation variable was included in the rent in equation but it became insignificantly different from zero. One explanation that we can suggest for this contradictory result is that some of the basic theoretical assumptions are violated (constant returns to scale). The other explanation is the rationing going on in the market such that primarily well-established households that have oxen and that also tend to have larger farm sizes are those that have access to the market as tenants. It is possible that there are economies of scale in production at such small farm sizes due to the lumpiness of oxen and the essential role oxen play in land cultivation. A certain farm size may be necessary to meet the feed requirement of the oxen, making it harder for relatively smaller farms to keep oxen and thus forcing them to rent out their land. Furthermore, female-headed households, who are female-headed because the women lost their husbands or got divorced, may have less land than male-headed households due to the rule of dividing the land such that the widow or divorced woman keeps only half of the land the couple had before. This may cause a correlation between farm size, sex of household head, and ownership of oxen.

We found that households with more female labor rented in significantly more land (significant at 10% level in only one of the model specifications). We also found in two of the models that tenants with better access to major markets (*mkt1* = 1) were found to rent in significantly less land than those with poorer market access. This may be due to more lucrative employment opportunities in locations with better market access.

Market Symmetry and Rationing

A market symmetry test[2] has been made using the approach of Skoufias (1995), applying it on the censored tobit models. The results of the various Wald tests conducted to test market symmetry (the equality of coefficients on both sides of the market) are presented in Table 4-7 and show that the hypothesis of sym-

TABLE 4-7 Tests of market symmetry on intensity of participation in LRM (Wald test of equality of coefficients)

Hypotheses Tested	*Wald Statistic (R)*
Simultaneous equality of variables ($r = 19$)	132.36***
Individual variable tests ($r = 1$ for each)	
Fs	*16.21****
Road	*2.33*
Adulteq2	*3.85**
Kins3	*6.18***
Hhhsex	*1.53*
Hhhage	*5.25***
Hhheduc	*0.04*
Totmale	*4.89***
Totfemale	*0.82*
Ox	*0.3*
Tluox	*0.81*
Ofdist	*1.17*
Irrlp1	*0.63*
Offma	*0.56*
Lrmpb	*10.84****
Irrg	*2.81*
Mkt1	*7.91****
Vgini	*8.75****
Pop1	*1.99*

Note: r = number of restrictions

* significant at 10%; ** significant at 5%; and *** significant at 1%

metric resource adjustment in both sides (the demand and supply sides) of the tenancy market had to be rejected. This is also supported by the clear differences we found between the Heckman models on the two sides of the market as well as the descriptive analysis demonstrating rationing on the tenant side of the market. As shown in Table 4-7, the simultaneous test for the equality of all coefficients in the two sides of the market is rejected with a Wald statistic of 132.36, significant at 1% level. To identify the factors (variables) responsible for the asymmetric nature of the tenancy market, individual variable tests were made, comparing the two sides of the market. As the results reported in Table 4-7 show, variables like owned farm size (*fs*), village access to market (*Mkt1*), village distribution of land (*vgini*), previous participation in the tenancy market (*Lrmpbfd*), age of household head (*hhha*), and total male labor (*Totmal*) have been identified as variables with significant asymmetric roles in the market. These results suggest that the households' marginal responsiveness toward adjusting the desired cultivated area for these variables is not the same in each side of the market. Great caution, however, has to be exercised, since the results (tests) are entirely based on the censored tobit model that fails to capture the sample selection problem (selection bias) which was found significant especially on the tenant side of the market.

The preceding results, together with the findings in Tables 4-4, 4-5, and 4-6, give ample evidence that the tenancy market does not clear given a market clearing price. Rather, access and degree of access appear to depend on the resource endowments of households—especially oxen, earlier participation in the market, and the household's access to kin partners; and the level of trust and reputation that may depend on earlier market participation. Newcomers may face entry barriers, especially if they do not have oxen. Kinship relationships may help reduce these entry barriers. Finally, we should make the reservation that our analysis does not control for unobservable household and farm heterogeneity (Holden et al. 2007a). This is an area for future research using household panel data.

Conclusions

Our study in Tigray region in northern Ethiopia has demonstrated the importance of land rental markets for factor ratio adjustment in agriculture because of imperfections in nonland factor markets. The continued prohibition of land sales markets makes land rental transactions important for more long-term as well as short-term adjustments of factors of production across households. Ownership of oxen turned out to be a key determinant of land rental market participation as households without oxen typically failed to cultivate their land themselves and instead rented it out to households with oxen. The market for plowing services by oxen is severely constrained due to the lumpiness of oxen; the synchronized demand for oxen in rain-fed agriculture, which is particularly important in semi-arid areas where rainfall is very limited and erratic; the crucial need of exact timing of farm operations relative to rainfall; and the moral hazard problem related to renting out oxen without the owner as a driver.

The household relative endowments of male and female labor were found to have weak effects on the likelihood of participation in both the demand and supply sides of the tenancy market. This may not be due to a well-functioning labor market, as off-farm labor opportunities are very scarce, but rather may be due to the relative abundance of household labor.

We found signs of significant transaction costs limiting the access to and the degree of adjustment in the land rental market. Although a high ratio of kinship contracts in the communities appeared to be associated with better functioning land rental markets, households that previously participated in the market appeared to face lower transaction costs in the market. Even though tenants were found to be wealthier than landlords, landlords controlled the market by selecting tenants such that many tenants and potential tenants perceived themselves to be rationed in the market, as a large share of them managed to rent in less land than they desired at the going contract conditions. The dominance of sharecropping may also help explain the lack of market clearing through a market clearing price.

The growing landlessness due to continued population growth, increasing land scarcity, and limited opportunity to further subdivide farms among children create an increasing demand for alternative employment opportunities and an increasing pressure on the land rental market. Our findings indicate, however, that there may be limited prospects for poor landless households to utilize the tenancy ladder as a way out of poverty since households without oxen and other farm endowments are likely to be rationed out of the market for tenancies. On the other hand, households that are poor in nonland endowments but that have land may benefit from the land rental market, for it enables them to get a higher income by renting out the land than they would have obtained by farming the land themselves. Many female-headed households belong to this category, and the land rental market has become an important source of livelihood and a safety net for these poor landlords. Thus, the salient conclusion of this chapter is that although the land rental market is imperfect, it works for the benefit of the poor and contributes to more efficient allocation of land among farm households.

Notes

1. Due to incompleteness of data and respondent dropouts, the entire analysis in this chapter considers 372 sample respondents only.
2. Testing for the equality on the two sides of the market with respect to the signs and size of parameters.

CHAPTER 5

Efficiency and Equity Impacts of Land Markets in Kenya

TAKASHI YAMANO, FRANK M. PLACE, WILFRED NYANGENA, JULLIET WANJIKU, AND KEIJIRO OTSUKA

Population continues to grow rapidly in sub-Saharan Africa and has exerted strong pressure on limited cultivable land areas. Thus, how to allocate scarce land resources among farm households efficiently and equitably is a major policy issue in many parts of the continent (see Chapters 1 and 2). This issue is particularly relevant for poverty reduction, as the poor usually have limited access to land. Due to its increasing scarcity, new forms of accessing land have emerged in Africa, including market-based mechanisms. As is reported in various chapters in this book, in some countries, such as Ethiopia, land rental markets have developed; in others, such as Uganda, land sales markets have developed; and in still other countries, such as Kenya, both types of land markets have developed. In all these countries, such market transactions date back a few decades, but they are becoming increasingly common in response to the growing scarcity of land.

This chapter explores the operation of both land sales and rental markets and their consequences in Kenya, with particular focus on the identification of types of participants in these markets and the efficiency impacts of market transactions. Although Kenya has attempted a more comprehensive formalization of land administration than in most countries, its actual practice is *ad hoc* and operates on an outdated legal framework (Ministry of Lands and Settlement 2004). For example, although there are numerous laws and statutes governing the leasing of land—such as the Indian Transfer of Property Act (1882), the Registered Land Act (Cap 300), the Crown Land Ordinances of 1902 and 1915, the Lands and Titles Ordinance of 1908 (currently Cap 208), and the Conversion of Leases Regulation and Rules of 1960 (Okoth-Ogendo 1999)—many farmers in rural Kenya cannot afford to engage in the complex legal leasing arrangements. Consequently, the majority of farmers are involved in informal land rental arrangements. Similarly, many farmers find that the costs involved in updating the land registry following land subdivisions or other transfers exceed the potential benefits of doing so.

The empirical analysis relies primarily on survey data from households located across 15 districts in Kenya in 2004. First, by using such information, we examine the efficiency of the land sales and rental markets by estimating the land adjustment model first proposed by Bliss and Stern (1982) and later elaborated by Skoufias (1995).[1] For a second dataset of households obtained in the Laikipia area, we examine the factors associated with both the renting in and renting out of land. If the land market fails to equalize land-labor ratios among farm households, the inverse relationship between farm size and productivity is likely to arise. This is expected because labor-abundant small farmers cultivate their limited land more intensively than large land-abundant farmers in the absence of capital-intensive farming systems (Otsuka et al. 1992; Otsuka 2007). Second, we examine the existence of the inverse relationship between the farm size and farm management performance indicators, such as maize yield and net returns to maize.

Conceptual Framework and Hypotheses

A conceptual framework of the impacts of land tenure institutions on land management and productivity is shown in Figure 5-1. Let us assume that the original farm size distribution before the market transactions is exogenously given, which actually may have been determined by customary land tenure systems and passed down by parents of farmers in the current generation. Data from Africa show that the number of persons per household varies much less than do land holdings, so that large landholders also have higher land–family labor ratios. The potential for equating land-labor ratios through the labor market is dim because of the difficulty in monitoring hired workers on spatially dispersed and ecologically diverse farm land; therefore, farm labor markets seldom function efficiently (Otsuka et al. 1992). Given the unequal original farm size distribution, marginal products of land are high on small farms, and this provides incentives for smaller farmers to buy or rent in land from larger farmers. Land markets, including both land sales and renting markets, however, may not function competitively for a number of reasons.

First, prevailing land tenure institutions may not allow land sales or renting due to traditional land access and transfer rules and weak land rights systems (Otsuka and Place 2001). Second, because of the imperfect credit markets, it is possible that small farmers cannot afford to acquire land, particularly through land purchase. Although the credit requirement is less important in the case of renting, the practice of advance fixed-rent payment may prevent very poor farmers from entering tenancy contracts. Credit constraints may have more pervasive effects on other factor markets as well, such as input and labor markets. Third, the cost of monitoring tenants may be very high for landowners,

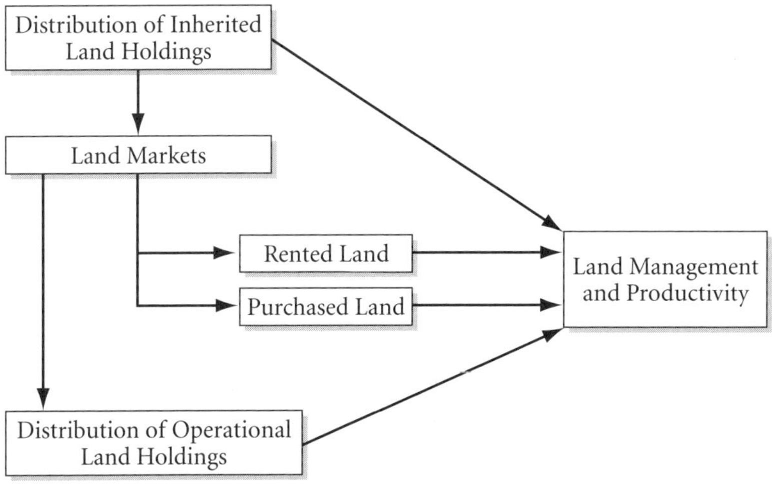

FIGURE 5-1 Conceptual framework of land tenure institutions and land management

which reduces the attractiveness of renting out land. Such cost is particularly high under share tenancy, because the sharing of output reduces tenants' incentives to work. Thus, Otsuka (2007) argues that unless landlords and tenants are tied by kinship or friendship with sufficiently strong trust, share tenancy will not be chosen, indicating that the choice of fixed-rent leasehold tenancy itself is a sign of high monitoring cost. Fourth, to the extent that the optimum contract length is short, tenants may not be interested in maintaining soil fertility and long-term land investments, such as tree planting. This reduces attractiveness of renting out land by land-rich households. Fifth, in areas where animal plowing is common, imperfections in the rental markets of draft animals may lead to rented land moving only to farmers who can afford to own draft animals. We believe that rental markets of draft animals do not work in sub-Saharan Africa primarily because of the simultaneous demands for draft animals by almost all farmers under rain-fed conditions, as the optimum timing of plowing is determined by rain.

Based on these arguments, we will postulate several empirically testable hypotheses that are relevant to Kenya. First, land markets do act to improve equality of land holdings across households. However, our second hypothesis is that land markets, and land sales in particular, are inefficient due to high transaction costs and credit constraints. Credit constraints may also reduce the area of land renting due to the advance payment requirement of fixed rent tenancies. Third, because of less than socially optimum area of land transferred through market transactions, the land-labor ratio, as well as other factor ratios, will not be equalized across farm households even after accounting for land market transfers. An inverse relationship between farm size and management intensity will arise, as revealed in lower use of inputs and, hence, lower crop yield on larger farms—that is, the inverse relationship between farm size and farm performance.

The Data

The main dataset was collected from a survey of over 943 households in 100 communities in Kenya as part of the RePEAT project.[2] A second dataset used was collected from 320 households in the Laikipia district in the Rift Valley province just to the west of Mt. Kenya. A brief description of that dataset is presented later, along with the empirical results. The RePEAT households were distributed across 15 districts in five provinces: eastern, central, Rift Valley, Nyanza, and western. The sampling scheme followed that of an earlier study by the International Livestock Research Institute (ILRI), which was stratified down to sublocation level in the medium- to high-potential areas, followed by a random selection of households. The household survey instrument emphasizes agricultural activities such as cropping, raising livestock, and growing trees. It included questions on land acquisition and rights, endowment of family labor, and access to capital. The survey also measured soil quality through the sampling and analysis of soils collected from the major cereal field, which was maize in all cases.[3] As for measures of poverty, data were collected on household assets, consumption, expenditures, and income. Finally, there are also significant sections on household demographics, education, and health. As a companion to the household surveys, the project also administered a community survey to assess the prevalence of livelihood and farming practices, land market activity, and prevailing prices and wages at the community level.

All the sites are located in medium- to high-potential areas: Owing to relatively high rainfall of more than 800 mm (31 inches) a year in two rainy seasons, these areas are potentially favorable for growing maize, and indeed maize is grown by all of our sample farmers. All farmers in our survey areas plant maize in the long rain season, whereas a majority of sample farmers, but not all, plant maize in the short rain season. In addition to maize, a large number of annual and perennial crops are grown. The perennials are most common in the higher elevation areas, with coffee and tea frequently observed in a few of the communities. Beans, cassava, plantain, and a variety of vegetables and fruits are also grown. The majority of farmers in our sample are also dairy farmers, usually with zero grazing units. In addition to milk production, this system has the advantage of generating manure for crop farming, including maize production. Farm sizes are generally between 1 and 3 hectares, with smaller farms found in the more densely populated areas of central, Nyanza and western provinces, and larger farms in the Rift Valley province. Farmers hold between one and three parcels of land, but rarely more than that.

Land Holdings and Land Markets

In Table 5-1, we find that the average size of inherited land per farm is 1.2 hectares. In the Rift Valley province, the average size of inherited land is about 1.6 hectares, which is 0.6 hectare larger than the average sizes in central and western/Nyanza provinces. The average size of purchased land is 0.6 hectare, and about 34% of the sample households own at least some purchased land. Fewer than 10% of parcels were purchased in two of the districts, and more than 30% were purchased in four of the districts. The highest was in Nakuru District, where 52% of parcels were purchased. About two-thirds of those who purchased land have no other type of land and the rest own both inherited and purchased land. The majority of those who have purchased land have done so only once. As a result, about 10% of all the sampled households own both inherited and purchased land, 24% only purchased land, and 66% only inherited land. The average years since acquisition is about 23 to 24 years for inherited and purchased plots, respectively. Thus, few (if any) sample farmers were able to make frequent adjustments to their land size through the land sales market.

Regarding the land rental market, we find that 18% of the households cultivate some rented in land.[4] Less than 10% of parcels had been rented in 8 of the 15 districts, but in 3, the percentage of rented parcels was greater than 20%. Almost all contracts are the fixed-rent type, in which rents are paid just before the beginning of the cropping season. Sharecropping is very rare in Kenya. The size of rented in land is quite small: The average size is about 0.1 hectare and total rented area is about 4% of the total cultivated area. Again, in the Rift Valley province, the size is slightly larger, at 0.2 hectare and the percentage of households with rented in land is higher. The average length of a land rental is 4 years, with more than half being 2 years or less. At the other end, just over 10% of rented parcels were operated by the tenant for 10 years or longer.

TABLE 5-1 Land size and rights by province

		Land Size			*Households with*	
Province	*Number of Households (1)*	*Inherited Land (2)*	*Purchased Land (3)*	*Rented in Land (4)*	*Purchased Land (5)*	*Rented in Land (6)*
	Number	Ha	Ha	Ha	%	%
Central	394	1.0	0.5	0.1	30.2	13.7
Rift Valley	231	1.6	1.0	0.2	56.9	26.3
Western & Nyanza	299	1.0	0.3	0.1	21.4	17.1
All	924	1.2	0.6	0.1	34.1	17.9

To compare the equity effects of land markets, Gini coefficients were calculated for three land holding variables: inherited land, owned land (inherited plus purchased), and total land (owned land plus rented land). The Gini coefficient for inherited land is high: .64. This is partly explained by the fact that as many as 21% of households reported that they do not operate any inherited land. The Gini coefficient reduces considerably following the inclusion of land purchases to .53. The percentage of land controlled by the largest 10% of farms likewise drops from 48.3% to 41.0% in comparing inherited land and owned land. Rental markets also play an equalizing role, as shown by the further reduction of the Gini coefficient to .51. The reduction is small because rented parcels are significantly smaller than purchased and inherited parcels.

Respondents were asked about the rental value per acre of each of their parcels. The range was between $4.30 and $210 per acre, with a mean of $33. The current rental value of rented land was estimated to be the same as that for purchased land, but both were greater than the estimated rental value of inherited parcels (mean of about $36 compared to $30 for inherited parcels). The stated rental values on rented parcels are obviously informed and strongly correlated with actual prices paid.

Interestingly, the percentage of purchased parcels in households currently headed by females is the same as that in males. This does not mean that women are active buyers of land, however, since many of the purchased plots were acquired two decades ago when the husbands may still have resided in the household. On the other hand, the percentage of rented plots was only 10% in female-headed households as compared to 15% for male-headed households, suggesting that it is easier for males to participate as tenants in land rental markets. Finally, the descriptive statistics shown in Table 5-2 indicate that inherited plots are the more fertile, purchased plots second, and rented plots the lowest score. This suggests that due to information asymmetry, parcels that are transacted through markets tend to be of inferior quality than others and that land is exploited by short-term tenants who are not interested in maintaining soil quality (Otsuka et al. 1992).

Land Markets and Tenure Security

We measure the degree of individualization of land rights by the ability to sell plots without permission from extended families or authorities (permission from spouses or children was not used to distinguish degree of individualization). According to Table 5-2, households have stronger individualized land rights on purchased plots than inherited plots. This is expected, because inherited land can be sold only with the consent of extended family members under many customary land tenure systems, and this is enforced through a requirement that land sales transactions be witnessed by village and extended family leaders. Although some tenants claim that they have individualized land

TABLE 5-2 Description of land income security and productivity

	All Plots	Tenure		
		Inherited	Purchased	Rented
	(1)	(2)	(3)	(4)
Plot Characteristics				
Plot Size (Ha)	0.9	1.2	1.3	0.4*
Years since acquired	21.9	23.3	24.1	11.7*
Title deed (%)	72.4	83.4	100	0.0*
Land Rights				
Concerned conflicts (%)	17.5	17.6	14.8	22.8
Rights to sell (%)[a]	55.5	56.0	76.4*	11.7**
Rights to plant timber (%)	72.0	78.6	86.0*	14.7**
Maize Plots				
Soil fertility index	0.36	0.50	0.20*	–0.36**
Yield (kg/ha)	1,549	1,436	1,789*	1,902*
Value Production (Kshs/ha)	14,253	12,722	17,579*	18,885*

Note: A blank, * or ** denotes similarity between numbers/values in the same row. Those with the same symbol (or lack of one) are not statistically different from each other while those with different symbols are.

[a]Individualized rights to sell land indicate that the households can sell land without approvals from extended families or local authorities.

rights, this could be because the landlords are their fathers, children, or other immediate family members.

In terms of titling land, we will focus on differences in titling between purchased and inherited land.[5] The differences between the two are striking. All purchased parcels were noted to be titled, whereas about 17% of inherited parcels are said to have no title. Furthermore, of the titled purchased parcels, 84% were titled in the name of the current household head or spouse. On the other hand, only 47% of inherited parcels were under title of the current household head or spouse. Almost 35% were still titled under one of the parents' names. This suggests that there is stronger incentive to update the registration and title status for purchased parcels. However, whether a current title confers stronger tenure security is an empirical issue; it may be that the lack of updating land titles on inherited land is because the holders already feel sufficiently secure. One particular concern is the tenure security of women in female-headed households. Across all inherited and purchased parcels in female-headed households, a worrying 22% are titled in the name of the deceased husband.

Estimation Models and Variables

For the analysis of land renting and purchase, we follow the models of Bliss and Stern (1982) and Skoufias (1995), which assumes that each household transacts land (i.e., selling, buying, renting in, or renting out) toward the most desirable size of cultivated land in the absence of transaction costs, which is determined by the endowment of nontradable, nonland resources such as family labor (see a theoretical discussion of this in Chapter 2). It is also assumed that a household is only able to adjust its farm size by a portion of the difference between the desired farm size and actual size because of transaction costs (along with credit constraints). For example, the leader of the extended family may want to confiscate the land rented out by his family member who possesses the cultivation right, and to reallocate it to some other members, because renting may be regarded as a sign of excessive family land holdings. In such a case, land renting is suppressed. We treat inherited and purchased land as exogenous or predetermined in the land rental regression. Following Skoufias (1995) and Tibako and Holden (2004), we estimate the following model, comparing tobit and double-hurdle methods[6]:

$$R_{ij} = -\theta \overline{A}_{ij} + \theta A^{*}_{ij}(H_{ij}, V_{j}) + \varepsilon_{ij} \tag{5.1}$$

where R_{ij} is the size of net rented in area of household i in community j.[7] Actual landholding is denoted by $\overline{A}_{ij}$, which is either inherited land or alternatively the sum of inherited and purchased land, and the desirable size of cultivated land by A^{*}_{ij}. Note that essentially the same estimation model will be applied to the determination of purchased land areas, while using inherited land to represent $\overline{A}_{ij}$. A^{*}_{ij} is not observable and so it is assumed to be a function of a set of household characteristics, H_{ij}, and community characteristics, V_j. To control for these characteristics, we include a set of household characteristics into the model explicitly as we will discuss later and also include community dummies.

We estimate θ from the coefficient of $\overline{A}_{ij}$, and test if it is significantly smaller than unity. If θ is less than unity but greater than zero, farmers who want to rent out land actually cultivate more land than desired, given their nontradable resources, and will therefore have lower land productivity than the optimal level, whereas farmers who want to rent in land will achieve higher land productivity because they devote more nontradable resources than the optimal level to their limited land. As a result of this allocative inefficiency, an inverse relationship between farm size and farm productivity is likely to emerge. This is probably the most important reason for observing an inverse relationship between farm size and productivity (Carter 1984; Newell et al. 1997; Benjamin and Brandt 2002). However, there appears to be at least two additional explanations for the inverse relationship discussed in the literature.

The first one is that small farmers produce more high-value crops than large farmers. The second alternative explanation is that small farmers' land quality is better than that of large farmers. Some previous studies have shown that after controlling for the soil quality, the observed inverse relationship between the farm size and performance disappeared (Newell et al. 1997; Benjamin 1995).[8]

In order to measure the degree of allocative inefficiency (through the inverse relationship between farm size and agricultural productivity), we estimated maize yield and maize income models. Although one possible reason for observing the inverse relationship is the crop mix, as discussed earlier, we excluded this possibility from our analysis by focusing on parcels with maize production. The yield and income models are specified as

$$Y_{sij} = \delta(\overline{A}_{ij} + R_{ij}) + \beta_Q Q_{sij} + \beta_X X_{sij} + \beta_H H_{ij} + \beta_V V_j + \eta_{sij} \quad (5.2)$$

where Y_{sij} is either log of the maize yield in kilograms per hectare or log of maize income expressed in Kenyan Shillings per hectare, on plot *s* of household *i*; Q_{sij} is the soil fertility index of plot *s*; and the other variables are the same as defined before. We estimated these models using community fixed effects, which can control for the effects of factor and product prices and agroclimate that are common within a local community. In order to examine the importance of controlling for the effects of soil fertility, we first estimated these models without the soil fertility index, followed by the estimation with such index. If the soil fertility significantly contributes to the inverse relationship between the farm size and performance, the inverse relationship may disappear when we insert the soil fertility index into the yield and income models.

Explanatory Variables

Since only a small proportion of households in our countrywide sample are found to rent out land, we examined only the rented in land area function, but not rented-out land.[9] In the less densely populated Laikipia case study, we examined determinants of area rented in and area rented out. A set of plot characteristics include the log of walking time in minutes from the homestead to the plot, the log of years since acquisitions, and a dummy variable for a steep plot. A set of household characteristics includes the log of owned area, a female-headed household dummy, the maximum education levels of male and female adults, the number of male adults, the number of female adults, the number of oxen, the log value of nonland assets, and a dummy for the second cropping season (the November to January harvest). We expect that owned area will have a negative effect on renting in of land and a positive effect of renting out.[10] With supervision costs making hired labor more expensive than household labor, the household labor supply will positively affect the decision to rent in land. As was explained, soil samples from the largest maize field for each household were used to develop a soil fertility index. We expect that the soil fertility variable will have a positive effect on maize yield and income.

Estimation Results

Determinants of Land Purchase and Renting In

We compared a double-hurdle model to a tobit specification for both the land purchasing and renting in. Although a likelihood ratio test revealed that the less restricted double-hurdle approach is the preferred specification, we have included the results from both in Table 5-3.[11] We will focus our discussion on the double-hurdle results but augment those with results on key variables from the tobit regression.

In the land purchase analysis we characterized inherited land by whether any was received (a dummy variable) and the amount received. This was done because of the unusually high number of households who operate no inherited land. In the land rental analysis, we combined inherited and purchased land together to form "owned land." In the land purchase participation regression (first column), the coefficient on inherited land is not significant but the coefficient on the dummy indicating no inherited land is positive and highly significant, showing that land purchase markets are a means for those without inherited land to acquire some land. The same pattern holds for the truncated model, suggesting that the size of land purchased is higher among those with no inherited land, but does not vary for different inherited landholding sizes among those with some inherited land. Hence, contrary to the premise of the Bliss-Stern and Skoufias models, purchases are not used so much to "adjust" holding size as they are used to secure land when inheritance is not an option. According to our selected interviews, young sons who did not receive inherited land worked outside the village to acquire capital and, if successful, returned later on to purchase land to initiate farming. We further ran tobit and truncated models dropping the dummy variable for noninheritance in order to assess the efficiency of the land markets. The coefficient on inherited land size was insignificant in the truncated regression, which is not altogether surprising since there are a large number of zero cases of inherited land. In the tobit model, the coefficient of –.82 was significantly different from zero, but its marginal effect is –.48 after transforming it from its unconditional value. Both estimates suggest that the land purchase market is inefficient in removing imbalances in inherited land holdings.

The education of the head of household and the value of assets also have positive impacts on market participation. This is sensible as land prices in many rural areas are high and education and asset value indicate that the more wealthy households are able to purchase land. The age of household

TABLE 5-3 Determinants of purchased land and land rent-in (Two step probit and truncated regressions with district fixed effects)

	Land Purchased (Ha)			*Land Rent-in (Ha)*		
	Probit Model	*Truncated Model*	*Tobit*	*Probit Model*	*Truncated Model*	*Tobit*
Total Land Owned (ha)				–0.0692 (2.42)**	.1328 (0.96)	–.1107 (2.27)**
Land Inherited (ha)[a]	–0.0083 (0.22)	1.8791 (1.09)	.0188 (0.17)			
No Inheritance Land (=1)	3.5191 (9.69)***	18.3220 (1.86)*	10.0931 (10.92)			
Household Characteristics						
Head's age	0.0071 (1.08)	0.8895 (3.18)***	.06777 (3.16)***	–0.0038 (0.79)	–0.0421 (1.50)	–.0112 (1.05)
Female Head (= 1)	–0.2580 (1.24)	1.7872 (0.32)	–.8229 (1.27)	–.1689 (1.08)	2.0915 (1.88)*	–.2168 (0.75)
Education of Head	0.0572 (3.22)***	0.3041 (0.55)	.1546 (2.29)**	0.0120 (1.46)	0.0770 (0.97)	.0496 (1.60)
ln (Asset values)	0.2267 (3.47)***	2.1688 (1.28)	.6945 (3.21)***	0.0760 (1.56)	0.3214 (1.16)	.1946 (1.81)*
Number of men	–.0226 (0.42)	2.9498 (1.46)	.0679 (0.31)	0.0132 (0.31)	1.0771 (2.90)***	.0878 (0.90)
Number of women	.1173 (2.47)**	.2736 (0.20)	.3078 (1.72)*	0.0737 (1.87)*	–0.0719 (0.27)	.1354 (1.48)
Number of oxen				0.2142 (1.98)**	–0.1685 (0.38)	.3611 (2.05)**
Constant	–4.7821 (6.10)***	–127.3689 (2.77)***	–19.1023 (5.94)***	–1.1406 (2.08)**	–7.1789 (1.87)*	–3.1461 (2.46)**
Observations	896	260	896	896	171	896

Absolute value of t-statistics are in parentheses and *, **, and *** indicate significance at the 10, 5, and 1%, respectively.

[a]If the No Inherited Land dummy is removed the coefficient on inherited land size is –.8249 and is statistically significant with a t-stat of 2.66.

head is positively related to the size of purchased land, which likely shows that available size of land offered for sale has decreased over time with population growth. The final significant result is that the number of women is positively related to participation in the purchase market, but no other household size variables are significant. If the land purchase markets functioned to adjust holdings to available labor, one would expect to see more significance among the household size variables. In this case, it is not apparent why the number of women is more important than the number of men.

The estimation results from the renting-in participation model shown in the fourth column shows that owned land size (inherited plus purchased) is highly negatively related to the renting in of land. Just as with the purchase regression, this shows a tendency for rental markets to reallocate land from the relatively land-rich to the land-poor. The amount of owned land did not influence the amount of land rented in, however (column 5), which shows that the land rental market is not efficient in equalizing land holdings (which would be indicated by a coefficient of –1.0). The tobit model, although obtaining a significant negative coefficient on owned land size, reveals a small coefficient that also supports the conclusion that land rental markets display allocative inefficiencies (column 6). Finally, in the last column, we have included inherited land size in hectares instead of owned land size, because the owned land size includes purchased land, which could be considered an endogenous variable. For instance, a household may decide to rent in land when it finds it difficult to purchase land. The result also suggests that the inherited land size has a negative and small coefficient, suggesting some, but not perfect, adjustments of land in the land rental market.

There are several important findings about the effects of household characteristics. We find that the number of oxen is significantly related to participation in rental markets, indicating that those households (relatively few in number) seek and are able to increase their land holdings to take advantage of improved land preparation capability. If instead there were efficient oxen-leasing markets, the coefficient would have been nonsignificant. The number of adult women is positively related to participation in the rental market, whereas the number of adult men is positively related to the size of rented land. This indicates that rental markets are to some extent improving factor ratio imbalances by moving land to households with more labor. It also suggests that labor markets are not functioning smoothly. The final result is that larger rented areas are associated with female-headed households. Thus, there appear to be two types of households accessing more land through rental markets: the female-headed households and the male-headed households with more adult males present. The former result may reflect the fact that female-headed households, some of which are widows, have small land holdings (male-headed households own 33% more land than female-headed households). Many women are unable to participate in rental markets, but those who do, tend to fare well.

The result on the number of male adults reflects the need for adjusting land sizes to meet the needs of more household members.

Determinants of Land Rented In and Rented Out in Laikipia District

A total of 320 households were sampled for interviews during 2002.[12] Agriculture is the main activity in the district and contributes 75% of household income in Laikipia. Nearly 79% of farmers in Laikipia practice food crop farming, suggesting that subsistence farming is the dominant activity. Cut flowers and french beans are grown for the export market. Livestock tending is also common, but less important than agriculture.

Laikipia district was designated as "White highlands" and had large-scale farms and ranches, group or individually owned lands. Land-buying cooperatives were formed to buy these large-scale farms left behind by departing British settlers. A number of cooperatives and land-buying companies still operate in the district, subdividing the farms and settling members on the plots. There has been continuous immigration into the district by the new settlers. Faced with increasing demand for land and government restrictions on subdivisions, many have resorted to land rentals to increase their land holdings. An important feature of the land rental market is that rental agreements are temporary. In the sample of 302 households, 102 had rented in land, and 86 had rented out land. The average size of inherited land was slightly below 2 hectares (5.2 acres), with the average inherited area for tenants being just 0.004 hectare (0.01 acre) and for landlords 3 hectares (8.3 acres).

Results of tobit regression models of amount of land rented in and rented out are presented in Tables 5-4 and 5-5. First, we estimate the model with the amount of land owned, and, second, we replace land owned with the amount of land inherited and the amount of land purchased. In Table 5-4, the coefficient on the land-owned variable confirms the hypothesis that households owning more land are less likely to rent in land. In column 2, we drop the land-owned variable but our results remain robust to the alternative specification. More important, they confirm the hypothesis that initial land endowment of the household is negatively correlated with renting in land. We also find that the purchased land size has a negative coefficient, suggesting that the land sales market partially substitutes for the land rental market. Households that purchased land have less need to rent in land.

Estimation results for the renting out of land are reported in Table 5-5. The sign on the coefficient for land owned is positive and significant, which indicates that large landholdings increase the likelihood of renting out land, keeping other factors constant. Possession of a land title has a positive and significant effect on renting out land. One argument is that property rights remain quite insecure in the eyes of landholders without the title and hence reluctance to rent out.

TABLE 5-4 Parameter estimates for tobit with amount of land rented in

	Land Rented in (Ha) (Tobit)	
Amount of Land Owned	–0. 243 (3.02) ***	
Land Inherited		–0.356 (1.31)
Area Purchased		–0.318 (1.67)
Age of Head	0.212 (2.94) ***	0.411 (2.26)**
Age of Head Squared	–0.035 (2.68) ***	–0.019 (2.19)**
Education of Head	–0.312 (1.94)*	–0.232 (2.36)***
Education of Head Squared	0. 179 (2.52) **	0.731 (2.21)**
Household Size	0.195 (2.75)***	0.154 (2.63) ***
Nr. Active Adults (15–55)	1.015 (2.25)**	0.916 (2.24)**
Livestock Value	–0.002 (0.512)	–0.001 (0.392)
Member Cooperative	0.954 (0.356)	0.845 (1.145)
Partner Land Buying	1.563 (2.456) **	1.235 (2.52)***
Distance to Road	–0.089 (1.89) *	–0.041 (1.96)**
Lumuria	2.152 (2.25) **	2.141 (2.26)**
Ng'arau	–2.123 (1.112)	–2.117 (1.102)
Rumuruti	0.604 (2.106) *	0.343 (2.28)**
Constant	–4.345 (6.065) ***	–4.342 (2.56)***
Nr. of Observations	102	

Absolute values of t statistics are in parenthesis. *, **, and *** indicate significance levels at the 10%, 5%, and 1% level.

We test for the presence of transaction costs on the land rental market and we test the joint hypothesis of the land owned parameters being equal to –1. This hypothesis is significantly rejected on both markets, which implies that adjustment realized by the land rental market remains imperfect. Next, we address the symmetry of adjustment when land is rented in or out. If transac-

TABLE 5-5 Tobit model of amount of land rented out in Laikipia

	Land Rented Out (Ha) (Tobit)	
Amount of Land Owned	0.057 (2.87)***	
Land Inherited		0.068 (1.67)
Area Purchased		0.121 (1.98)**
Age of Head	0.035 (1.82)*	0.054 (1.89)*
Age of Head Squared	–0.0002 (1.165)	–0.002 (1.23)
Education of Head	0.170 (0.324)	0.214 (0.324)
Education of Head Squared	–0.019 (0.352)	–0.024 (0.207)
Household Size	–0.012 (1.79)*	–0.142 (2.25)**
Nr. Active Adults (15–55)	–0.245 (3.25)***	–0.211 (2.34)**
Livestock Value	–2.320 (3.31)***	–1.984 (2.68)***
Member Cooperative	1.564 (2.566)***	1.502 (2.615)***
Partner Land Buying	0.075 (1.376)	0.091 (1.321)
Distance to Road	–0.244 (0.189)	–0.241 (0.203)
Lumuria	–0.472 (0.825)	–0.442 (0.685)
Ng'arau	2.543 (2.212)**	2.218 (2.19)**
Rumuruti	–0.815 (0.196)	–0.692 (0.231)
Constant	–3.345 (5.061)***	–3.131 (2.793)***
Nr. of Observations	86	

Absolute values of t statistics are in parenthesis. *, **, and *** indicate significance levels at the 10%, 5%, and 1% level.

tion costs affect the renting out of land in a similar manner as the renting in of land, we would expect equality of coefficients in both sign and magnitude. Recall that the land owned by households negatively affects the land area rented in and positively affects the land area rented out. We conducted a Wald test, in which the hypothesis of parameter equality cannot be rejected. These results show that land rental market transaction costs do not appear to be noticeably greater for tenants or landlords, so that improvements in land market efficiency would need to address both sides of the market.

Allocative Efficiency of Land Markets

In the bottom portion of Table 5-2, we provided information about maize yield and the value of maize production according to mode of land acquisition. Despite the lower soil fertility, the maize yield, as well as the value of maize production per hectare, is higher on rented in plots than on inherited plots partly because a larger amount of chemical fertilizer is applied on rented-in plots. However, after deducting paid-out costs for inputs and labor, the net revenue or income from rented in plots is not significantly different from that of inherited or purchased plots. Note that the income thus calculated is expected to represent returns to family labor, land, and management ability.

For the analysis of yield and income, we focus on one single crop (maize), which is grown by all households during the long rainy season. According to the maize yield model without the soil fertility index (see the first column in Table 5-6), we find a strong inverse relationship between farm size and maize yield. However, when we include the soil fertility index, we find that the inverse relationship is no longer significant (see the second column). Next, we estimate the maize income model without and with the soil fertility index (third and fourth columns in Table 5-6). Again, we find a large negative coefficient on the farm size in third column. Similar to the yield model, the size of the negative coefficient reduces its size, and is no longer statistically significant when the soil fertility index is added in the fourth column.

These findings suggest that small farms are more fertile, which is partly responsible for the inverse relationship between the farm size and maize yield. Indeed, Figure 5-2 shows a negative relationship between the soil fertility index and farm size, which is created by using a nonparametric regression model. Small farmers apply more inputs per unit of land than do large farmers and thus the soil fertility of small farms improves or at least does not deteriorate as much as large farms. (See Chapter 13 for more analysis of land investments.) Thus, the nonsignificance of farm size when the soil fertility index is used as an additional explanatory variable reinforces the inverse relationship between farm size and performance. Indeed, it seems reasonable to conclude that the imperfection in the land market leads to a significant maize production inefficiency in land allocation across households.

It is important to emphasize that both maize yield and income are significantly lower on rented in plots than on inherited plots, except in the last column. The last result, however, is not contradictory, because soil fertility is lower on the rented in plots (recall the data in Table 5-2). There is no sign of significant difference in maize yields and income between purchased and inherited land, which does not support our expectation.

TABLE 5-6 Determinants of maize yield and income per hectare

	Maize Yield		*Maize Income*	
	Basic Model	*Add Fertility*	*Basic Model*	*Add Fertility*
	(1)	*(2)*	*(3)*	*(4)*
Farm Size				
ln (Farm Size in ha)	–0.123 (1.82)*	–0.072 (0.80)	–1,001 (1.83)*	–706.1 (1.01)
Plot Characteristics				
Purchased Plot (= 1)	–0.192 (1.54)	–0.307 (1.81)*	–907.5 (0.90)	–1,849 (1.39)
Rented In Plot (= 1)	–0.422 (2.17)**	–0.557 (1.68)*	–2,726 (1.74)*	–973.5 (0.38)
Soil Fertility Index		13.33 (1.18)		165,571 (1.87)*
ln (Walking time)	–0.096 (2.27)**	–0.162 (2.31)**	293.1 (0.86)	–216.2 (0.39)
ln (Years since acquired)	–0.055 (1.00)	–0.140 (1.75)*	–433.8 (0.97)	–555.0 (0.89)
Steep Plot (= 1)	0.040 (0.34)	0.104 (0.64)	–1,320 (1.39)	–588.9 (0.46)
Household Characteristics				
Female Headed (= 1)	–0.370 (3.05)**	–0.271 (1.70)*	–2,498 (2.55)**	–2,236 (1.79)*
Education men	0.014 (0.87)	0.017 (0.79)	108.3 (0.84)	233.8 (1.43)
Education women	–0.011 (0.64)	–0.003 (0.12)	–229.2 (1.71)*	–267.6 (1.61)
Men	–0.030 (0.74)	–0.018 (0.33)	–208.1 (0.64)	–38.78 (0.09)
Women	0.106 (2.58)**	0.132 (2.41)**	645.3 (1.95)*	1,059 (2.48)**
Oxen	0.106 (1.66)*	0.133 (1.59)	528.0 (1.02)	427.7 (0.65)
ln (value assets)	0.018 (0.37)	–0.002 (0.03)	–827.3 (2.11)**	–1,068 (2.02)**
Second Season (= 1)	0.430 (4.25)**	0.308 (2.28)**	–1,207 (1.48)	–2,685 (2.55)**
Constant	6.055 (13.06)**	6.161 (9.48)**	17,321 (4.62)**	16,518 (3.26)**
Observations	1,923	1,180	1,923	1,180

* and ** indicate the 10 and 5% significance level, respectively.

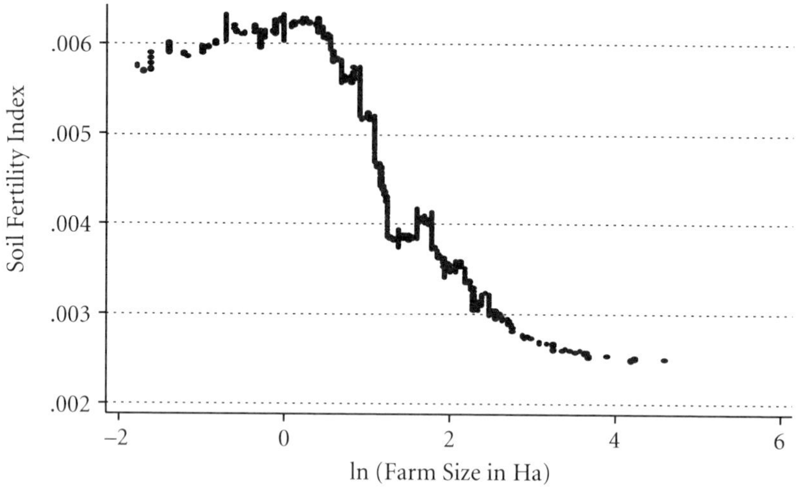

FIGURE 5-2 Soil fertility index and farm size in ha on maize plots

Concluding Remarks

Evidence from rural Kenya shows that both the land sales and rental markets promote the movement of land from the land-rich to the land-poor. As such, they both improve the equity of landholdings. Thus, fear that land sales might lead to increased concentration of landholdings is not supported by our analyses of smallholder farming communities in Kenya. Land sales markets are particularly important in providing land to those who were unable to inherit land, but they are not used for minor adjustments of landholdings. Rental markets, on the other hand, do tend to be used for short-term adjustments in factor ratios, but the size of land transacted is relatively small. Both markets together are unable to bring about the socially optimum distribution of landholdings because of the high transaction costs. In other words, land markets tend to transfer land from land-rich to land-poor households only to some extent. The resulting distribution still generates inefficiencies in production as the smaller farms have higher maize productivity than the larger ones. In this restricted sense, land markets contribute to both efficiency and equity of land allocations among farm households in Kenya.

The most critical remaining question is why land markets do not function more effectively in Kenya, despite the most thorough implementation of land titling programs in sub-Saharan Africa. This may be partly due to poorly functioning credit markets whereby households that would like to increase their landholdings are unable to raise significant financial resources to do so. Land prices are also very high in rural Kenya because of the severe land scarcity. Consequently, land plays many roles in society, due to inefficiencies in other markets and in government social insurance programs. This tends to increase the sales price of land beyond the level justified by potential productivity. If these other markets and programs were in place, one would expect the price of land to fall and more land market activity. The inverse relationship between farm size and farm productivity also arises from the poor functioning of land rental markets. Potential landlords may be reluctant to rent out land for fear of losing land rights to tenants or other claimants from within their extended families. Thus, we suspect that the implementation of a land titling program alone is not sufficient to strengthen land rights. We will take up this issue again in Chapter 13, when we analyze the land conflicts in Kenya.

Notes

1. The model has been estimated recently by Tikabo et al. (2008) in Pender and Fafchamps (2006) in Ethiopia, and Kimura et al. (2007) in China; see also Chapters 3, 4, 6, and 7 for other estimates in Africa. See also Chapter 2 for a further elaboration and extension of the model for the case with nonlinear transaction costs.
2. RePEAT is Research on Poverty, Environment and Agricultural Technology and has been launched to generate panel datasets in Kenya, Uganda, and Ethiopia. The project is funded and coordinated by GRIPS and FASID of Tokyo (Yamano et al., 2005).
3. A small number of samples were spoiled or lost, reducing the number of parcels that could be included in analyses using the soil fertility variable.
4. Among the sample households we find that less than 3% of the households have rented-out land. We might have failed to capture information about rented-out land. Therefore, we examine determinants only of rented in land in this study.
5. In principle, rented parcels cannot be titled by the operator.
6. As noted earlier, because the data on renting out of land is poor, the model captures only the renting in of land which is truncated at zero.
7. We estimate a similar function for rented out land using the Laikipia dataset.
8. Many of the studies on the inverse relationship come from Asian countries, but only a few studies have examined this issue in Africa (Pender and Fafchamps, 2006).
9. We are not sure why there are far fewer households renting out land than renting in. We suspect that we failed to obtain accurate data on renting out land.
10. An exception would be where land ownership increased access to credit markets in which case it may have a positive correlation with both renting in and purchasing of land.
11. The double-hurdle approach is captured by the coupling of a probit model on participation in the markets and a truncated regression model on amount of land acquired (conditional on having participated).
12. The data were collected as part of the International Food Policy Research Institute (IFPRI) eastern Africa Food Policy Network Studies (Obunde 2004).

CHAPTER 6

Land Rental Market Participation and Tenure Security in Malawi

RODNEY LUNDUKA, STEIN T. HOLDEN, AND RAGNAR ØYGARD

Security of property rights for agricultural land has been argued to have important effects on investments, access to credit, and the functioning of land sales and rental markets (Feder and Feeny 1991; Besley 1995; de Soto 2000). In many parts of sub-Saharan Africa, the system of communal property rights on cultivated agricultural fields has moved in direction of more individualized rights (Bruce and Migot-Adholla 1993; Otsuka and Place 2001). Under individualized property rights and immobility of land, when partners move together to make a household, the moving partner gains only limited rights to that land, and may be less tenure secure than the partner. For example, in patrilocal societies women often obtain usufruct rights to family land, but do not possess inheritance rights (Lastarria-Cornhiel 2001). When, or if, the marriage is terminated through death or divorce, the partner who brings land into the family may have more secure rights to the land than the other partner (see Chapter 9). Other sources of insecurity recognized in literature are land conflict, unclarified land rights, and attempts by tenants to squat in rented land (Deininger 2003; Chapter 13).

Several studies have accessed the impact of tenure security on investments (Deininger and Jin 2006; Amsalu and de Graaff 2006; Place and Hazell 1993; Place and Otsuka 2002; Braselle et al. 2002; Feder and Feeny 1991; Platteau 1996) and have found diverging results. Few studies have looked at tenure security and land rental market participation in Africa (see Chapters 3, 4, 5, and 7 for details). In examining the challenges facing land rental markets in the Europe and central Asia region, Vranken and Swinnen (2006) indicated that security of ownership is a condition for the efficient operation of land rental markets. Where land rights are insecure, landlords are reluctant to rent out for longer periods, are less likely to use formal contracts, and restrict renting to farms from the same ethnic or social group. Macours et al. (2004) found that in the Dominican Republic insecure property rights not only reduce the level of activity on land rentals but they also induce segmentation. Landlords who have reasons to fear losing their land will restrict renting to narrow local circles of confidence. Despite tenure security being a very highly debated issue in Africa, few studies have examined the relationship between emerging land markets and tenure security. In this chapter we focus on the link between tenure security and participation in the emerging land rental markets in Malawi.

Holden et al. (2006) provided evidence of emerging land rental markets in Malawi through which land-poor households are able to access land through the markets. However, the inheritance system provides differing tenure security to individuals in the household depending on the amount of land they bring into marriage and their residential area. The amount of land brought into marriage may affect the household decision either to rent in more land or rent out excess land. The bargaining power of the individuals plays an important role in the final decision on use of land. Quisumbing and Maluccio (2003) found that land brought into marriage by either spouse influenced decisions regarding the use of household resources.

This chapter uses random effects instrumental probit and tobit models to analyze how intrahousehold land rights and tenure security affect decisions to participate in the land rental market as landlords or as tenants, and by how much to participate. We propose that individuals who reside in the home area of their spouse tend to be more insecure and they have weaker rights to make decisions on the land—and therefore are less able to participate in land rental markets. Based on the assumption that insecure individual rights to land at the beginning of marriage reduces the bargaining position of an individual in a household, this study tests whether the inheritance system (matrilineal or patrilineal) and residence (patrilocal or matrilocal) affect land renting decisions. It is hypothesized that men are more active in the land rental market if they possess stronger land rights.

Land Tenure System in Malawi

There are three legally recognized tenure regimes operating in Malawi: the public tenure system,[1] the private tenure system,[2] and the customary tenure system.[3] The customary tenure system is by far the largest in terms of land area and the number of people working. Customary tenure systems in the late 1970s operated on about 80% of the total arable land, and by 1997 less than 10% had been converted to leasehold.

Within the customary tenure sector, the methods of land transfer differ principally according to descent practices—namely, matrilineal or patrilineal—and to residency practices—namely, patrilocal (wife residing in the husband's village) or matrilocal or uxorilocal (husband residing in wife's village) (Place and Otsuka 2001a). In the matrilineal system, as Peters (2002) reports, descent, succession, and inheritance run through the matriline. Marriage is overwhelmingly uxorilocal, so that men marry into a village and use the land belonging to their wives' families. Land is passed matrilineally, but almost all passes go to daughters and not to sons. This is because sons leave; that is, they get married and are then expected to use land belonging to their wives (Peters 2002). Therefore, in such a system women have more secure land rights than men. Men married under an uxorilocal system may experience insecurity, as their continued enjoyment of land rights in the matrimonial village depends on the longevity of the marriage (Place and Otsuka 2001a). A divorced man or one who has been predeceased by his wife is expected to return to his own village. The children remain in their mother's village because they are deemed to belong to their mother's matrilineage. Young men still living at home or who return for periods between divorce and remarriage (which is common) may be given a field to use by their mothers or sisters. However, it is very rare for a married man to have full rights to a field belonging to his matrilineage. In some cases married sons who have received fields are in families where there is no daughter who can inherit, or who have large amounts of land—more than sufficient for the claims of daughters.

The other system, common in the northern part of the country, is both patrilineal (receiving land through the father's side) and patrilocal (the couple living in the husband's village) (Place and Otsuka 2001a). Land is passed from the father's side to sons, or in some very few cases (e.g., if there are no sons or if there is abundant land) to daughters. The son continues to reside in the father's village, normally bringing in a wife from another village. In patrilineal systems a divorced woman is expected to return to her own village and relinquish all rights she had to her husband's land. A widow would normally remain in her deceased husband's village and continue to enjoy cultivation rights, provided a bride price was paid for her; but she may be forced out if her deceased husband's kin believe that she somehow contributed to her husband's death, or the marriage produced no offspring, or she refused to be inherited by a kinsman of the dead as required by custom, or she is otherwise considered to be of bad character.

Divorce, or death of a spouse under both residential arrangements, effectively renders the nonlocal partner landless. When such persons return to their own villages, allocations of land made to them are understood to be temporary, as those entitled to the land under customary rules may claim it at any time. To avoid losing land rights in the circumstances discussed, men in matrilineal social systems may opt for neolocal[4] residence, which would give them more control of the land rights, as opposed to their wives. Therefore insecurity of land tenure to an individual is reduced if he or she moves away from his or her original home. Figure 6-1 shows that the residence of the household determines the individual's land tenure security.

As indicated earlier, the matrilocal system is slowly being replaced by the patrilocal system. Men now prefer to take their wives to their village in order

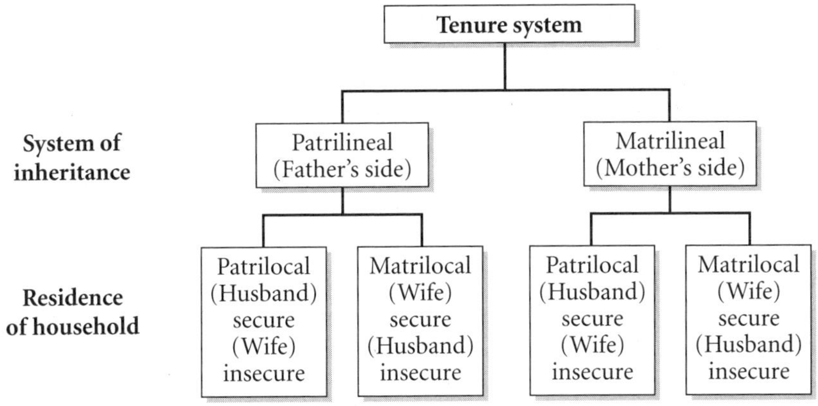

FIGURE 6-1 Source of individual land insecurity

to increase their tenure security, as they there have full rights over the land resources. However, this requires payment of a bride price that in itself is determined by the resource endowments the man has before getting marriage. If the man has more resources, he is able to pay the bride price, but if he has less endowments, he is forced to move to the wife's village. Therefore, better-off men would prefer to increase their security by paying the bride price.

Theoretical Framework

Rural peasant farm households typically face a number of imperfections in markets for factors of production (Sadoulet and de Janvry 1995; Eswaran and Kotwal 1985; Bell and Sussangkarn 1988; Bliss and Stern 1982; Feder 1985; Chapters 2, 3, 4, and 5). High transaction costs in factor services such as labor lead to market imperfections. Thus, households having, for example, surplus labor in relation to their landholdings rent in land, whereas households with excess land in relation to their factor endowments rent out land. In light of this, we postulate that participation in the land rental market in Malawi is a result of imperfections in the markets for labor (including management) and credit.

Households' Decisions to Participate in Land Markets

A household with endowments of land, labor, and initial wealth can derive income from agricultural production on its own farm and from off-farm wage employment. Agricultural production is based on the operational land used by the household, family labor is the effective labor input on the farm, and the initial wealth is used to purchase inputs (seeds and fertilizer) and pay rent, if the household rents in land. In Malawi, fixed rent contracts are dominant (Holden et al. 2006), and hence our theoretical framework is developed on the assumption that farmers have only fixed rent contracts.

We refer to the Bliss and Stern (1982) model as presented in Chapter 2. Bliss and Stern argued that participation in the land market is an attempt to make up the difference between desired cultivated area (*DCA*) and owned land (L) and this difference is the net land rented in (L_j). The household can either adjust nonland endowments (e.g., labor), the land rented in or out, or other inputs. In our case, however, where rights and tenure security may be different for the husband and wife depending on inheritance system and residence location, they may also each have a different "desired cultivated area" that will depend on their relative bargaining power and control over resources. Still, if we deviate from the unitary household model and replace it with a bargaining model, also such households face similar fixed and variable transaction costs in the land rental market, and the basic hypotheses that are derived from the model are still relevant to test.

We will therefore test the basic hypotheses derived from the Bliss and Stern model, assuming that the labor endowment is the most important non-land resource endowment in Malawi:

H1: Land-rich households will rent out land, whereas land-poor households will rent in land.

H2: Households with less labor endowments will rent out land, whereas households with more labor endowment will rent in land.

The extent of non-participation in the land rental market may be an indicator of the fixed and variable transaction costs in that market. Nonparticipants may include households that are completely rationed out of the market (Bell and Sussangkarn 1988; Skoufias 1995), and rationing may be explained by other factors than those explaining the degree of market participation.

If we relax the assumptions of the unitary household model, it is possible that intrahousehold characteristics related to inheritance rights and location of residence may affect land rental market participation. Inheritance rights and location of residence may matter for the degree of control and relative bargaining power of husband and wife. Rights to land are normally held primarily by the partner who actually brought the land into marriage and this also typically determines location of residence, R ($R \in$ (*patrilocal*, *matrilocal*)). Residential location is also influenced by the inheritance system, I (i.e., $I \in$ (*patrilineal*, *matrilineal*)) which also affects the rights of individuals within the household.

With these complications we resort to estimating a reduced form model where we assess whether net land rented in is affected by inheritance system and residential location in addition to the nonland resources and inherited land.

$$L_j = f(N_h, L^i, I, R(L^i, Z_h)) \tag{6.1}$$

L_j = Net land rented in
L^i = Inherited land by husband or wife
R = Residence, $R \in (1 = matrilocal, 0 = patrilocal)$
I = Inheritance system, $I \in$ (1 = matrilineal, 0 = patrilineal)
N = Nonland factors of production
Z_h = Household characteristics
$j \in \{s = \text{landlord}, b = \text{tenant}, * = \text{non-participant}\}$
$i \in$ (husband, wife)

Land is immobile, and this has important consequences. For example, if a husband moves to live in the wife's village, he cannot move with his land. This may imply that he loses his rights to land in his village. He can only gain usufructory rights to land in the wife's village. He does not have rights to sell, subdivide, or even rent out the land. A similar situation arises for a wife moving to her husband's village. Therefore, the decision to rent out will depend more on the spouse with such rights or the bargained rights to land gained in the marriage. However, the husband living in his wife's village may have a stronger influence on the decision to rent in land, since the man is typically the head of household. For this reason, the wife living in the village of her husband would have a weak influence on the decision to rent in land. The stronger position of the husband as the household head may imply that the matrilineal and patrilineal systems do not represent "mirror images," as the husband's position in matrilineal communities in relation to household decision making over land is likely to be stronger than the wife's position in patrilineal communities.

The total effect of a change in land transacted in the rental markets may be decomposed into a land endowment effect, $\partial L/\partial L^i$ (households with more inherited land will rent in less land and rent out more, and households with less inherited land will rent in more land and rent out less) and a shift in the rights or security effect, $\partial f/\partial R * \partial R/\partial L^i$, resulting from a change in the residence, which is itself affected by the landowner and inheritance system (patrilineal or matrilineal). That is

$$\partial L_j \,|\, \partial L^i = -\partial f/\partial L^i \,|_R + \partial f/\partial R * \partial R/\partial L^i \qquad (6.2)$$

We may hypothesize that an increase in security of tenure of (potential) landlord households increases rental market participation. Thus, hypothesis H1 must be qualified to the extent that the residence affects the tenure security.

The decisions on participation in land markets may follow the rights on the inherited land. This means that if we assume symmetry in rights to land by the husband and wife in the household, then residential location should not matter for land renting decisions. However, Peters (2002) and Place and Otsuka (2001a) reported that under the matrilocal system, in which men bring less or no land into marriage, it is still primarily men who make the major farm management decisions, including decisions to invest in land improvement. This may also imply that they dominate land rental decisions when it comes to renting in decisions, while they may have less influence over renting out decisions. We therefore hypothesize that

> H3: Men in matrilineal-dominated society have less security and control over land and will hence rent in more and rent out less land than households in predominantly patrilineal societies.

Residential location may be determined by the inheritance system and the assets, including land that the spouses potentially have control over. The chosen location is likely to be where the married couple gains better access to resources. The consequence of the choice of residential location may, however, also affect the relative bargaining power of husband and wife in terms of controlling the resources they brought into the marriage. We assume that only user rights are given to the spouse but rights to sell and rights to subdivide remain with the person who inherited the land. It is very rare for a married man to have full rights to a field belonging to his matrilineage (Peters, 2002). These rights are exogenous and can be used to determine residential area. Land rights are treated separately from security of tenure, which is defined as the probability or likelihood of losing land as used by Holden and Yohannes (2002), Sjaastad and Bromley (1997), and Alemu (1999). We asked the households the probability of losing the user rights to the land due to any other external factors such as death of the spouse, divorce, or land being grabbed away by other people. We perceived this to indicate their security on the land. The security of tenure is exogenously treated in our analysis and we assume that it is determined by culture and traditions (e.g., the risk of being evicted after divorce or death of spouse).

Data and Econometric Estimation

Sampling, Data, and Descriptive Statistics

The data used in this study are from six districts (Thyolo, Chiradzulu, Zomba, Machinga, Kasungu, and Lilongwe) with a total of 462 households. Table 6-1 shows that 62% of the households were in matrilineal societies and of those, 47% lived in the wife's village and 15% lived in the husband's village. Of the 42% that were in a patrilineal society, 12% lived in the wife's village and 26% lived in the husband's village. The average landholding size was highest for patrilocal-patrilineal households and lowest for matrilineal-matrilocal households. This could be because the majority of the matrilineal households are in the south, which generally has a higher population density and smaller landholding size than the patrilineal-dominated area of central Malawi. However, there is a systematic pattern such that patrilocal households both in matrilineal and patrilineal societies have larger land sizes as compared to their counterparts in matrilocal societies (see Table 6-1).

TABLE 6-1 Number of households by inheritance system, residential area, land rental market participation in 2005/06 cropping season, and household land endowment

Inheritance System and Residential Area	*Number of Households*	*Percentage of Households*	*% Renting In*	*% Renting Out*	*Average Land Endowments, Ha*
Matrilineal					
matrilocal	199	47	21	6	0.784
patrilocal	64	15	14	14	0.8
Patrilineal					
matrilocal	53	12	23	4	1.21
patrilocal	109	26	26	13	1.476
Total	425	100	20.3	7.56	1.038

TABLE 6-2 Description, mean, and standard deviation of the variables used in the regression

Variable	*Mean*	*Standard Deviation*	*Gini Coefficient*	*N*
Renting in Dummy *(1 = yes 0 = no)*	0.207	0.406		425
Amount of land rented in (ha)	0.439	0.266		95
Amount of land rented out (ha)	0.423	0.240		36
Total land inherited (ha)	1.038	2.153	0.4613	425
Total operated land (ha)	1.106		0.4342	425
Age household head	41.426	16.202		425
Education of household head *(years in school)*	6.998	3.916		
Per capita labor supply of household per year *(hours)*	767.935	838.226		425
Value of assets (MK)	5830	18160		425

Of the 462 households, 28% participated in the land rental market in the 2005–2006 growing season. Participation in land rental markets was observed in all four types of societies. Renting-in land was more common among patrilineal-patrilocal households (26% of the households) and less common among matrilineal-patrilocal households. Renting out was more common among patrilocal households in both patrilineal and matrilineal societies.

Table 6-2 shows that the average amounts of land rented in and out per household were 0.439 hectare and 0.423 hectare, respectively. Average total operational land per household in 2005–2006 growing season was 1.106 hectares. The Gini-coefficients for inherited land and operated land were 0.46 and 0.43, respectively. This indicates that the land rental markets contributed to

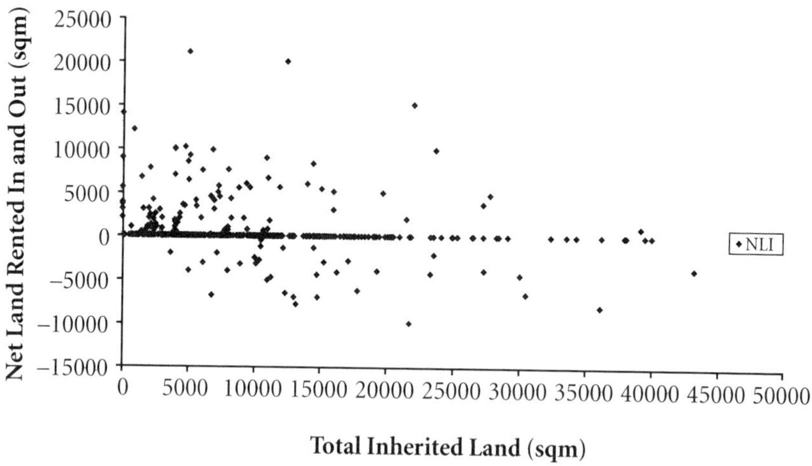

FIGURE 6-2 Net rented land vs. total inherited land (sqm)

equalizing the land distribution. This is further illustrated in Figure 6-2, relating own farm size to net land rented in or out. One can see that land-poor households and even landless households can access land through the land rental market. The land rental market to some extent reallocates land from land-rich to land-poor households, leading to a more egalitarian land distribution.

Note from Table 6-3 that tenants were land-poor households with the lowest average inherited land of 0.69 hectare and landlords were relatively land-rich with average land size of 1.56 hectares. This was similar in matrilineal and patrilineal areas, but patrilineal areas have larger land sizes for both tenant and landlord households. Tenants had higher asset values than landlords in all districts, but when breaking up in patrilineal and matrilineal areas, we see that this was due to the large difference in patrilineal areas, while this was not the case in matrilineal areas. In matrilineal areas, landlords had higher asset values than tenants, and nonparticipants were the poorest in terms of assets. Tenants also tended to have higher labor endowments and household sizes than landlords. Tenants (heads of households) were also on average younger than landlords and nonparticipating households.

In Figure 6-3 households are ordered by the size of net land rented in, illustrating the relative size of the three categories of households relative to the land sizes rented in or out. Relatively few households rented out land, and the maximum land rented out is about 1 hectare. It was also more common to rent in very small pieces of land than to rent out small pieces of land. This may indicate something about the nature of the transaction costs in the market. The willingness to rent in small pieces of land may indicate that there are no economies of scale and entry barriers due to lumpiness of some inputs. This may be due to hoe-based cultivation, unlike in Ethiopia where oxen are used for plowing (see Chapter 4). The land rental market may therefore be a pathway for land-poor and landless households to access land. The graph may also

TABLE 6-3 Description, mean of the variables used in the regression

Variable	All Districts			Matrilineal			Patrilineal		
	Tenants	*Non-participants*	*Landlords*	*Tenants*	*Non-participants*	*Landlords*	*Tenants*	*Non-participants*	*Landlords*
Average inherited land (ha)	0.69ab	0.97ac	1.56 bc	0.53ab	0.82ac	1.3bc	0.92ab	1.2ac	1.7bc
Land rented in	0.460	0	0	0.34	0	0	0.581	0	0
Land rented out	0	0	0.413	0	0	0.382	0	0	0.445
Assets value (MK,000)	10.42a	5.22a	6.22	7.7a	3.9ac	8.9c	14.2ab	7a	4.4b
Total labor	0.94ab	0.8a	0.7b	0.92a	0.75a	0.79	0.95b	0.869	0.70b
Male labor	0.34ab	0.27a	0.25b	0.33a	0.25a	0.30	0.35b	0.31	0.22b
Female labor	0.31ab	0.27a	0.25b	0.31b	0.26	0.24b	0.30	0.29	0.26
Household size	5.4	5.3	4.6	5.3b	5.1	4.2b	5.6	5.6	4.9
Age	37.5ab	43.3a	44.3b	37.5ab	44a	43.1b	37.6ab	42.4a	45.1b
CW-ratio	1.36	1.4	1.28	1.2	1.47	1.3	1.5	1.3	1.2
Education	7.6ab	6.8a	6.5b	8.3a	6.7a	6.1	6.7	6.9	6.8
Rights									
Sell land	0.46a	0.35a	0.37	0.34a	0.24a	0.14	0.64ab	0.53a	0.52b
Sub-divide	0.57ab	0.39a	0.37b	0.52a	0.3a	0.21	0.64a	0.52a	0.57
Security									
Probability not to stop cultivate	0.22ab	0.73a	0.74b	0.25ab	0.73a	0.85b	0.17ab	0.07a	0.66b
Conflict	0.1ab	0.051a	0.085b	0.02ab	0.07a	0.07b	0b	0.02	0.09b
Number of observations	96	331	35	55	200	14	41	132	21

Note: a,b,c are significant levels at 10%. a = tenants different from nonparticipant, b = tenant different from landlords, and c = landlord different from nonparticipant.

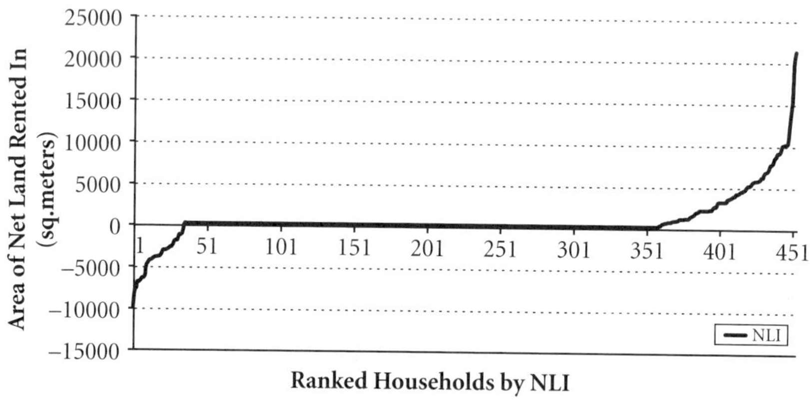

FIGURE 6-3 Net land rented in

imply a need for assessing the two sides of the markets separately, as different factors may be at play on the two sides. Tenure insecurity, customary rules, and poverty may also be at play and restrict adjustment on the landlord side, as focus group discussions revealed that the land rental market may serve as insurance during times of emergency (distress rentals), such as for cash to pay hospital bills, buy food, or even buy inputs.

The high share of nonparticipating households (the flat part in Figure 6-3) may indicate that there are significant transaction costs in the market. This may also be due to its recent occurrence (Fafchamps 2004) as market participation may grow gradually with expansion of trust and experience with such transactions.

Econometric Estimation

A simple linear specification of the reduced form model was chosen. This can be written in its econometric form as:

$$L_j = \alpha_j + \beta_j N + \beta_j I + \delta_j R - \xi_j L^i + \mu_j \qquad (6.3)$$

This equation can be estimated econometrically to study the leasing behavior of households participating in the land rental market, although estimation is not straightforward. Leasing behavior involves two decisions: the decision to participate and how much to participate, given participation on each side of the tenancy market (Tikabo et al. 2008). To investigate the decision to participate, we used an ordered probit model by categorizing households into tenants, nonparticipants, and landlords. We let the categories of farmers be the latent variable $J \in (b, *, s)$.

$$J = \beta_i N + \beta_{jL} + \delta_j \hat{R} - \xi_j L^i + e_j \qquad e \mid h, L, R, Z \sim Normal\ (0,1) \qquad (6.4)$$

The error term *e* is assumed normally distributed and the cumulative density function is Φ. We let α_1 and α_2 be cut-off points and we define: $J = -1$ if $L_j \leq \alpha_1$ $J = 0$ if $\alpha_1 < L_j \leq \alpha_2$ $J = 1$ if $L_j > \alpha_2$.

To investigate how much to participate and the effects of security on participation, we use an instrumental tobit model on area rented in or out:

$$L_j = \max(0, \alpha_j + \beta_j N + \beta_j I + \xi_j L^i + \delta_j(\beta_{jh} L^i + \beta_{jI} I + \theta Z_j + v) + \mu_j \qquad (6.5)$$

$$u \mid N, L, Z \sim normal(0, \sigma^2 + \delta^2 \tau^2$$

$$R = \gamma_j + \beta_{jh} L^i + \beta_{jL} I - \xi_j Z + v \qquad v \mid N, L, Z \sim Normal(0, \tau^2) \qquad (6.6)$$

Using probit to predict $\hat{R}$ and get the predicted error term $\hat{v}$ and using the *R* and the predicted error term in a tobit model of the net rental in land on household and plot variables gives consistent estimators (Wooldridge, 2002). We therefore estimated the following model with robust standard errors:

$$L_j = Max(0, \alpha_j + \beta_j N + \beta_j I - \xi_j L^i + \delta_j R + \theta \hat{v} + e_j) \qquad (6.7)$$

Using the ordered probit model, we tested the hypothesis that participation in land markets is a result of households adjusting available cultivable land to the labor endowments (i.e., households with more labor relative to land will rent in land, whereas households with less labor relative to land endowment will rent out land. Using the instrumental probit model, we further investigated if the security and residential location influence the decision for the households to be either a landlord or tenant. We estimated the determinants of renting by including the residential location variable, which is a proxy for tenure security. However, residential location is endogenous as it is a function of other variables that also affect renting in for inherited land rights. As a remedy for the endogeneity, residential area was instrumented by rights to sell land, rights to subdivide land, and initial inherited land. These are assumed to be exogenous as they come before the residential decision. A wife or husband will have acquired the rights to land even before getting married. These instruments are used in the instrumental tobit model.

Findings

Participation in Land Markets

As indicated earlier, 28% of the farmers interviewed participated in land renting either as landlords or tenants. Using an ordered probit model for the households that participated in rental markets by either renting in or renting out and those that did not participate, we found that having more inherited land had a negative effect on the rented in land for all the districts and separately for matrilineal-dominated and patrilineal-dominated areas. Table 6-4 provides the results, which indicate that farmers were using the land market to adjust for imperfections in other factor markets like labor. Tikabo and Holden (2003) demonstrated the importance of the land rental market for adjustment to nontradable or semi-tradable nonland household endowments in Eritrea. Ghebru and Holden (Chapter 4) demonstrate the same in Ethiopia.

Labor is another factor that is important for households to be able to get into the land market. Households with less labor were unable to increase their operational holding, hence both male and female labor were positively correlated with leasing in land. Male-headed households preferred to rent in land when in matrilineal societies, as indicated by a significant positive coefficient

TABLE 6-4 Parameter estimates for ordered probit regression on decision to participate in land markets

	All districts *b/se*	*Matrilineal dominant* *b/se*	*Patrilineal dominant* *b/se*
Inherited land	–0.405** (0.17)	–0.495** (0.25)	–0.354 (0.24)
Sex of hhh	0.107 (0.17)	0.394* (0.23)	–0.119**** (0.02)
Male labor	0.290 (0.50)	–0.450 (0.39)	1.024** (0.41)
Female labor	0.786*** (0.27)	1.202** (0.58)	0.393**** (0.08)
Cwratio	0.040 (0.05)	0.122** (0.05)	–0.018 (0.06)
Educ of hhh	0.005 (0.03)	0.037 (0.03)	–0.014 (0.02)
Asset value	0.013 (0.01)	0.004 (0.03)	0.012 (0.01)
Age of hhh	–0.006 (0.01)	–0.006 (0.01)	–0.005 (0.01)
Security	–0.905**** (0.15)	–1.063**** (0.20)	–0.811**** (0.20)
Cut1_cons	–2.220**** (0.50)	–2.175** (0.85)	–2.240*** (0.72)
Cut2_cons	0.311 (0.45)	0.676 (0.67)	0.086 (0.60)
Prob > chi^2	0.000	0.000	0.012
Number of obs.	394	225	169

Robust standard errors. The superscripts ****, ***, ** and * indicate the 0.1% 1%, 5%, and 10% levels of confidence, respectively.

of sex of household head in matrilineal districts. As they could not bargain for more inherited land from the wife's relatives, the land rental market becomes an option to access more land. This could encourage men to prefer rented in land while in an insecure situation. Female-headed households, on the other hand, preferred to rent out in patrilineal-dominated societies. This may be an indication of resource poverty and distress, as land rentals in Malawi are used as a safety net in times of emergency. Evidence in other countries in Africa suggests that landlords are poor and often female-headed households with insufficient labor to operate land themselves. They therefore rent it out to less poor tenants who have the necessary resources (Tikabo and Holden 2003; Holden and Ghebre 2006; Kassie and Holden 2006; Chapters 3, 4, 8, 9, 10).

Separating the landlords and tenants and assessing the location of residence (by instrumenting for it) in an instrumental probit model, we found that patrilocal households participate more in land markets as tenants than matrilocal households do in both patrilineal and matrilineal societies (see Table 6-5). Land-poor and labor-rich (female labor) households have a high probability of participating in land markets as tenants in both matrilineal and patrilineal societies.

There was a significant difference between patrilocal and matrilocal households when it came to the renting-out decision (for all districts and in matrilineal societies separately), as can be seen from Table 6-6. Households with more land endowment rented out significantly more land in the total sample model and in both matrilineal and patrilineal areas. Similarly, households with less land endowment were more likely to rent in land in the total sample model and in matrilineal and patrilineal areas (Table 6-5). This indicates that land-poor and even landless households (Figure 6-2) access land through the land rental market, thereby leading to a more egalitarian distribution of land.

How Much to Participate

The results from instrumental variable tobit models are presented in Table 6-7 for area rented in and in Table 6-8 for area rented out. Notice in Table 6-7 that area rented in was significantly higher for land-poor households in the total sample and in matrilineal and patrilineal areas separately. It appears, therefore, that the land rental market is "pro-poor" in both patrilineal and matrilineal areas. Patrilocal residence was associated with significantly larger areas rented in the total sample and in the matrilineal areas. Land-rich households tended to rent out land, whereas households poor in land tended to rent in land. Households with more assets were more able to rent in land in the overall sample and patrilineal society. Households with access to more capital may be able to rent in land and intensify land use by using higher levels of purchased inputs such as labor and fertilizer (Holden et al. 2006), which is indicated by their level of assets. Holden et al. (2006) also observed that, in central and southern Malawi,

TABLE 6-5 Instrumental variable probit models for land rented in

	All districts *b/se*	*Matrilineal dominant* *b/se*	*Patrilineal dominant* *b/se*
Patrilocal	1.295**** (0.37)	1.240** (0.52)	2.013**** (0.55)
Inherited land	–0.470*** (0.14)	–0.589** (0.26)	–0.426*** (0.16)
Asset value	0.019 (0.01)	0.006 (0.01)	0.036* (0.02)
Sex of hhh	0.226 (0.33)	0.910* (0.53)	–0.561 (0.57)
Male labor	0.068 (0.40)	–0.214 (0.51)	0.125 (0.72)
Female labor	1.129** (0.45)	1.262* (0.67)	1.352* (0.73)
Cwratio	0.021 (0.09)	0.064 (0.12)	–0.026 (0.10)
Age	–0.001 (0.01)	–0.010 (0.01)	0.013 (0.01)
Educ. of hhh	0.013 (0.02)	0.065* (0.04)	–0.005 (0.01)
Security	–1.164**** (0.20)	–1.149**** (0.25)	–0.922* (0.48)
Patrilineal	–0.541** (0.23)		
Constant	–0.761 (0.49)	–1.321* (0.80)	–1.614** (0.66)
Prob > chi^2	0.000	0.000	0.000
Number of obs.	337	205	132

Robust standard errors. The superscripts ****, ***, **, and * indicate the 0.1%, 1%, 5%, and 10% levels of confidence, respectively.

land-scarce households with sufficient cash and nonland factors of production rented in land, whereas land-rich and cash- and labor-poor households were more likely to rent out their land. Female labor was positively correlated with the area rented in (Table 6-5) and negatively correlated with area rented out at 10% significance level (Table 6-6).

Effects of Security on Land Market Participation

By incorporating the security variable, we relaxed the assumption that all households have secure tenure and take into account the rights on the land

TABLE 6-6 Instrumental variable probit models for land rented out, with household random effects

	All Districts	*Matrilineal Dominant*	*Patrilineal Dominant*
Patrilocal	–1.115**	–1.278*	–0.883
	(0.51)	(0.73)	(0.69)
Inherited land	0.461****	0.461*	0.474***
	(0.13)	(0.25)	(0.15)
Asset value	0.004	0.034***	–0.012
	(0.00)	(0.01)	(0.03)
Sex of hhh	0.371	0.270	0.938*
	(0.33)	(0.56)	(0.49)
Male labor	0.272	1.004	–1.362*
	(0.49)	(0.66)	(0.81)
Female labor	–1.031*	–1.147	–0.735
	(0.54)	(1.01)	(0.77)
Cwratio	–0.082	–0.323*	0.017
	(0.08)	(0.18)	(0.09)
Age	0.002	–0.004	0.007
	(0.01)	(0.01)	(0.01)
Educ. of hhh	–0.027	–0.062	–0.014
	(0.03)	(0.04)	(0.03)
Security	0.221	0.486*	0.115
	(0.20)	(0.28)	(0.29)
Patrilineal	0.701***		
	(0.26)		
Constant	–1.622***	–1.145	–1.629*
	(0.53)	(0.84)	(0.86)
Prob > chi^2	0.001	0.007	0.076
Number of obs.	337	205	132

Robust standard errors. The superscripts ****, ***, **, and * indicate the 0.1%, 1%, 5%, and 10% levels of confidence, respectively.

that one needs to have in order to make a decision to rent in or rent out land. Table 6-7 shows that the security variable was significant and negatively correlated with renting in for the whole sample as well as for the subsamples. As indicated earlier, renting-in and renting-out decisions are made based on the bundle of rights one has. A husband in a matrilocal residence will have only user rights to the land, which are gained at marriage. Matrilocal residence gives men (the decisionmakers according to Place and Ostuka 2001a; Peters 2002) only user rights to land, and likewise a wife in a patrilocal residence will have only user rights while the husband has extra rights such as rights to sell, subdivide, rent out, and borrow land.

TABLE 6-7 Instrumental variable Tobit models for land rented in

	All districts	*Matrilineal Dominant*	*Patrilineal Dominant*
Patrilocal	0.787***	0.714**	1.230*
	(0.28)	(0.35)	(0.66)
Inherited land	–0.224***	–0.278**	–0.243**
	(0.07)	(0.12)	(0.10)
Asset value	0.008****	0.015	0.007****
	(0.00)	(0.01)	(0.00)
Sex of hhh	0.150	0.406	–0.159
	(0.18)	(0.25)	(0.47)
Male labor	–0.011	–0.133	0.037
	(0.20)	(0.24)	(0.42)
Female labor	0.664***	0.497*	1.089*
	(0.25)	(0.30)	(0.60)
Cwratio	0.018	0.011	–0.000
	(0.06)	(0.06)	(0.08)
Age of hhh	0.000	–0.003	0.011
	(0.00)	(0.00)	(0.01)
Educ. of hhh	0.011	0.034*	–0.004
	(0.01)	(0.02)	(0.01)
Security	–0.616****	–0.480****	–0.751****
	(0.10)	(0.12)	(0.21)
Patrilineal	–0.275*		
	(0.16)		
Constant	–0.653**	–0.819**	–1.377**
	(0.31)	(0.41)	(0.64)
Prob > chi^2	0.000	0.000	0.000
Number of obs.	337	205	132

Robust standard errors. The superscripts ****, ***, **, and * indicate the 0.1%, 1%, 5%, and 10% levels of confidence, respectively

The land rental market is localized, and land market participation is affected by social norms. We observe in Table 6-8 (total sample) significantly (at 5% level) larger areas rented out in patrilineal areas, after controlling for inherited land size, and this may be due to higher land supply by households who can make decisions to rent out without fear of losing the land. The security variable was insignificant in Table 6-8 for area rented out and was significant (at 10% level only) in matrilineal areas in Table 6-6 for the probability of renting out. This was surprising, as we would have expected that security would have been more important for renting out decisions.

TABLE 6-8 Instrumental variable Tobit models for land rented out

	All Districts	*Matrilineal Dominant*	*Patrilineal Dominant*
Patrilocal	–0.980** (0.49)	–0.881 (0.63)	–0.763 (0.58)
Inherited land	0.384**** (0.11)	0.384** (0.18)	0.346**** (0.10)
Asset value	0.003 (0.00)	0.021*** (0.01)	–0.003 (0.02)
Sex of hhh	0.293 (0.28)	0.106 (0.44)	0.796** (0.40)
Male labor	0.202 (0.40)	0.699 (0.43)	–1.043* (0.61)
Female labor	–0.712* (0.42)	–0.635 (0.62)	–0.522 (0.60)
Cwratio	–0.067 (0.07)	–0.253** (0.12)	0.014 (0.07)
Age of hhh	0.001 (0.00)	–0.003 (0.01)	0.005 (0.01)
Educ.of hhh	–0.018 (0.02)	–0.041 (0.03)	–0.007 (0.02)
Security	0.148 (0.16)	0.329 (0.21)	0.044 (0.22)
Patrilineal	0.602** (0.25)		
Constant	–1.333*** (0.44)	–0.871 (0.72)	–1.272** (0.61)
Prob > chi^2	0.002	0.000	0.049
Number of obs.	337	205	132

Robust standard errors. The superscripts ****, ***, **, and * indicate 0.1%, 1%, 5%, and 10% levels of confidence, respectively

Conclusions

There is strong evidence that land rental markets are developing in Malawi. Land-poor households are increasingly able to access land through the rental market. We found evidence that emerging land rental markets in Malawi redistribute land from more land-rich to more land-poor households, as seen by reduction in the Gini-coeffient for operated land as compared to for own landholding. The endowment of nonland factors, such as labor, has a significant effect on land market participation. Households with more female labor endowments were more likely to rent in land, which is a sign of imperfections in the labor market.

Due to the dual land tenure system, security over land and the bundle of rights that one acquires vary, also affecting decisions on land rental market participation. The security of men who are traditionally the main decision-maker in the household (head of household) has implications for household participation in the land rental market. We found that households in patrilineal areas were more likely to rent out and rented out significantly more land than in matrilineal areas. Households in matrilineal societies do not offer men full land rights, and hence men are not the main decisionmakers with regard to inherited land, and this may have affected land renting-out decisions. We also found that patrilocal households rented in significantly more land than matrilocal households overall and separately in matrilineal and patrilineal areas. In addition, we found that less tenure-secure households rented in more land. Additional renting in of land may thus be a response to such tenure insecurity, yet the only evidence that tenure insecurity causes landlords to rent out less land was that households in patrilineal areas rented out more land than households in matrilineal areas.

It is a challenge to introduce a land law and land reform in a country with two such markedly different land tenure and inheritance systems as in Malawi. In contemplating the land reform in this country, it must be clearly recognized that the emerging land rental markets contribute to both efficiency and equity of land allocation by transferring land from land-abundant to labor-scarce households, even though such markets are subject to high transaction costs and are constrained by tenure insecurity due to the traditional inheritance and residence systems. Thus, land reform must be designed to reduce transaction costs and to strengthen tenure security, including tenure security of women, if it aims to promote equitable and efficient land allocation among farm households.

NOTES

1. Public land is defined as that which is occupied, used, or acquired by the government and any other not being customary or private, and includes settlement schemes, national parks, forest reserves, and lapsed leaseholds.
2. Private land is defined as that which is held or owned under freehold or leasehold title, Certificate of Claim, or is registered under the Registered Land Act.
3. Customary land is that which is held or used under customary law.
4. Neutral residential arrangement in which land is from a chief in neither husband's village nor wife's village. However, it is becoming difficult to acquire such land as chiefs do not have any more free land to distribute.

CHAPTER 7

Land Markets in Uganda

What Is Their Impact and Who Benefits?

Klaus Deininger and Paul Mpuga[1]

In situations of high risk, imperfect credit markets, and in which nonagricultural uses affect demand for land purchases, rural land sales markets may not always lead to productivity-enhancing outcomes (Binswanger et al. 1995). Fear of such efficiency-reducing outcomes has led a number of countries (e.g., Ethiopia, South Africa, India, Mexico, and China) to impose restrictions on the operation of land sales or rental markets and is prompting researchers to caution against premature liberalization of land markets (Lastarria-Cornhiel 1997; Platteau 1996) or even to warn about outright negative consequences of doing so (Manji 2003). Whereas rural markets in Africa suffer from many imperfections, available empirical evidence that would be required to justify interventions in this area suffers from two shortcomings.

First, although one would expect land sales and rental markets to work quite differently and that informal mechanisms for transacting land can work quite differently, little distinction is made between the different types of land transfers, with few exceptions (Baland et al. 2003). Temporary exchange of land offers not only greater flexibility but are also less affected by market imperfections than sales markets and can play an important role in many developing as well as developed countries (de Janvry et al. 2001). Therefore, even in situations in which, as a result of high risk, imperfect markets for credit and insurance, limited ability to use land as a collateral to obtain credit financing, and nonagricultural demands for land, agricultural sales markets may lead to land transfers that would not be socially optimal. Informal exchange or formal land rental can do much to facilitate temporary land transfers that significantly increase productive efficiency and equity. Using data from Uganda, this chapter provides evidence on the relative performance of land sales and rental markets.

Second, although a fundamental objective of land markets is to enhance productivity, data limitations may reduce the scope for inferences on productivity impacts of market transactions. In-depth analysis of land market outcomes in Africa is a relatively recent phenomenon (Lavigne Delville et al. 2002). Available evidence provides interesting insights into the mechanisms at work but, because it often relies on case studies or small samples, makes it difficult to generalize, conduct statistical tests, and draw policy conclusions from the results obtained. Empirical evidence on the level of activity and productivity impact of land markets is of policy relevance as, with greater development of the off-farm economy, the scope for productivity-enhancing land transfers, and thus the losses associated with administrative prohibition or malfunctioning of such markets are likely to increase significantly.

In this chapter we address both of these issues by separately providing evidence on functioning of land rental and sales markets for Uganda using a large nationally representative household survey. In line with the literature on the subject, we assume that land markets are driven by unobserved ability and imperfections in labor and capital markets. Households' unobserved agricultural ability becomes more relevant as opportunities in the off-farm sector emerge. In addition, both rental and sales markets are differentially affected by labor market imperfections in the form of supervision constraints and capital market imperfections that are widespread in rural areas. Specifically, the scope for dealing with capital market imperfections is much higher in rental than in sales markets.

We use panel data to provide an estimate of households' agricultural ability, a variable that can then be used to make inferences on the extent to which markets transfer land to more efficient producers. Together with information on households' asset endowment, this can be used to make inferences on the extent to which land markets would make a contribution to productivity as well as equity. At a descriptive level, we assess whether land sales or rental markets have contributed to concentration of land in the hands of the better off, and whether prices are in line with returns from agricultural cultivation or overvalued due to nonagricultural demand—something that might put land out of the reach for the poor. Econometrically, we are able to identify not only determinants of market participation but also check the presence of systematic differences between rental and sales markets.

We find that land rental markets transfer land to more efficient and land-poor or landless households. By comparison, participation in land sales markets requires some initial asset endowment and, as predicted, there is less evidence of these markets transferring land to more productive households. At the same time, even though land prices are higher than the present value of agricultural profits, there is no sign of land ownership becoming more concentrated over time, and land sales markets as well as informal gifts appear to provide access to young households. Availability of data from before and after economic reform and agricultural market liberalization also allows us to make inferences on how this policy, together with broader economic growth, has affected performance of land rental markets. Results point toward a large increase in such transactions. There is no evidence to support the claim that land market activity has put the poor at a disadvantage; to the contrary, the expansion of formal and informal channels for land transactions seems to have greatly improved the ability of those with low land endowment and limited off-farm opportunities to access land and make more productive use of their labor.

The chapter is structured as follows. The next section reviews the literature on land markets in Africa and presents the analytical framework and the estimation strategy. This is followed by a discussion on data, descriptive evidence, and variable definitions. Then econometric results are presented. The chapter concludes with a number of policy implications.

Background and Analytical Framework

Chapters 1 and 2 have already provided evidence on incidence and performance of land rental and sales markets in Africa, allowing us to directly introduce the conceptual model, the hypotheses to be tested empirically, and the data to be used in doing so.

Conceptual Background

With differences in ability across households, it will be advantageous to transfer land from less able to more skilled producers so as to enhance overall output, holding other things constant. In addition, labor market imperfections would lead land markets to transfer land from large producers who would have to rely on wage labor to "poor but efficient" small producers who cannot fully utilize available family labor just by cultivating their endowment. Capital market imperfections would, however, counteract this advantage and reduce or eliminate the labor cost advantage of small producers due to lack of access to other inputs. After briefly reviewing the basic factors, we assess the differential implications of these two market imperfections on sales and rental markets separately.

In a world of perfect information and complete markets, with zero transaction costs, the distribution of land ownership will affect households' welfare but will not matter for efficiency outcomes, and everybody will operate their optimum farm size (Feder 1985). Imperfections in rural labor markets are mainly due to the cost of supervision that arises from the fact that, except in very limited circumstances, a wage worker's true effort is not easily observable. This implies that wage workers will have limited incentives to exert effort and either need to be supervised at a cost[2] or be offered contracts that provide higher incentives. Family members have higher incentives to provide effort than do hired laborers, implying that it would be advantageous for those who do not have enough land to fully utilize their family labor endowment to rent in land, or for those who are relatively land abundant to rent out, rather than engaging in labor market transactions that incur supervision costs. Land markets would thus have a positive impact on improving land access by land-poor households.

Indeed, a large literature has demonstrated that many agricultural activities do not exhibit true economies of scale in production, with the exception of highly specialized machinery, livestock production, or plantation crops in which economies of scale are transmitted from the marketing to the production stage. A number of studies found a negative farm-size productivity relationship for all but the smallest farm size classes (Berry and Cline 1979; Carter 1984; Newell et al. 1997; Kutcher and Scandizzo 1981; Burgess 2001), or have not been able to reject the hypothesis of constant returns to scale in agricultural production (Lanjouw 1999; Feder et al. 1992; Wan and Cheng 2001; Olinto 1995).

Credit market imperfections can, however, offset or even eliminate supervision cost advantages of family farmers. For example, if there is a need for up-front working capital (e.g., to acquire inputs in addition to land and labor) and there is no credit market or access to capital depends on initial wealth, the optimal size of the operational holding would vary systematically with the size of owned holdings even if land rental markets operate perfectly. This will give rise to a positive relationship between farm size and productivity, as is indeed confirmed by studies from environments where credit markets are important, such as in Sudan (Kevane 1996) or in Malawi (Dorward 1999).

Thus, empirically, the extent to which land sales or rental markets will transfer land to more productive (and small) producers will depend on a host of factors, including the functioning of other markets. Given the multiplicity of imperfections in various factor and credit markets, the direction of transactions in rental markets can be from large owners to smaller operators or the other way around, depending on the nature of the dominant market imperfections, although it will differ systematically between rental and sales markets.

In rental markets, share contracts provide scope for overcoming capital market imperfections at relatively low cost (Basu 1992; Otsuka et al. 1992; Ghatak and Pandey 2000; Ray and Singh 2001). A large number of studies provide support to the hypothesis that share tenancy provides a second best arrangement which, in any given environment, is difficult to improve on unless overall circumstances change so that the operation of factor and credit markets improves (Otsuka et al. 1992; Sharma and Dreze 1996; Sadoulet et al. 1994; Quisumbing 2001b; Otsuka 2002). In fact, systematic variation of contract parameters depending on tenants' wealth provides empirical confirmation for this (Quibria and Rashid 1986; Shetty 1988; Shaban 1991; Laffont and Matoussi 1995). We therefore hypothesize that, although credit market imperfections will affect the nature of contracts adopted, it will have less impact on the tendency of rental markets to transfer land to more productive producers.

Compared to rental markets, in which adjustments within the rental contract can be made in the contractual structure to overcome the impact of capital market imperfections, land sales markets will be significantly affected by credit market imperfections. Where agricultural production is very risky and insurance is not available, farmers may be forced to sell land off at very low prices in distress sales (Cain 1981; Bidinger et al. 1991; Kranton and Swamy 1999) and may be unable to obtain it again once conditions have returned to normal (Carter and Zimmerman 2000). Also, to the extent that land is valued as an asset, in addition to its use as an input into agricultural production, nonagricultural land demand, tax and subsidy policies, and macroeconomic conditions can increase land values over and above the present value of profits from agricultural production for specific groups but not for others (Robison et al. 1985; Brandao and Rezende 1992; Gunjal et al. 1996), making it more difficult for productive farmers to acquire land. Lack of long-term credit can further exacerbate the ability of small and productive farmers to acquire land through the sales market.

In environments where credit markets do not work well, land sales markets are more likely to lead to undesirable outcomes. Therefore, market imperfections or other distortions in other markets could give rise to the emergence of efficiency-reducing outcomes such as speculative purchases, distress sales, and artificially inflated land values that reduce the access to land by low income and landless buyers. This has often led governments to intervene in the functioning of land sales markets. Land sales markets would be even more constrained from transferring land to producers based solely on considerations of productivity in cases in which absence of long-term finance makes it more difficult for the latter to finance such acquisition. All of this leads us to expect that rental markets will be more able than sales markets to transfer land to more productive producers.

Conceptual Model

Let household i be endowed with fixed amounts of labor ($\bar{L}_i$) and land ($\bar{A}_i$), and a given level of agricultural ability (α_i). Relative land abundance is assumed to make farming based on hired labor infeasible and households can allocate their labor endowment between farming their own land and off-farm employment at an exogenous wage (w_i). Renting incurs transaction costs (T) proportional to the amount of land transferred and we assume that working capital is not binding. With this, household i will choose l^{a*}, l^{o*} as well as A^* by solving the maximization problem:

$$\max_{l^a, A} p\alpha_i f(l_i^a, A_i) + wl_i^o - I^{in}(A_i - \bar{A}_i)(r+T) + I^{out}(\bar{A} - A_i)(r-T) \quad (7.1)$$

where p is the price of agricultural goods, l^o is the amount of time allocated to off-farm labor ($= \bar{L}_i - l_i^a$), I^{in} is an indicator for rent in (= 1 for rent in, and 0 otherwise), I^{out} is an indicator for rent out (= 1 for rent out, and 0 otherwise), and all other variables in (7.1) are as defined above. The optimal choices of l_i^{a*}, l_i^{o*} and A_i^* will solve the first-order conditions (FOC) of problem (P)—that is

$$p\alpha_i f_{l_i^a}(l_i^a, A_i) = w \tag{7.2}$$

plus, for households who rent in,

$$p\alpha_i f_{A_i}(l_i^a, A_i) = r + T \tag{7.3}$$

or for households who rent out,

$$p\alpha_i f_{A_i}(l_i^a, A_i) = r - T \tag{7.3a}$$

and for autarkic households,

$$r - T < p\alpha_i f_{A_i}(l_i^a, A_i) < r + T \tag{7.3b}$$

The first-order conditions allow to derive three empirically testable hypotheses:[3]

Hypothesis 1: The amount of land rented in is strictly increasing in ability, α, and strictly decreasing in their land endowment $\overline{A}$. Rental markets will thus transfer land to "poor but efficient" producers.

Hypothesis 2: Transaction costs drive a wedge between those renting in and those renting out with any increase in T decreasing α_l and increasing α_u, thereby expanding the range of producers who remain in autarky, reducing the number of households who participate in rental markets, as well as the amount of land transacted through rental markets. Reduction in transaction cost will increase social welfare.

Hypothesis 3: Increases of the wage for off-farm employment will increase the amount of land transacted in rental markets and overall welfare. This will be associated with a decrease in the equilibrium rental rate and, in a risk-free environment, will make everybody better off.

Concerning land sales markets, we follow the literature on intertemporal asset accumulation. In general, household *i*'s choices of consumption, land accumulation, and saving in each period solves the utility maximization over the life time horizon:

$$\text{Max} \qquad V = E_t \sum_{k=0}^{T-t} (1+\delta)^{-k} U(c_{t+k}) \tag{7.4}$$

$$\text{Subject to} \quad c_{t+k} + (\overline{A}_{t+k} - \overline{A}_{t+k-1}) p_{t+k} + (S_{t+k} - S_{t+k-1}) \le Y_{t+k}(\overline{A}_{t+k}) \tag{7.4a}$$

and $$\bar{A}_{t+k} \geq 0 \tag{7.4b}$$

where $U(c_t)$ is one period utility function, c_t is the level of consumption, δ is the rate of time preference, $\bar{A}_t$ is the amount of land owned in time t, S_t is the amount of saving in time t, $Y_t(\bar{A}_t)$ is the amount of income household i generates in time t, which itself is a function of the land household i owns in that period (assume household self-cultivates all the land endowment). Equation (7.4b) is a form of borrowing constraints, restricting the end stock of tradable assets to be nonnegative in each period.

If income is stochastic, an analytical solution to this problem cannot be derived (Zeldes 1989) and the literature has resorted to simulation to obtain solutions under general assumptions (Deaton 1991; Zeldes 1989; Rosenzweig and Wolpin 1993; Carter and Zimmerman 2000). One key result in this context is that the supply of land to the sales market may be driven by distress sales whereby households have to liquidate land at low values in situations of emergency. A second insight of relevance is that, in environments where capital markets that could provide long-term financing for land acquisition are either not available or available only to the wealthy, the wealth bias may overwhelm the productivity advantage of farms based on family labor. This would imply that land sales markets might lead to accumulation of land in the hands of households who are not necessarily the ones to make the most productive use of land. Although this could, in principle, justify restrictions on the functioning of land sales markets, such restrictions normally have undesirable side effects and should therefore be adopted only if there is empirical evidence that such distortions are of relevance and if other, less distortionary measures are not available.

The fact that wealth effects may dominate in the land sales market, but are likely to be of limited importance for land rental, provides the basis for a number of testable predictions. First, we expect that initial wealth is more important as a determinant of participation in sales than in rental markets. Second, even though we would expect households' ability to increase their demand for cultivable land, the variable would be less important in sales markets. Finally, we expect a positive time trend in rental markets due to the fact that higher levels of off-farm activity, possibly together with lower transaction costs, make it more desirable for households to rent out their land.

Estimation Strategy

To obtain an estimate of households' agricultural ability, we use the panel households to estimate a production function with household fixed effects. Let technology be represented by the Cobb-Douglas production function:

$$Q_{ijt} = \exp(\alpha_i + \alpha_j) A_{ijt}^{\theta_1} L_{ijt}^{\theta_2} K_{ijt}^{\theta_3} \tag{7.5}$$

where Q_{ijt} is agricultural output produced by household i in village j in year t; A_{ijt}, L_{ijt} and K_{ijt} are the land, labor, and capital used to produce this output Q, with technical coefficients θ_1, θ_2, and θ_3, and $exp(\alpha_i + \alpha_j)$ is the efficiency parameter that consists of a household- and a village-specific element reflecting access to infrastructure and markets, soil quality, climate, and so on. Taking logs of both sides of equation (7.2) and adding a time trend and an iid >error term, and letting $\alpha_{ij} = \alpha_i + \alpha_j$, we obtain an estimable equation for production by household i in village j at time t:

$$q_{ijt} = \alpha_{ij} + \theta_1 \alpha_{ijt} + \theta_2 l_{ijt} + \theta_3 k_{ijt} + \phi t + \varepsilon_{ijt} \tag{7.6}$$

where lower case letters refer to the logarithm of the quantities referred to above. Availability of multiple observations per household in the panel allows us to estimate this equation using household fixed effects.

$$q_{ijt} - \bar{q}_{ij} = \alpha_{ij} - \bar{\alpha}_{ij} + \theta(\mathbf{Z}_{ijt} - \bar{Z}_{ij}) + \phi(t - \bar{t}) + \varepsilon_{ijt} - \bar{\varepsilon}_{ij} \tag{7.7}$$

where **Z** is a vector consisting of a, l, and k, and θ is a coefficient vector. This equation allows us to recover the composite efficiency parameter α_{ij}. As this variable still includes unobservable characteristics at the village level, we apply a similar procedure at the village level to obtain α_j the subtraction from which provides an estimate of α_i, the pure idiosyncratic effect for each producer in the sample. The estimation results are reported in Annex Table 7A1.

To determine factors that caused households to participate in land purchase markets or received land through either inheritance or as a gift in the 1992 to 1999 period, we estimate probit and tobit regressions with either a zero/one participation dummy or the amount of land purchased as dependent variables. Right-hand side variables include a vector of household characteristics and initial endowments, whether or not the household migrated, and for households where this information is available, agricultural ability. Formally,

$$P_i = \beta_0 + \beta_1 \alpha_i + \beta_2 \mathbf{X}_i + \beta_3 \mathbf{E}_i + \beta_5 \mathbf{M}_i + \varepsilon_i \tag{7.8}$$

where P_i is either a dummy or the area purchased as described above; α_i is agricultural ability, derived as described above; $\mathbf{X}_i$ is a vector of household characteristics including, age, sex, and the head's education; $\mathbf{E}_i$ is a vector of (initial) endowments; and $\mathbf{M}_i$ is a vector of migration dummies, indicting whether the household migrated either before or after 1992. Proper identification of the land purchase equation requires information on initial land endowments (i.e., in 1992), participation in rental markets is decided at any given point in time. The estimating equation for land renting is

$$L_i = \beta_0 + \beta_1 \alpha_i + \beta_2 \mathbf{X}_i + \beta_3 \mathbf{E}_i + \beta_5 \mathbf{M}_i + \mu_i \tag{7.9}$$

where L_i is either a dummy for renting in (or renting out) land or the log of the area rented. Other variables are similar to what was discussed in equation

(7.8), with the only difference being that $\mathbf{E}_i$ includes only contemporaneous land ownership.

To assess changes in land market performance over time, we use the pooled 1999 and 1992 sample and estimate a model with a time-interactive dummy for land owned, value of household and agricultural assets in which we had observations for 1999 and 1992, and a pure trend of the form

$$C_{it} = \beta_0 + \beta_1 \mathbf{X}_{it} + \beta_1 \mathbf{E}_{it} + \beta_3 \mathbf{M}_{it} + \delta_1 \mathrm{E}_{it} T + \delta_2 R_{it} T + \alpha_3 T + \mu_{it} \qquad (7.10)$$

where the dependent variable C_{it} is a categorical variable for leasing in or leasing out land; $\mathbf{X}_{it}$ is a vector of contemporaneous household characteristics; $\mathbf{E}_{it}$ is a vector of endowments including education, land area owned, and value of assets; and $\mathbf{M}_{it}$ is dummy for current migration, μ_{it} is an iid error term, and T is a year dummy to capture the time trend in rental markets. To explore whether the relevance of certain endowment variables has shifted over time, we add an interaction between the time trend and endowment variables $\mathbf{E}_{it}T$.

Description of the Data Used

Uganda's national household surveys provide an excellent opportunity to test the hypotheses stated earlier. We use the 1999–2000 Uganda national household survey (UNHS), which covered 10,696 households, as well as the 1992–1993 integrated household survey (IHS) covering 9,600 households. Both are multipurpose household surveys providing rich information on household characteristics, assets, and consumption. The agricultural module in the UNHS obtained data on asset and landholdings, including total area owned, cultivated, rented in, and rented out, both for the survey year (1999) and, through recall, for 1992.[4] This allows us to recover, for 7,324 UNHS households who had existed in 1992, the amount of land bought or sold during the period, thereby avoiding the need to resort to information on whether the household *ever* purchased any land. In addition, pooling the two datasets allows us to make inferences on the extent to which rental markets have changed over time. Moreover, the fact that the UNHS contains a panel of about 1,300 households that had already been included in the IHS allows us to use a fixed effects production function in order to derive a measure of households' agricultural ability, which will be included as an independent variable in the land market equations.

Household Characteristics

Table 7-1 reports characteristics of the UNHS sample. With a mean education level of slightly below 5 years, 26% of the households were female headed and 12% widowed. About one-fourth of household heads had migrated to their present location before 1992 and an additional 15% did so after 1992.

TABLE 7-1 Descriptive statistics

Variable description	*All Uganda*	*Central*	*East*	*North*	*West*
Household characteristics					
Age of head (years)	43.32	42.62	43.47	43.01	44.1
Education of head (years)	4.91	5.37	5.00	4.30	4.71
Female head (%)	26.32	29.68	24.22	30.41	22.27
Non-widow (%)	14.34	17.59	12.84	17.15	10.62
Widow (%)	11.98	12.09	11.38	13.26	11.65
Migrated before 1992 (%)	26.14	34.05	23.39	20.48	23.91
Migrated after 1992 (%)	15.14	19.84	16.2	12.87	10.48
Non-farm activity (%)	34.06	38.39	35.25	28.91	31.45
Asset ownership					
Value of hh assets in 99 (US $)	281.56	346.96	247.7	113.18	348.89
Value of hh assets in 92 ($)	209.7	258.22	179.19	75.1	270.88
Value of agric assets in 99 ($)	103.58	85.88	109.28	110.65	112.47
Value of agric assets in 92 ($)	71.41	54.94	71.16	93.54	75.57
Iron sheet roof (%)	61.54	83.05	53.58	15.21	75.03
Land market participation					
Cost of renting in (US $)[a]	14.82	12.83	11.74	11.74	16.3
Profit per acre ($)	49.44	46.39	82.58	22.92	38.04
Price of land per acre - rural, 99 ($)	1,098.40	1,270.49	982.6	70.81	1,051.55
Rented in 99 (%)	19.48	15.74	27.09	28.36	10.62
Rented in 92 (%)	8.92	7.2	10.3	17.87	3.88
Length of contract in 99 (years)	4.3	6.25	3.47	5.48	4.34
Length of contract in 92 (years)	2.6	3.4	2.51	2.5	2.38
Household purchased land (%)	34.63	50.12	30.89	2.62	43.98
Household inherited land (%)	27.04	14.8	30.21	31.33	31.74
Household received free land (%)	37.54	45.73	34.6	38.23	33.04
No of. observations	10,696	3,110	2,865	1,802	2,919

Source: Own analysis based on the 1999/2000 UNHS.

Note: All US $ values in terms of 1999 US $.

[a]Cost of renting and length of contract only for households who rented in land.

Nonfarm activities were pursued by about one-third. The total value of assets owned by households in the sample remained low, although, since 1992, it is estimated to have increased by between 30 and 40%.[5] This increase was greatest in percentage terms in the North, although the very low basis implies that the value of household assets in this region still remained at less than half of any of the other regions. In most African countries, the fact of having an iron roof is a clear measure of wealth that is highly correlated with other housing and welfare indicators. Although the variable is not available for 1992, the

ANNEX TABLE 7-A1 Production function for panel households

Dependent Variable	*Fixed Effects Model*
Education of head	0.013 (0.81)
Age of head	0.004 (0.91)
Female head (non-widow)	–0.151 (0.90)
Female head (widow)	–0.181 (1.14)
Iron sheet roof	0.029 (0.31)
Age group 6–14 years (log)	0.107* (1.79)
Age group 15–64 years (log)	0.116 (1.23)
Labor cost (log)	0.054** (2.19)
Input costs (log)	0.434*** (11.63)
Uses manure	0.047 (0.34)
Uses pesticides	–0.045 (0.28)
Value of hh assets (log)	0.073*** (3.28)
Value of agric assets (log)	0.068*** (3.23)
Area cultivated in (log)	0.321*** (5.40)
Urban area	0.888 (0.90)
Year	–0.767*** (6.43)
Constant	1.800*** (6.07)
No. of observations	2,365
R^2	0.42
No. of panel households	1,265

Note: Absolute value of t statistics in parentheses

* Significant at 10%; ** significant at 5%; *** significant at 1%

cross-sectional variation in 2000 reinforces the relative backwardness of the North as compared to other regions; about 60% of households nationwide have a roof with iron sheets, but this share is highest in the Center with 83%, only 15% do so in the North.

The bottom panel of Table 7-1 describes land market participation and agricultural productivity. The return to fixed factors and management by the land owner per acre of land, abbreviated as profit in the table, is arrived at by subtracting the value of purchased inputs from the total output obtained from a plot. Comparing this figure with households' self-assessed land values provides an indication of the extent to which land is "overvalued" (i.e., land prices are above the net present value of profits at the current rate of interest).[6] Results from doing so illustrate that self-reported land values are only about 3 times the annual profits in the North and amount to about 12 times this figure in the East. Land values amount to more than 20 times the value of annual profits in the Center and the West, suggesting that, possibly because of the lack of attractive assets or stores of wealth, speculative demand pushes land prices in these regions beyond what would be justified by the productive value of the land. This would make it more difficult for efficient agricultural producers, and in particular the poor, to engage in mortgage-based land acquisition. However, the fact that almost two-thirds of transactions in rental markets in 1999 did not involve compensation (see Table 7-2) and that, even in the cases in which money is involved, rental rates are significantly less than one-third of profits, suggests that there is little spill-over from such speculative demand for land on rental markets and that the latter can be equity enhancing. This illustrates not only that land rental can have a substantial welfare impact but also that barriers to land rental continue to be very low.

Two other findings that are of interest in the context of rental markets are that the share of households in the cross-section that had rented in land more than doubled between the two periods under concern, from 9% in 1992 to 19% in 1999. At the same time, the average length of existing rental contracts also increased, implying that the average time period for which a rented plot had been under the possession of the current occupant increased from 2.6 to 4.3 years. Concerning the modalities under which households access land, we note the high degree of informality in the sense that free receipt is, with 37% in the overall sample, the most common way of land acquisition, followed by purchase (34%) and inheritance (27%). These national aggregates hide considerable regional variation; for example, the share of those who purchased land ranges from virtually zero in the North (2.6%) to 50% in the Center. Also, the fact that somebody in the past had acquired land through the purchase market is useful only if other initial conditions can be recovered and if social and economic conditions have not changed significantly in the interim. Even if this is the case, it is not possible to compare the 35% who had purchased land at some point to those who are currently renting and, on this basis, conclude that land sales markets are more active than rental markets. Instead, doing so requires one to choose a cut-off point in time.

TABLE 7-2 Land market participation for households that existed in 1992 and 1999

	All Uganda	Center	East	North	West
	Between 1992 and 1999				
Households who purchased land (%)	11.86	16.35	12.07	1.06	13.08
Price of land per acre ($)	864.77	848.62	611.94	44.24	1,100.24
Households who inherited land (%)	5.43	3.75	6.21	6.70	5.60
Households who received land free (%)	14.10	21.31	15.03	12.57	8.09
	1999				
Total area owned (cultivable; acres)	4.20	5.11	4.06	4.11	3.63
Total area cultivated (acres)	2.29	2.59	2.45	1.58	2.30
purchased (%)	27.55	40.74	23.19	2.65	34.39
received free (%)	41.09	42.51	37.16	62.37	31.33
inherited (%)	23.96	12.22	26.65	32.96	26.30
rented (%)	7.41	4.53	12.99	2.01	7.98
Share of households renting in (%)	35.91	36.20	43.42	35.17	29.61
paid rent	14.45	5.50	24.80	5.58	17.64
free	21.46	30.70	18.62	29.59	11.97
Rental price per acre ($)	10.14	4.34	10.65	11.01	14.32
Mean area rented (acres)	0.49	0.36	0.77	0.71	0.24
Share of households renting out (%)	8.90	10.67	12.86	6.22	5.54
Landless (%)	10.01	6.37	11.99	23.84	3.61
	1992				
Total area owned (cultivable; acres)	3.43	4.24	3.26	3.58	2.84
Total area cultivated (acres)	1.99	2.18	2.03	2.08	1.76
purchased (%)	28.59	41.90	24.45	2.80	34.83
received free (%)	39.77	40.98	35.22	61.85	30.95
inherited (%)	24.44	12.57	27.55	33.52	26.74
rented (%)	7.20	4.54	12.78	1.82	7.48
Share of households renting in (%)	12.58	13.23	13.57	19.78	7.22
paid rent	3.43	1.25	6.63	1.44	3.52
free	9.15	11.98	6.94	18.34	3.70
Rental price per acre ($)	7.80	7.06	6.61	3.51	14.43
Mean area rented (acres)	0.30	0.29	0.34	0.58	0.13
Share of households renting out (%)	2.32	2.35	4.41	1.76	0.89
Landless (%)	27.46	22.16	29.54	38.28	23.98
No. of observations[a]	7,324	1,837	1,960	1,254	2,273

Source: data analysis by author based on 1999/2000 UNHS survey.

Note: All asset values are in 1999 US $

[a]Households who existed in 1992 based on recall.

Land Market Participation between 1992 and 1999

Patterns of land market participation for households that had existed in 1992 at two periods in time are illustrated in Table 7-2, based on recall data gathered in 1999.[7] The top panel, which provides information on participation in land sales markets between the two periods, illustrates that use of cross-sectional evidence on a household ever having purchased land will not necessarily give an accurate picture of land sales market activity. Although 35% of households had, at some point in their lives, purchased land (Table 7-1), we find that only about 12% of households had done so over the entire 7-year interval covered. This implies that, on average, about 1.7% of households participate in land purchase markets every year. Comparing this with the 36% who were engaged in some rental in 1999 (or even the 14% who rented for payment in the same period) suggests that land rental markets are significantly more active than land sales markets. Also, the mean rental contract in 1999 has been in existence for 4.2 years as compared to 2.6 years in 1992. The top panel of Table 7-2 shows that the self-assessed land prices reported in Table 7-1 are (with the exception of the West) somewhat higher than what had been realized in actual market transactions, although the difference could of course also be due to unobserved differences in land quality.

The middle and bottom panels of Table 7-2 provide information on land ownership and rental market participation in 1999 and 1992, respectively. These figures point toward a strong increase in rental market activity between the two periods. Only about 13% of households were engaged in land rental in 1992, but 36% were so in 1999. The share of households renting out also increased from 2% to 9% while the mean area rented increased from 0.3 to 0.5 acres. Although there has been a modest decline, from 75% to about 60%, in the share of those who rented in land for free, the absolute number of households who were able to complement their own endowment with land that was given to them without payment has more than doubled, from 9% to 21% of the sample.

To give a first indication of the extent to which households who entered the rental market on either the demand or the supply side differ from the average or from each other, Table 7-3 provides evidence on household characteristics as well as land ownership by households that had rented in, rented out, or purchased land, in any of the periods. We note that households leasing in are younger (age 41 as compared to age 45), have lower levels of household assets than the average (U.S. $240 as compared to U.S. $274; 53% with iron roofs compared to 57%), and have significantly lower land endowments (2 as compared to 4.6 acres). By comparison, those leasing out are older (age 49), richer (U.S. $337 of household assets and 63% with iron roofs), and more land abundant (10 acres). Those who purchased land were also younger than the average but at the same time more educated (5.32 years of education) and

TABLE 7-3 Characteristics of households who participated in land market transactions

	All		*Rented In*		*Rented Out*		*Purchased*	*Inherited*	*Received Free*
	1999	*1992*	*1999*	*1992*	*1999*	*1992*	*1992–99*	*1992–99*	*1992–99*
Household characteristics									
Age of head (yrs)	45.04	46.24	41.08	42.86	49.47	53.86	40.32	37.57	39.17
Female headed (%)	25.24	25.82	26.89	32.63	21.94	19.35	18.51	22.06	26.94
of which widow (%)	13.22	13.94	11.88	13.97	14.10	12.90	7.32	12.55	12.43
of which non-widow (%)	12.02	11.88	15.01	18.66	7.83	6.45	11.19	9.51	14.51
Education of head (yrs)	4.42	4.35	4.43	4.00	4.84	4.84	5.32	4.97	4.69
Household assets in 92 ($)		228.28		165.55		322.30	288.58	180.73	167.05
Household assets in 99 ($)	274.11		240.03		337.19		413.61	263.66	229.42
Agric assets in 92 ($)	83.10	90.04	70.91	100.58	122.10	186.15	96.65	68.42	50.03
Agric assets in 99 ($)	120.02		104.54		160.78		160.65	124.02	84.42
Roof sheets (%)	57.21	57.63	53.74	47.01	63.68	66.67	73.95	52.43	53.61
Land ownership and use									
Owned area in 1992 (acres)	4.26	3.99	1.47	1.03	8.51	11.97	2.79	2.38	1.64
Owned area in 1999 (acres)	4.64		2.04	1.84	10.13	12.29	5.63	4.54	2.66
Cultivated in 1992 (acres)	2.35	1.99	1.05	1.02	3.15	3.96	1.54	1.28	1.29
Cultivated in 1999 (acres)	2.24		1.26	0.85	3.18	3.61	3.14	2.27	1.73
No. of observations	8,137	7,324	3,031	1,002	702	186	956	494	1,440

Note: All asset values are in 1999 US $.

less likely to be headed by a female (18%). They also had somewhat higher levels of assets (U.S. $288 as compared to U.S. $228) and also differ from those households who either received land for free or inherited it during the period, suggesting that initial wealth could be a precondition for land purchases, as will be tested in detail in this chapter.

Econometric Results

The fact that information is available on different modalities of land access allows us to identify and compare determinants of households' land access through rental and purchase markets. In line with theoretical expectations, land sales markets are found to be less efficiency enhancing than land rental markets, possibly due to the existence of barriers to entry in terms of preexisting wealth. Rental markets are both efficiency enhancing and pro-poor (i.e., provide access to those who do not own land as well as to migrants). The evidence of a strong increase in rental market activity over time suggests that, with economic liberalization, poor rural households' opportunities to access land have improved considerably.

Incidence and Impact of Land Sales Markets

Concerning participation in land sales markets as compared to receipt of land through inheritance and gift,[8] we find in Table 7-4 that, in line with predictions, young and better-educated households are more likely to purchase land (i.e., land sales market participation is partly a life cycle phenomenon). Also, large households are more likely to have the labor force necessary and are thus more likely to purchase land. At the same time, lack of land and lower agricultural or household asset endowments in 1992 reduced the scope for land purchases, contrary to inheritance or gift of land.[9] Moreover, female-headed households, in particular widows, seem to be discriminated against in the land purchase markets; being a widow reduces the probability of having purchased land by 3.3%.

The regressions suggest that, in addition to human capital assets, some initial wealth is needed to enter the land purchase market; those who did so had both higher levels of household and agricultural assets in 1992. At the same time, migration, especially after 1992, has a significant and very positive impact, suggesting that, in view of large regional differences in availability of land, migration and subsequent purchase (or gift) of land may be a strategy employed by a larger number of young Ugandan households.[10] Consistent with the need for some initial wealth in terms of physical or human capital, we find that it is more difficult for the landless, those with few assets, and for households headed by widows, to acquire land through purchase markets, although the landless are marginally more likely to receive land as a gift.

To proxy for households' relative productivity in the absence of the direct measure of agricultural ability, which is available only for panel households, we included the deviation of profits per unit area from the village mean. As Table 7-4 illustrates, this variable is insignificant throughout. The same is true for our measure of "true" agricultural ability for the 1,178 panel households (column 4 of Table 7-4). This suggests that sales markets do not transfer land to more productive producers and that other considerations are relevant for producers' decisions in this respect. At the same time, and contrary to fears repeatedly raised in the literature that economic liberalization and the associated "commodification" of land would lead to a wave of land sales by the poor, even operation of land sales markets does not seem to have led to land concentration. In fact, the regression suggests that households who had less land in 1992 were more likely to subsequently acquire land through purchase markets. A neutral overall effect of land sales markets is confirmed by the fact that the Gini coefficient for land ownership distribution was 0.55 in both 1992 and in 1999; in other words, there was no increase of land concentration between the two periods. This is consistent with the insignificant coefficient on dispersion of land ownership at the village level.

We conclude that land sales and purchase markets, although providing an avenue for young households with some means to access land and contributing less to concentration than often feared, also do not seem to make a strong contribution to enhancing productive efficiency. Finally, the regional dummies suggest that land purchase markets are most active in the Center, followed by the West and the East, and least active in the North where inheritance and gifts are more important. Comparing purchases to other modalities of land acquisition during the same period illustrates that gifts include a more redistributive element than land sales and purchase transactions.

Incidence and Impact of Land Rental Markets

Contrary to the sales equation in which most right-hand side variables refer to initial conditions and limited data on land sales force us to consider only the demand side of the market, better data availability for rental markets allows us to estimate equations for those renting in and those renting out, using contemporaneous household characteristics as independent variables.[11] The discussion here will focus on differences to land sales markets and the extent to which these correspond to what would be predicted by theory.[12]

Table 7-5 highlights that both the coefficients on age and household size are larger than in the land sales equation, suggesting that rental markets are more effective than sales markets in transferring land to larger households with young heads, thus providing an opportunity to employ relatively abundant family labor. This is reinforced by the magnitude of the coefficient on the landless dummy (60%, or four times as large as what was observed in the sales

TABLE 7-4 Determinants of land purchases in the 1992 to 1999 period

	Cross Section			*Panel*
	Purchased	*Inherited*	*Gift*	*Purchased*
Agricultural ability				0.006 (0.68)
Age of head	–0.002*** (9.76)	–0.002*** (9.08)	–0.004*** (11.45)	–0.002*** (4.42)
Education of head	0.003*** (3.50)	0.002** (2.15)	0.003** (2.30)	0.001 (0.49)
Female head (non-widow)	–.015 (1.50)	–.009 (1.19)	–.008 (0.61)	–.036 (1.60)
Female head (widow)	–0.033*** (3.22)	0.037*** (3.81)	0.016 (1.18)	–0.018 (1.02)
Household size (log)	0.032*** (4.95)	–0.002 (0.46)	–0.016** (2.03)	0.015 (1.11)
Area owned in 1992 (log)	–0.045*** (7.29)	–0.007 (1.36)	–0.091*** (9.70)	–0.021 (1.54)
Landless in 1992	–0.021** (2.00)	0.015* (1.66)	0.019 (1.24)	0.003 (0.13)
Land dispersion at village level (log)	0.005 (1.04)			0.003 (0.29)
Value of hh assets 92 (log)	0.007** (2.46)	–0.003 (1.54)	–0.007* (1.95)	–0.004 (0.62)
Value of agric assets 92 (log)	0.007*** (3.47)	0.001 (0.69)	–0.007*** (2.58)	–0.002 (0.46)
Migrated before 1992	0.019** (2.33)	–0.036*** (5.68)	–0.002 (0.16)	0.019 (1.17)
Migrated after 1992	0.171*** (9.75)	–0.014 (1.43)	0.050*** (2.67)	–0.034 (0.58)
Urban area	–0.032*** (2.91)	–0.026*** (2.79)	0.010 (0.59)	–0.042 (1.49)
Profits per acre less village mean	–0.00001 (0.93)	–0.000002 (0.31)	–0.000006 (0.43)	
East	–0.041*** (5.15)	0.028*** (3.24)	–0.082*** (7.30)	–0.034* (1.91)
North	–0.120*** (10.93)	0.024** (2.17)	–0.093*** (6.81)	–0.098*** (3.43)
West	–0.027*** (3.48)	0.038*** (4.62)	–0.152*** (14.31)	–0.024 (1.43)
Constant	–3.468*** (8.63)	–1.567*** (2.86)	–0.076 (0.43)	–1.717* (1.79)
Observations	7,294	7,294	7,294	1,178
Pseudo R^2	0.14	0.08	0.14	0.11
Log likelihood	–3,101.21	–1,901.32	–3,843.06	–321.56

Note: Absolute value of z statistics in parentheses.

* significant at 10%; ** significant at 5%; *** significant at 1%.

TABLE 7-5 Determinants of participation in land rental markets in 1999

	Full Sample		Panel Only	
Dependent Variable	*Rent In*	*Rent Out*	*Rent In*	*Rent Out*
Household's agricultural ability			0.053**	0.014
			(2.41)	(1.56)
Age of head	–0.004***	0.000*	–0.005***	0.001
	(8.96)	(1.83)	(4.38)	(1.44)
Education of head	–0.000	0.000	0.009*	–0.001
	(0.07)	(0.43)	(1.85)	(0.30)
Female head (non-widow)	–0.018	–0.013	–0.024	–0.039*
	(0.85)	(1.43)	(0.42)	(1.81)
Female head (widow)	–0.026	0.003	–0.034	–0.016
	(1.28)	(0.31)	(0.70)	(0.84)
Household size (log)	0.093***	–0.019***	0.117***	–0.013
	(6.39)	(3.07)	(3.32)	(0.94)
Iron sheet roof	0.002	0.018**	0.025	0.011
	(0.14)	(2.24)	(0.61)	(0.59)
Land dispersion at village level (log)	0.025**	–0.014***	0.014	–0.020
	(2.54)	(2.90)	(0.57)	(1.63)
Area owned (log)	–0.294***	0.074***	–0.297***	0.078***
	(23.89)	(17.54)	(9.33)	(7.12)
Value of hh assets (log)	0.019**	–0.004	0.013	–0.006
	(2.26)	(1.06)	(0.91)	(1.09)
Value of agric assets (log)	0.012***	–0.003	–0.019*	0.003
	(2.76)	(1.57)	(1.75)	(0.73)
Landless	0.608***		0.661***	
	(14.77)		(5.30)	
Migrated before 1992	0.029*	0.010	–0.028	0.018
	(1.89)	(1.51)	(0.76)	(1.13)
Migrated after 1992	0.095***	0.008	0.187	0.135
	(3.21)	(0.59)	(1.14)	(1.50)
Urban area	–0.072***	–0.005	–0.019	0.024
	(2.79)	(0.39)	(0.26)	(0.68)
Profits per acre less village mean	0.000	0.000		
	(0.95)	(0.92)		
East	0.043**	0.031***	0.043	0.018
	(2.29)	(3.65)	(0.92)	(0.93)
North	–0.095***	–0.028***	–0.173***	–0.034
	(3.74)	(2.86)	(2.67)	(1.62)
West	–0.079***	–0.037***	–0.037	–0.071***
	(4.73)	(4.99)	(0.88)	(4.15)
Observations	7,293	7,293	1,088	1,006
Pseudo R^2	0.265	0.122	0.262	0.185
Log likelihood	–3,502.73	–1,922.68	–523.27	–247.15

Note: Absolute value of z statistics in parentheses.

* significant at 10%; ** significant at 5%; *** significant at 1%.

market equation), which suggests that it is much easier for landless households to gain access to land through rental than through sales markets. The coefficient on area owned is also highly significant and negative, implying that land rental markets transfer land to those with limited endowments and do so much more than markets for sales. Finally, the coefficient on land dispersion in the village is positive and significant at 5%, implying that rental markets in contrast to sales markets are more active in locations where, other things equal, there is greater scope for productivity gains from equalizing endowments across producers because endowments are distributed more unequally. There is no evidence for discrimination against widows, contrary to what was observed earlier. All this suggests that rental markets provide opportunities to the poor beyond what is available through sales markets.

The advantages that rental markets provide for the poor are reinforced by the absence of wealth effects. The coefficient on the main indicator of wealth, having an iron roof, is insignificant. The same is true for education, the lack of significance of which suggests that entry requirements for rental are lower than for sales markets. At the same time, the significant coefficients on household and agricultural assets indicate that some assets will still be required to access land through rental. Even though the regression suggests that migrants are more likely to rely on land rental markets to gain access to land than those who did not migrate, the size of the coefficient is smaller than for land purchase markets. This would suggest that, even though rental is one option through which migrants can access land, it is less for migrants to establish themselves solely based on land rental, making purchase the more desired option.

Shifting from the whole sample to panel households for whom information on agricultural ability is available, we note that able producers are more likely to make use of land rental markets.[13] The magnitude of the coefficient is large; compared to the least productive producer in the sample, the most productive one has a 40% higher probability of accessing land through the rental market. It is of particular interest to observe that the magnitude of the productivity effect, as well as the coefficients of other variables, are clearly large enough to easily outweigh any wealth effect.[14]

Turning from the demand to the supply side, we note that many of the coefficients are indeed opposite from those observed for demand. Most important, land is supplied to the rental market by households who own more land, are wealthier (as indicated by them having an iron roof), older, and smaller, all in line with theoretical predications. Although agricultural ability is insignificant, implying that improvements in productivity are not the driving forces underlying supply to the rental market, the positive and highly significant coefficient on wealth (iron roof) could imply that households with off-farm employment are more likely to supply land to the rental market. All of this provides strong evidence in favor of the hypothesis that, in Uganda, land rental markets transfer land from large and older land owners to smaller and more productive young and often landless producers and have hardly any wealth bias.

Evolution of Land Rental Market Activity

Results based on the pooled sample of 16,200 households in both periods as indicated in Table 7-6 illustrate the significant increase of renting in land through the market by 16% between 1992 and 1999, which is likely to be due to increased levels of economic activity and associated reductions in transaction costs. Although land rental market activity is estimated to have increased in all regions, there was a relative decrease in the North. The combined sample also suggests that rental markets transfer land to less educated people and that female-headed households may have an advantage in gaining access to land through such markets. The redistributive effect of rental markets is supported by the negative and highly significant coefficient on area owned; in fact, we cannot reject the fact that the coefficient on land owned was the same in 1992 and 1999. At the same time, the importance of nonagricultural assets as a determinant of land rental has decreased over time; it is impossible to reject the hypothesis that, in 1999, the coefficient on such assets was equal to zero, implying that in 1999, possession of nonagricultural household assets no longer affected the probability of participating in land rental markets. By comparison, agricultural assets have become more important between the two periods, a finding that we take as an indication that land rental markets have, over time, shifted land to more productive producers.

On the supply side to rental markets, we find a similarly significant, although smaller, increase of market activity with time. Furthermore, and corresponding to greater demand by the land-poor, it is the relatively land-rich who are likely to rent out land, and the predictive power of our regression is quite high.

Conclusions and Implications

We started this chapter by noting that, especially for Africa, policy advice regarding the operation and impact of land markets is often based on scant empirical evidence that fails to do justice to a complex set of factors. Such factors are likely to affect the behavior of participants in formal and informal land markets, and thus the outcomes resulting from them, including differences between different (market and nonmarket) modalities of acquiring land. Our results demonstrate that not only do modest additions to standard household surveys facilitate a rather thorough exploration of these issues but also that doing so provides evidence of a positive role of land rental markets and illustrates the variety of options for land access open to rural households.

TABLE 7-6 Determinants of rental market participation: Pooled 1992/1999 sample

Dependent Variable	1992 and 1999 Pooled Sample	
	Rent In	*Rent Out*
Age of head	–0.002*** (8.73)	0.000*** (3.79)
Education of head	–0.005*** (5.17)	–0.000 (0.71)
Female head (non-widow)	0.024** (2.53)	–0.004 (1.17)
Female head (widow)	0.011 (1.12)	0.004 (1.04)
Area owned (log)	–0.141*** (29.14)	0.035*** (20.74)
Year*area owned	0.001* (1.77)	–0.000** (2.29)
Value of hh assets (log)	0.019*** (4.43)	–0.001 (0.42)
Year*hh asset value	–0.014** (2.54)	0.000 (0.11)
Value of agric assets (ϕ_1)	0.010*** (3.25)	–0.000 (0.10)
Year*agric asset value (ϕ_2)	0.019*** (5.02)	–0.001 (0.60)
Land dispersion at village level	–0.000 (0.30)	–0.000** (2.17)
Year*land dispersion	–0.000 (0.68)	0.000 (1.46)
Migrated	–0.002 (0.17)	0.001 (0.30)
Urban area	–0.123*** (14.47)	–0.007* (1.74)
East	–0.015 (1.05)	0.007 (1.23)
North	0.067*** (3.75)	–0.020*** (3.41)
West	–0.100*** (7.25)	–0.022*** (4.07)
Year	0.161*** (5.75)	0.032*** (2.79)
East*year	0.080*** (4.13)	0.004 (0.56)
North*year	–0.042** (2.15)	0.009 (0.85)
West*year	0.123*** (6.01)	0.006 (0.77)
Observations	16,200	16,200
Pseudo R^2	0.136	0.171
Log likelihood	–7,347.07	–2,711.81

Note: Absolute value of z statistics in parentheses.

* significant at 10%; ** significant at 5%; *** significant at 1%.

We find that, in the case of Uganda, land rental markets improved productivity in a pro-poor way and helped to provide land access to those in need, especially landless households. In line with theory that would predict the scope for land market transactions to increase with overall economic development, the level of land market participation has increased considerably and at a rapid pace, and there is evidence that wealth constraints have become less important. Sales markets are increasing less in productivity but nonetheless, and despite the fact that land prices are above the level that would be justified based on the present value of agricultural profits in some regions, we fail to find any evidence for concentration of land through sales markets. This is in line with what was found in other studies (Platteau and Baland 2001) and suggests that, especially in situations characterized by egalitarian access to land, land rental markets will have a key role to ensure productive use of land, similar to what has been found in China (Deininger and Jin 2002).

From a policy perspective, this suggests that, in Uganda, where endowments of land are distributed fairly equitably across the population, there is little justification for imposing restrictions on the functioning of land markets—but at the same time there is no great need for policy interventions to improve the functioning of land (rental) markets.[15] Countries with an egalitarian land ownership distribution (e.g., Ethiopia) that continue to restrict the functioning of land rental markets may incur significant efficiency losses and might consider ways to eliminate such restrictions. Even in settings where land reform is considered to be necessary to bring about changes of the land ownership distribution, measures taken in this context should not undermine the functioning of rental markets. Instead, devising mechanisms of land redistribution that complement rental markets, rather than try to substitute for them, would be of importance well beyond Africa. Furthermore, to the extent that ill-functioning land markets can lead to the build-up of conflict, as in Rwanda (Andre and Platteau 1998), establishing a regulatory framework within which households can freely exchange land in formal or informal ways will be important to facilitate adjustment to changed economic circumstances.

Although our analysis provides a less pessimistic picture about the operation of land markets than is often maintained, and suggests that rental markets can respond quite rapidly to changed economic conditions, it is only a first step toward a better understanding of the dynamics of land markets in an African context. In fact, the large extent of informal land transactions suggests that, similar to what has been reported from other African countries, exchange of land in Uganda is embedded within a complex network of social relationships. More specific investigation of the specific contractual arrangements underlying informal land transfers, especially the implicit and explicit obligations associated with them and the way in which these have changed over time with economic development and changes in the scarcity of different factors, could go a long way toward improving our understanding of the working and evolution of land markets. This would be critical to assess the impact of land markets on household welfare in the short term and to better assess the dynamics of rural factor markets and their long-term effect on investment, welfare, and productivity in rural areas.

NOTES

1. We thank the Uganda Bureau of Statistics, especially James Muwonge and Tom Emwanu, for making the data available. Support from the African Statistical Capacity Building Initiative and the Norwegian ESSD Trust Fund is gratefully acknowledged. The views expressed in the chapter are those of the authors and do not necessarily reflect those of the World Bank, its Board of Executive Directors, or the countries they represent.
2. In agricultural production, supervision is particularly difficult or costly due to the spatial dispersion of the production process, and the vagaries of nature imply a need to constantly adjust to microvariations of the natural environment.
3. A detailed derivation of the model is available from the authors on request.
4. The fact that this module was applied only to households that are currently engaged in crop production implies that supply of land in both the land rental and sales market are likely to be truncated. As there are many households that still own considerable amounts of land in urban areas, this may be a problem. We desist from estimation of the supply side of the sales market.
5. Since enumeration of assets in 1992 was incomplete, retrospective information on the 1992 value of about 25 types of household and enterprise assets was collected in the 1999 survey. A discrete ranking with five levels of values was used to indicate whether, for each of these, the value of the asset under concern was about equal, somewhat more, somewhat less, or much more than what is available now, or whether the household did not own any of this type of asset. To transform this into a value for the household's asset stock in 1992, the rankings for each asset were transformed into percentage increase (0, +25%, –25%, +50%, and –100%, respectively), allowing to compute the total value of different categories of assets owned by the household. The different categories were aggregated to yield a total asset value in each period, allowing to make inferences on changes in investment between the two periods.
6. An assessment of the estimated monetary value is given for each plot. Note that, because there was no "market test" households that did not buy their land may overvalue their land, a hypothesis that receives some support from the fact that the prices recorded for lands that were subject to actual purchase transactions in Table 7-2 are about 20% lower than the self-assessed ones for all plots described in Table 7-1, although with regional variations. In-depth investigation of the differences between subjective land values and actual market prices would transcend the scope of this chapter; however, the general point of overvaluation of land in the West and the Center is valid even without such a detailed investigation.
7. For panel households, we can compare the information they reported in the 1992 IHS with the recall data available from the 1999–2000 UNHS on selected variables in order to assess data quality. Doing so for the case of land owned in 1992 reveals that, for almost 75% of the sample, the information matches exactly, pointing toward very high data quality.

8. The discussion is limited to results from probit regressions for land participation; tobit models using the amount of area transferred as dependent variables are consistent with this evidence and are available upon request.
9. The fact that widows, but not non-widows, have a higher probability of having inherited land suggests that, as is indeed noted by observers, local inheritance practices are slowly changing, partly in response to the HIV/AIDS epidemic.
10. Although a large-scale household survey cannot provide details on the often complex transactions involved in "gifts" of land, using smaller-scale follow-ups to do so may be of interest.
11. The fact that landless rural households will not act as suppliers of land requires us to estimate two separate equations for supply and demand of land.
12. Tobit estimates of the area transacted are consistent with the probit regressions presented here and again are available upon request.
13. The fact that individual producers' deviation from village-level profits is insignificant suggests that, at least for the data at hand, this variable is not a very good proxy for households' agricultural ability.
14. Note that the mean of the log of agricultural and household assets is less than 10 in each case.
15. From an analytical point of view, it would be of interest to explore further the extent to which improved access to alternative savings instruments is associated with a reduction in the premium paid for land over and above what is warranted by its productive value.

PART III

Contract Choice, Poverty, and Efficiency of Land Use

CHAPTER 8

Contract Choice and Poverty in Southern Highlands of Ethiopia

Million Tadesse, Stein T. Holden, and Ragnar Øygard

Sharecropping is a land lease contract in which a tenant and a landlord share the final output as compensation for the labor supplied by the tenant and the land supplied by the landlord. In addition to labor and land, farmers in many rural areas form informal cost-sharing contracts for market-purchased inputs. The fact that land is state-owned property under the current Ethiopian constitution implies that land acquisition is via membership in a particular peasant association (PA), allowing only parental transfer of user rights to the heirs and participation in land rental markets (sharecropping and fixed rent contracts), while formal land sales are prohibited. The 1974 Ethiopian land reform implied a radical redistribution of land from feudal landlords to all community members. Land was then allocated based on household size. Land renting as well as hiring of labor was prohibited, whereas frequent land redistributions were implemented to maintain the egalitarian land distribution and allocate land to new households. With the change in government in 1991, a more market-friendly policy was introduced and short-term land renting was again allowed. Despite the policy barriers related to access and transfers of land, Ethiopian farmers have a long history of renting land through sharecropping and fixed rent contracts. This has continued even after the egalitarian land distribution was introduced. Our study revealed a particularly rich variation in cost-sharing sharecropping arrangements in southern Ethiopia, yet there have been no studies trying to explain the rationale of this variation within and across local communities. This is what we try to do in this chapter.

In other parts of the world there have been many studies and researchers trying to explain the rationale of sharecropping and to assess its efficiency implications (Marshall 1890; Cheung 1968, Cheung 1969; Stiglitz 1974; Shaban 1987; Hayami and Otsuka 1993b). Various authors have provided alternative theoretical explanations for the existence of alternative contracts. Screening models (Hallagan 1978; Allen 1982; Shetty 1988) assumed that landlords have no information on the ability of tenants and that they use self-selection among contractual options as a device to prevent tenants from defaulting on rent payments. However, the assumption of nonobservable tenant quality may be unrealistic when landlords and tenants live together in the same community and can obtain knowledge on ability and land quality through direct observation (Bardhan 1984; Singh 1989; Hayami and Otsuka 1993b). Braverman and Stiglitz (1986) argued that cost sharing applies only to inputs that can be monitored and enforced by the landlord at low cost. If so, there is no incentive issue regarding the application of such inputs. Hayami and Otsuka (1993b) therefore suggested that inputs supplied under cost sharing can be regarded as *de facto* production loans. However, as shown by Kassie and Holden (2007) and in Chapter 10, tenants may have an incentive to use more fertilizer on their rented in land in order to increase the probability of contract renewal.

Bell and Zusman (1976) and Zusman and Bell (1989) formulated the problem of contract choice in a bargaining framework in which they argued that optimal contracts are the result of the relative bargaining power of the contracting parties. Sadoulet et al. (2001) argued that contracts take the form of resource pooling, which compensates for idiosyncratic market failures to which the parties are subjected. Tikabo and Holden (2003) proposed a model in which contract choice—including cost sharing, pure output sharing, and fixed rent contracts as the options—was endogenous. Imperfections in credit and insurance markets caused contract choice to be driven by risk aversion and liquidity scarcity. Cost-sharing contracts were therefore to be preferred by relatively cash-rich and less risk-averse landlords and by more cash-poor and risk-averse tenants.

Empirical studies investigating contract choices include Dubois (2002), who found that fixed rent contracts were preferred to sharecropping for the most fertile plots in the Philippines, but the latter is preferred when crop-enhancing land overuse, such as maize, are grown. In a study in Tunisia, Laffont and Matoussi (1995) found that a tenant's share of output is directly related to his or her working capital but inversely related to that of the landlord, which is in line with the capital constraint and resource-pooling hypothesis. Tikabo and Holden (2003) found that poor landlords and wealthy tenants were attracted to each other through a preference for fixed rent contracts, whereas less poor landlords and poorer tenants were attracted to each other through a preference for cost-sharing contracts. The intermediate wealth stage on both sides provides a preference for pure sharecropping contracts. Negative correlation between wealth and capital shortage, and between wealth and risk aversion, could possibly explain this variation in contracts.

We have not come across any empirical studies of variation in cost-sharing arrangements and factors explaining them. Our study is therefore unique, but we can build on the reviewed theoretical literature to assess alternative hypotheses to explain the observed variation in contracts in southern Ethiopia. In this chapter we develop a general theoretical model facilitating the testing of a number of hypotheses of why a diversity of cost-sharing contracts coexist within the same communities and across neighboring communities. We assess whether resource pooling or relative bargaining power can explain the variation, or whether the variation is primarily an adjustment to differences in soil quality across rented plots. With resource pooling, the wealthier landlords would provide more input cost, but if wealth and bargaining power are positively correlated, wealthy landlords would cover less of the input costs. With resource pooling, poorer tenants would also get more input costs covered by their landlords, whereas under the bargaining model, assuming that wealth and bargaining power are positively correlated, poorer tenants would have to cover more of the input costs.

Land Rental Contracts in Southern Ethiopia

The land rental contracts identified in our study areas are presented in Table 8-1. As can be seen from the table, we identified six different contract types. It was surprising, however, that output sharing was fixed at 50:50, since output shares are found to vary in other parts of Ethiopia. The fact that insurance

TABLE 8-1 Input cost sharing rules under crop-sharing and cash rent contracts in *Wolaita* and *Sidama* Zones, Ethiopia

Contracts (y^n_{obs})	*Description of Each Land Rental Contract*	*Frequency*	*%*
y_1	Landlord covers all input costs[a] and output is shared equally	18	2
y_2	Landlord covers all input costs but it is subtracted first before output is shared equally	53	6
y_3	Equal input cost share and equal output share	430	49
y_4	Tenant covers all input costs but it is subtracted first before output is shared equally	225	25
y_5	Tenant covers all input costs and output is shared equally	47	5
y_6	Fixed rent contract, tenant covers 100% of input costs and gets 100% of output	112	13
	Total	885	100

Note: Tenant preference ordering: $Y_1^t > Y_2^t > Y_3^t > Y_4^t > Y_5^t$ and landlord preference ordering: $Y_5^l > Y_4^l > Y_3^l > Y_2^l > Y_1^l$. It is only under distress rental condition that Y_6^l is preferred over the rest of the contracts by landlords. Observations are at plot level for each contract type for unmatched tenant and landlord households.

[a]Landlords contribute labor on rented out plots only in rare cases, except during harvesting. Otherwise, labor input is exclusively supplied by the tenant. Hence, our definition of input costs does not include labor and land.

or credit markets are imperfect in our study area may cause land rental markets to serve as a substitute to smooth consumption and mobilize resources as needed.

In Table 8-1 we have ranked the land rental contracts in terms of how favorable they are to the tenants, with the most favorable contract ranked first and assuming the opposite rank applies for the landlord. As marginal return to the tenant's work effort is 50% of marginal yield for all contracts, the work incentive provided by the contracts is the same in all cases, except for the fixed rent contract.

The tenants' expected return and utility from contracts decreases as we move from Y_1 to Y_5 but increases for the landlord as one moves from Y_1 to Y_5 (see Table 8-1). This is because contracts, all else being equal, at the top are less expensive for the tenant but more expensive for the landlord. In contract Y_1 the landlord covers all the cash input costs. In contract Y_2 the landlord covers all the up-front financial costs of the inputs but is compensated after harvest when the costs are shared equally (without interest). Contract Y_3 implies equal sharing of financial expenses for cash inputs. Contract Y_4 implies that the tenant covers the up-front financial expenses for cash inputs but is compensated after harvest by equal sharing of the costs (without interest). Contract Y_5 implies that the tenant carries all the cash input costs without compensation other than through equal sharing of the output, like in all the previous contracts. This is the standard pure sharecropping contract. Contract Y_6 is the fixed rent contract. The return to this contract and its ranking relative to the other contracts depends on the rental rate. Land rental market participation under cash rent is often an indication of financial distress by the landlord rather than a pure demand and supply-driven market condition. Those landlords who are really constrained by cash (liquid assets) for the purchase of inputs, or for meeting consumption demands such as expenses on school fees, social ceremonies, and medical treatment, may rent out their plots on cash rental basis. This will then be their most preferred contract, given their very high subjective discount rate. The average cash rental price in the study area was about 1,312 Ethiopian birr/hectare/year (1US$ = 8.8 birr), whereas the share of landlords' average net return from sharecropping contracts was 3,301 birr/hectare/year. The difference in mean returns between sharecropping and fixed rent contracts is statistically significant $t = 3.2064$; $p = 0.000$. In order to assess landlords' time preferences, we computed the discount rate at mean values of fixed rate and sharecropping contracts returns, assuming that the whole difference is due to discounting of future income. The result indicated that a landlord who prefers fixed rent contract from sharecropping has a discrete time discount rate of at least 150%. This must be seen as an upper bound on the discount rate because risk sharing is also implicit in the sharecropping contract. High discount rates have also been found for poor households elsewhere in Ethiopia and in other countries (Holden et al. 1998).

As can be seen from Table 8-1, the equal input cost sharing contract, Y_3, is the most common contract in our study area (49% of contracts), followed by Y_4 (25% of contracts) and Y_6, the fixed-rent contract (13% of contracts). The contract Y_2, in which the landlord advances for cash inputs but gets refunding at the time of harvest, before the remaining output is shared equally, and Y_5, in which the tenant pays for all inputs without compensation, and output is shared equally at harvest are about equally rare (6% and 5% of the contracts, respectively). Finally, the Y_1 contract, in which the landlord pays all the inputs without compensation at harvest while output is shared equally, is a rare contract that we found in only 2% of the cases.

A Model of Land Rental Contract Choice

We assume that there exists a set, S, of available contracts in an area. The size of the set is likely to increase with an expansion of the area, meaning that the set of contracts may be smaller in a sub-area, like a village, as compared to a district or region; therefore, the set of contracts in a village, $s^v \in S$. Furthermore, the set of available contracts to a specific household, h, in the village v may only be a subset of the village set of contracts, $s^{vh} \in s^v$. For households that are rationed out of the land rental market, $s^{vh} = \varnothing$. Contract characteristics are also likely to be affected by basic agroecological characteristics that affect the types of crops that can be grown, the profitability, and the risk. Specific plots that are offered in the rental market may therefore have a subset of contract alternatives that is a sub-set of the types of contracts available within a village, $s^p \in s^v$. The specific contracts available for a specific plot of a specific household in a specific village may be denoted $s^{vhp} \in s^{vh} \in s^v$, indicating also that contract choice may be a function of plot, household, and village characteristics.

We assume that households that are potential participants in the land rental market as landlords or as tenants have a preference ranking among the contracts that are available to them and that actual contract choice is a result of this preference ranking. However, landlords and tenants who agree on a contract may have conflicting interests with respect to contract choice, and the actual agreed contract may therefore depend on their bargaining power and be the best outcome of available choices. This will be the best offer from the contracting partner or possibly several potential contract partners. The actual observed contracts of landlords and tenants are a result of their constrained optimization. Thus, contract choice is likely to be a function of the characteristics of the landlords and tenants, the village characteristics, the agroecological and specific farm plot characteristics, and the market's characteristics.

$$Y_{obs}^{n} = Y_{obs}^{n}\left(z^{t}, z^{l}, z^{p}, z^{v}, p^{m}\right),$$

where Y^n_{obs} is observed contract choice of tenant (when $n = t$) and landlord (when $n = l$). z^t is tenant household characteristics, z^l is contract partner-landlord household characteristics, z^p is the plot and agroecologal characteristics, z^v are the village characteristics, and p^m is a vector of prices and market characteristics.

We develop the model for economies characterized by imperfect markets for land, labor, credit, and insurance. We assume that there is production risk but markets for cash inputs (fertilizer and seed) and outputs exist. Households are generally poor, face subsistence constraints, and have small farm sizes. Their endowments of nonland assets determine their relative poverty and their decisions on whether to participate in the land rental market or not. Households with more nonland resources relative to land are more likely to rent in land, whereas households with few nonland resources are more likely to rent out land. This is based on the assumption that transaction costs in the land rental market tend to be lower than in nonland factor markets (Tikabo et al. 2008; Holden and Ghebru 2006; Chapter 4). We will now introduce a number of hypotheses that may possibly explain observed contract choices of landlords and tenants:

H1. Bargaining Power Hypothesis

Observed contract choice depends on the relative bargaining power of landlords and tenants. Landlords and tenants with more bargaining power obtain better contracts than landlords and tenants with lower bargaining power.

Assumption: Landlords and tenants with more wealth have more bargaining power than landlords and tenants with less wealth.

Testable Implication: More wealthy landlords and tenants achieve more favorable land contracts.

H2. Resource-Pooling Hypothesis

This is the altruistic or more equal sharing of benefits model. Landlords and tenants with more resources contribute more of their resources and therefore accept less favorable contracts.

Assumption: Altruism and/or Pareto-efficiency gains can explain this behavior.

Testable Implications: Altruism may be associated with kinship contracts. Pareto-efficiency gains may be investigated through assessment of production efficiency in alternative contracts.

H3. Land Quality-Adjusted Contracts Hypothesis

Contract choice reflects differences in land quality. For good quality land the implicit rent will be higher (i.e., the tenant covers more of the input costs).

Assumption: Land quality is observable to both parties.

Testable Implication: Observable land quality characteristics are positively related to the rank obtained by landlords, implying that the tenant has to cover more of the input costs on better quality land.

H4. Duration-Adjusted Contracts Hypothesis

More long-term contracts are preferred by tenants and they are willing to cover more of the input costs for long-term contracts.

Testable Implication: Longer-term contracts are associated with tenants covering more and landlords less of the cash input costs.

H5. Distress Rental Hypothesis

Landlords facing emergencies rent out their land on fixed-rent basis due to lack of alternative sources of cash.

Assumption: Poor landlords lack access to credit and are cash-constrained.

Testable Implication: Landlords renting out land on fixed-rent contracts are particularly poor. Rent paid in fixed-rent contracts is low compared to return from sharecropping contracts.

H6. Spatial Variation Hypotheses

There is not one rental market for land, but due to the spatial distribution and immobility of land and the spatial dispersion of potential landlords and tenants, land rental markets are fragmented. Variation in contracts reflects local differences in demand for and supply of rental land.

Testable Implication: Location (village) dummies will explain the variation in land rental contracts.

H7. Reputation and Trust Hypothesis

Well-established tenants and landlords obtain more favorable contracts than new entrants in the market and entrants with poor reputation and low trust.

Testable Implication: Assess whether landlords and tenants that emphasize reputation and trust when selecting partners have given their partners more favorable contracts.

An Application to Southern Ethiopia

We will apply this theoretical framework to a case study in southern Ethiopia. Our data allow us to observe the range of contracts in a larger area containing variation in agroecological and market characteristics, variation in farm and household characteristics within villages, the actual contracts of individual tenant and landlord households, and the plots that are transacted. While we observed the set of contracts in Table 8-1, here they are formulated mathematically as seen from the tenant households' perspective:

$$S^{tp} = \begin{cases} \alpha\rho^t \sum_p \theta_p F_p^t\left(A_{rp1}^l, L_{rp1}^t, X_{rp1}^l\right) - \omega^t \sum_p L_{rp1}^t & :Y_1^t \\ \alpha\rho^t [\sum_p \theta_p F_p^t\left(A_{rp2}^l, L_{rp2}^t, X_{rp2}^l\right) - p_x \sum_p X_{rp2}^l] - \omega^t \sum_p L_{rp2}^t & :Y_2^t \\ \alpha\rho^t \sum_p \theta_p F_p^t\left(A_{rp3}^l, L_{rp3}^t, X_{rp3}^{lt}\right) - p_x \sum_p \gamma_p X_{rp3}^{lt} - \omega^t \sum_p L_{rp3}^t & :Y_3^t \\ \alpha[\rho^t \sum_p \theta_p F_p^t(A_{rp4}^l, L_{rp4}^t, X_{rp4}^t) - p_x \sum_p X_{rp4}^t] - \omega^t \sum_p L_{rp4}^t & :Y_4^t \\ \alpha\rho^t \sum_p \theta_p F_p^t(A_{rp5}^l, L_{rp5}^t, X_{rp5}^t) - p_x \sum_p X_{rp5}^t - \omega^t \sum_p L_{rp5}^t & :Y_5^t \\ \rho^t \sum_p \theta_p F_p^t(A_{rp6}^l, L_{rp6}^t, X_{rp6}^t) - p_x \sum_p X_{rp6}^t - \tau_p A_{rp6}^t - \omega^t \sum_p L_{rp6}^t & :Y_6^t \end{cases} \quad (8.1)$$

where $\rho^t = 1/1+\delta^t$ is the subjective discount factor, α is the output share (= 0.5), F_p^t denotes plot level concave production function that preserves the usual properties $F_p^{'t} > 0$ and $F_p^{''t} < 0$ and with positive cross-partial derivatives. Plot level output is a function of area, A_r, labor input per plot supplied by the tenant, L_{rp}^t, and purchased inputs, X_{rp}^l, either supplied (contributed) by the tenant (if superscript t), by landlord (if superscript l) or by both parties (if superscript lt), while subscripts 1, 2, 3, . . . ,6 denote available contract choices, Y_j^t ranging from 1 to 6. Output price is normalized to 1 and tenant's shadow wage rate, ω^t, and input price, p_x, are assumed constant across contracts. In addition, γ_p indicates input cost share of the tenant under equal cost-share contract choices ($\alpha = \gamma_p = 0.5$). The term τ_p denotes the fixed payment for cash rented plot, p.

The mirror image of this set of contracts as seen from the landlord is as follows:

$$S^{lp} = \begin{cases} (1-\alpha)\rho^l \sum_p \theta_p F_p^l (A_{rp1}^l, L_{rp1}^t, X_{rp1}^l) - p_x \sum_p X_{rp1}^l & :Y_1^l \\ (1-\alpha)[\rho^l \sum_p \theta_p F_p^l (A_{rp2}^l, L_{rp2}^t, X_{rp2}^l) - p_x \sum_p X_{rp2}^l] & :Y_2^l \\ (1-\alpha)\rho^l \sum_p \theta_p F_p^l (A_{rp3}^l, L_{rp3}^t, X_{rp3}^{lt}) - (1-\gamma_p) p_x \sum_p X_{rp3}^{lt} & :Y_3^l \\ (1-\alpha)\rho^l \sum_p [\theta_p F_p^l (A_{rp4}^l, L_{rp4}^t, X_{rp4}^t) - p_x \sum_p X_{rp4}^t] & :Y_4^l \\ (1-\alpha)\rho^l \sum_p \theta_p F_p^l (A_{rp5}^l, L_{rp5}^t, X_{rp5}^t) & :Y_5^l \\ \tau_p A_{rp6} & :Y_6^l \end{cases} \quad (8.2)$$

where $(1 - \alpha)$ and $(1 - \gamma_p)$ are the landlord's output and input shares, respectively with all other terms defined as before. Based on these sets we can now look at contract choice for individual tenants and landlords who are not likely to be able to choose among the full set of contract alternatives listed above.

Tenant Model

A typical tenant in our study area cultivates own plots, A_o and applies labor and purchased inputs, L_o and X_o, respectively, on own plots in addition to the available contract choices under the set s^{vt}. Hence, we can specify the tenant's discounted expected net income, denoted by y_j^t, from own plots and from the set of contract choices ranging from 1, 2, 3 . . . 6 in $\{s^{vtp}\}$, as:

$$y_j^t = \underbrace{\rho^t \sum_p \theta F_p^t (A_{op}^t, L_{op}^t, X_{op}^t)}_{\text{Income from own plots, where subscript } p \text{ is plots}} - \underbrace{\sum_p \left(p_x X_{op}^t + \omega^t L_{op}^t\right)}_{\text{Cost of inputs on own plots}} + \underbrace{\arg\max\left\{s^{vtp}\right\}}_{\text{Choice of best available contract}}$$

$$= \rho^t \sum_p \theta F_p^t (A_{op}^t, L_{op}^t, X_{op}^t) - \sum_p \left(p_x X_{op}^t + \omega^t L_{op}^t\right) + \pi\left\{Y_{obs}^t\left(z^t, z^l, z^p, z^v, p^m\right)\right\} \quad (8.3)$$

The expression $\sum_p \left(p_x X_{op}^t + \omega^t L_{op}^t\right)$ denotes the sum of input costs on own plots for purchased inputs and labor. The tenant's expected return from alternative contracts has the following preference ordering: $Y_1^t > Y_2^t > Y_3^t > Y_4^t > Y_5^t$. For a severely cash-constrained tenant the last contract, Y_6^t, will be the least preferred, while its rank is undetermined and depends on the rental price, relative risk of alternative contracts, the (subjective) discount rate and risk aversion of the tenant, in other cases. Based on this we are able to estimate a "reduced form" of the observed contract choices of tenants as an ordered rank model as follows:

$$Y_{obs}^t = Y_{obs}^t\left(A^t, L^t, z^t, z^l, z^p, z^v, p^m\right) \quad (8.4)$$

Since we do not have paired observations we assume that $z^l = z^l(z^v)$.

Landlord Model

A landlord in our setting usually cultivates own plots $(A_o - A_r)$ in addition to the plot currently rented out, either in the form of sharecropping or cash rent contracts. He or she lives in the same village as the tenant, or may in some cases reside in a neighboring village. The landlord's discounted expected net income, denoted by y_j^l, from own plots and from the choices of the best available contract, for each contract choice is specified as:

$$y_j^l = \underbrace{\rho^l \sum_p \theta F_p{}^l(A_{op}^l, L_{op}^l, X_{op}^l)}_{\text{Income from own plots, where subscript } p \text{ is plots}} - \underbrace{\sum_p \left(p_x X_{op}^l + \omega^l L_{op}^l\right)}_{\text{Cost of inputs on own plots}} + \underbrace{\arg\max\left\{s^{vlp}\right\}}_{\text{Choice of best available contract}} \tag{8.5}$$

$$= \rho^l \sum_p \theta F_p{}^l(A_{op}^t, L_{op}^l, X_{op}^l) - \sum_p \left(p_x X_{op}^l + \omega^l L_{op}^l\right) + \pi\left\{Y_{obs}^l\left(z^l, z^t, z^p, z^v, p^m\right)\right\}$$

The expression

$$\sum_p \left(p_x X_{op}^l + \omega^l L_{op}^l\right)$$

denotes the landlord's total costs (purchased inputs and labor) on own plots. Assuming that a cash rental contract is practiced only under distress condition (indication of the landlord's weak financial position), contracts have the following preference ordering: $Y_6^l > Y_5^l > Y_4^l > Y_3^l > Y_2^l > Y_1^l$. Other landlords not in a distress situation are likely to rank the fixed-rent contract lower. Based on this we may estimate the "reduced form" ordered response model for actual observed contacts of landlords as follows:

$$Y_{obs}^l = Y_{obs}^l\left(A^l, L^l, z^l, z^t, z^p, z^v, p^m\right) \tag{8.6}$$

allowing us to test our hypotheses, assuming that $z^t = z^t(z^v)$, since we do not have data on both landlord and household characteristics for each contract.

Econometric Specification and Estimation Issues

Ordered probit models were used in the econometric analysis. In order to capture unobserved household heterogeneity (such as tenant and landlord unobserved behavior), and also to adjust for within-cluster correlation, GLLAMM models (with 16 points of integration using ordinary quadrature points), following Rabe-Hesketh et al. (2004) were estimated. A two-level random intercept model taking only one plot specific explanatory variable, x_{pk} for simplicity, can be specified as

$$y_p^h = \eta_0^h + \beta_1 x_p^h + \varepsilon_p^h \tag{8.7}$$

where p denotes plots as level-1 units and households, h, as level-2 units, η_0^h are household-specific intercepts, β_1 is a regression coefficient, and ε_p^h is a plot level residual term. The η_0^h intercepts may be expressed as

$$\eta_0^h = \gamma_0^0 + \zeta_0^h \tag{8.8}$$

where γ_0^0 is the mean intercept and ζ_0^h is the deviation of the household-specific intercept η_0^h from the mean. Let $\theta \equiv \text{var}(\varepsilon_p^h)$ and $\psi \equiv \text{var}(\zeta_0^h)$. We assume household level clusters are independent and

$$\begin{aligned}
&\varepsilon_p^h \mid x_p^h \sim N(0,\theta),\\
&\text{Cov}(\varepsilon_p^h, \varepsilon_{p'}^h) = 0,\ p \neq p'\\
&\zeta_0^h \mid x_p^h \sim N(0,\theta),\\
&\text{Cov}(\zeta_0^h, \varepsilon_{p'}^h) = 0
\end{aligned}$$

The reduced form of the model is obtained by substituting equation (8.8) into equation (8.7) (Rabe-Hesketh et al. 2004):

$$y_p^h = \gamma_0^0 + \zeta_0^h + \beta_1 x_p^h + \varepsilon_p^h \tag{8.9}$$

The log likelihood-ratio test statistics favored GLLAMM models against the standard ordered probit models for both tenant and landlord sub-samples (1% level of significance), as zero variance restrictions were strongly rejected. This indicates that the two-level random intercept GLLAMM models are more suitable in analyzing the dataset and are therefore used in our subsequent discussion. Standard errors are also corrected for heteroskedasticity using the Huber/White/Sandwich estimator. One of the limitations of our analysis is that we do not have matched data for landlords and tenants. We therefore have to assess the contracts of landlords and tenants separately, causing possible omitted variable bias due to correlation of landlords' and tenants' characteristics. However, the use of the two-level random intercept model in GLLAMM (household being level 2 and plot being level 1) helps to control the unobserved heterogeneity of the overall responses of contract choices.

Instead of a constant term, the vector of explanatory variables includes threshold parameters (κ_0 to κ_6) that determine the six outcomes for observed responses.

$$Y_{obs}^{n} = \begin{cases} \overbrace{1 \Rightarrow Y_1}^{\text{Choices, } Y} & \text{if } \kappa_0 = -\infty \leq Y_j^* < \kappa_1 \\ 2 \Rightarrow Y_2 & \text{if } \kappa_1 \leq Y_j^* < \kappa_2 \\ 3 \Rightarrow Y_3 & \text{if } \kappa_2 \leq Y_j^* < \kappa_3 \\ 4 \Rightarrow Y_4 & \text{if } \kappa_3 \leq Y_j^* < \kappa_4 \\ 5 \Rightarrow Y_5 & \text{if } \kappa_4 \leq Y_j^* < \kappa_5 \\ 6 \Rightarrow Y_6 & \text{if } \kappa_5 \leq Y_j^* < \kappa_6 = \infty \quad \text{where } Y_j^* \text{ is unobserved contracts} \end{cases} \tag{8.10}$$

And can be estimated as

$$y_j^* = x'\beta + \varepsilon$$

Since the rank of the fixed-rent contract is not obvious and may vary with the discount rate (and risk preferences) of landlords and tenants, we have chosen to run the models with and without including the fixed-rent contracts (Table 8-5, shown later). We use this as a robustness test of the results as well and interpret results conservatively by saying that a hypothesis is supported only when the results are significant and in favor of the hypothesis in both model specifications.

Data

The dataset used for this study is based on data collected from the southern highlands of Ethiopia from September to December 2005. Six peasant associations (PA) were chosen to represent various agroclimatic zones in Wollaita and Sidama zones of the southern highlands of Ethiopia. The districts that are used in the present study are *Bolososore* and *Kindokoysha* in Wollaita zone and *Wondo Genet* area in Sidama zone. A total of 36 villages were included and households were randomly sampled from each village. Trained enumerators (who have diplomas and are qualified in agricultural sciences) were used to collect both farm plot and household level information under strict supervision by the first author of this chapter (Tadesse). Input use and outputs for the two rainy seasons (*meher* and *belg*) were collected from 278 households and 3,088 plots. Variable definitions and descriptive statistics of explanatory variables are presented in Tables 8-2, 8-3, and 8-4 for tenants and landlords.

As can been seen in Table 8-3, there are statistically significant differences between tenants and landlords for key wealth indicator variables. We found that tenants have significantly higher income, per capita farm size and male work force, contrary to other parts of the world where landlords are often richer than tenants. In addition, we also found that tenants are significantly younger, have smaller family size and lower dependency ratio compared to landlords. We also tested for significant correlation between

TABLE 8-2 Variables used in the models with summary statistics

		Tenants		*Landlords*	
Variables	*Description*	*Mean*	*SD*	*Mean*	*SD*
Dependent variable	Ordered responses for contract choice Y_j^n(j= 1, 2,3 . . . 6) and n = t or l				
Age	Age of household head in years	44	11.77	49	13.99
Sex	Sex of household head: 1 = male; 0 = female			0.82	0.38
Femaleworkf	Number of female work force per household	1.68	0.95	2.0	1.1
Maleworkf	Number of male work force per household	2.12	1.44	1.94	1.30
Cwratio	Consumer-worker ratio	0.89	0.56	0.92	0.74
Plotdist	Plot distance in meters	959	1317	394	1520
Farm size	Average farm size, ha	0.63	0.52	0.58	0.42
Pcfarmsz	Household's total farm size (sq. meters) divided by total family size	932	722	833	552
PA	5 dummy variables for 6 peasant associations				
Tlu	Tropical Livestock Unit excluding oxen units	1.61	1.1	1.59	1.04
Oxtlu	Oxen Tropical Livestock Unit	0.52	0.71	0.65	0.78
Durdum2	= 1 if contract duration ≥ 12 months but < 36 months ; = 0 if not	0.05	0.21	0.02	0.15
Durdum3	= 1 if contract duration ≥ 36 months; = 0 if not	0.13	0.34	0.14	0.34
Plotsucero	Plot exposure to erosion: 1 = high, 2 = medium, 3 = less, 4 = no				
Soildpth	2 dummy variables for 3 soil depth categories: 1 = shallow, 2 = medium; 3 = deep				
Reputation	If reputation is main criteria for partner selection: 1 = yes; 0 = no	0.14	0.31	0.10	0.28
Trust	If trust is main criteria for partner selection: 1 = yes; 0 = no	0.24	0.43	0.19	0.39
Accaltpart	Access to alternative partners: 1 = yes; 0 = no	0.40	0.48	0.35	0.48
Acccredit	= 1 if household has access to credit (applied and obtained credit); 0 if not	0.17	0.33	0.13	0.34
Iftenusecl	If you want, can you rent out your land on cash basis without PA approval? 1 = yes; 0 = no			0.49	0.01

contract choice and categorical variables using χ^2 test statistics (see the test results in Table 8-4). We see that access to credit, access to alternative partners, and reputation were significantly correlated with tenants' contract choice. However, credit access was not significantly correlated with contract choice for landlords.

TABLE 8-3 Mean comparison tests across major wealth indicator variables by household type

Continuous Variables	*Tenant Mean(SD)*	*Landlord Mean(SD)*	*Mean Diff. (t-Value)*
Income from crop and livestock per family size	666.1(512)	532.6(329)	133.5 *** ($t = 6.78$)
Per capita farm size (Sq. meters) (PCFARMSZ)	932.02(722)	832.8(552)	99.1 *** ($t = 3.37$)
Family size	7.06(2.47)	7.24(2.23)	–0.18* ($t = -1.65$)
Consumer worker ratio (CWRATIO)	0.89(0.56)	0.92(0.74)	–0.03 ($t = -0.92$)
Female workforce (FEMALEWRKF)	1.68(0.94)	2.01(1.1)	–0.33*** ($t = -6.93$)
Male workforce (MALEWRKF)	2.12(1.44)	1.94(1.29)	0.18*** ($t = 2.88$)
Livestock ownership (TLU) excluding oxen	1.61(1.03)	1.59(1.04)	0.019 ($t = 0.42$)
Age of household head	43.9(11)	48.4(14)	–4.5*** ($t = -7.66$)

***, **, * significance at 1, 5 and 10% level, respectively.

TABLE 8-4 Tests for significant correlation between contract choice and categorical variables

	Variations Across Contracts by Household Type	
Discrete Variables	*Tenant (Overall χ^2-test)*	*Landlord (Overall χ^2-test)*
Access to credit (ACCCREDIT)	32.54***	6.44
Access to alternative partner (ACCALPART)	42.14***	28.10***
Trust	9.96*	16.3***
Reputation	29.69***	9.10
Being male headed household (SEX2)	12.12*	33.69***
Freedom to rent out plots without PA approval (IFTENSEC1)	7.51	11.43*

***, **, * significance at 1, 5 and 10% level, respectively.

Results and Discussion

We present the results of our analysis of determinants of land rental contract choice from ordered probit models with household random intercepts (using the GLLAMM procedure in STATA) in Table 8-5 with models for tenants (M_1 and M_3) and landlord (M_2 and M_4)with and without the fixed-rent contracts. Robust standard errors for clustered data are reported in all the models. The Huber/White/Sandwich estimator of the covariance matrix of parameter estimates is used. We present the results with a focus on our main testable hypotheses in chronological order.

H1. Bargaining Power Hypothesis

If this hypothesis is correct, landlords and tenants with more bargaining power should obtain better contracts than landlords and tenants with lower bargaining power. If bargaining power and wealth are positively correlated (our assumption), more wealthy landlords and tenants should achieve more

TABLE 8-5 Determinants of land rental (cost sharing) contract choices, southern Ethiopia

	With Fixed Rent Contract		*Without Fixed Rent Contract*	
	Tenant (M_1)	*Landlord (M_2)*	*Tenant (M_3)*	*Landlord (M_4)*
	b/(se)	*b/(se)*	*b/(se)*	*b/(se)*
Fixed part				
Age	–0.046* (0.025)	0.102*** (0.030)	–0.075** (0.032)	0.019** (0.008)
Femalewrkf	–0.347 (0.333)	0.665 (0.492)	–1.185*** (0.408)	–0.275* (0.150)
Malewrkf	0.591*** (0.177)	1.085** (0.431)	1.469*** (0.377)	0.052 (0.131)
Cwratio	–1.689*** (0.351)	2.716*** (0.589)	–0.564 (0.619)	0.846*** (0.260)
Plotdista	0.000*** (0.000)	0.000 (0.000)	0.000*** (0.000)	0.000 (0.000)
Pcfarmsz	0.002*** (0.000)	0.003*** (0.001)	0.000* (0.000)	0.000 (0.000)
PA2	4.428*** (0.630)	3.662*** (0.939)	3.451*** (0.553)	–1.325*** (0.369)
PA3	–0.290 (0.421)	–0.026 (1.143)	–2.556*** (0.708)	–2.765*** (0.414)
PA4	–1.318** (0.559)	–0.330 (1.740)	–0.790 (0.665)	–0.401 (0.471)
PA5	–1.109 (0.863)	–4.476*** (1.262)	0.510 (1.436)	–0.913 (0.632)

(Continued)

TABLE 8-5 *(Cont.)*

	With Fixed Rent Contract		*Without Fixed Rent Contract*	
	Tenant (M_1)	*Landlord (M_2)*	*Tenant (M_3)*	*Landlord (M_4)*
	b/(se)	*b/(se)*	*b/(se)*	*b/(se)*
PA6	0.755 (0.796)	1.994*** (0.763)	1.160** (0.525)	0.233 (0.484)
Tlu	–0.150* (0.084)	0.616*** (0.217)	–0.011 (0.095)	0.185 (0.123)
Oxtlu	–0.127 (0.133)	–0.035 (0.167)	–0.267** (0.111)	0.113 (0.131)
Durdum2	3.351*** (0.560)	–1.583 (1.480)	2.438** (1.010)	0.052 (0.388)
Durdum3	1.673** (0.703)	8.163*** (2.692)	–0.928*** (0.302)	0.252 (0.255)
Plotsucero2	–1.279** (0.606)	0.358 (0.755)	0.169 (0.419)	–0.658* (0.376)
Plotsucero3	–2.405*** (0.522)	–2.237** (0.932)	–2.015** (0.941)	–1.567*** (0.395)
Plotsucero4	1.724*** (0.551)	5.055*** (1.299)	–0.509 (0.595)	–1.612*** (0.512)
Soildpth2	0.994*** (0.353)	1.180*** (0.397)	0.966** (0.482)	0.700*** (0.219)
Soildpth3	3.314*** (1.050)	2.549*** (0.695)	5.973*** (1.496)	1.672*** (0.454)
Reputation	–1.101 (0.860)	–0.018 (2.285)	–2.301* (1.259)	1.387* (0.731)
Trust	–0.063 (0.592)	–5.136*** (1.483)	0.497 (0.608)	0.928** (0.455)
Accaltpart	0.770 (0.906)	1.935* (1.151)	–0.052 (0.417)	–0.373 (0.418)
Acccredit	0.732*** (0.219)	–0.769 (0.827)	–1.124* (0.600)	0.031 (0.430)
Sex2		–1.277 (0.868)		0.480* (0.285)
Iftensec1		–4.400*** (1.131)		–0.302 (0.246)
Random part				
Variance(cov)				
HH level	14.22(6.1)	9.21(2.69)	13.90(2.03)	8.79(2.56)
N (level 1,plots)	333	254	303	231
N (level 2, HHID)	62	70	62	70
Log likelihood	–209.26	–157.53	–110.79	–125.25

Note: $^{*}p < 0.10$, $^{**}p < 0.05$, $^{***}p < 0.01$. Robust standard errors for household (HH) level clustered data are reported. The variance of HH level RE is 14.22 with SE 6.1 for M1 and so on.

favorable contracts. However, we see from Table 8-5 that there is little support for this hypothesis. Perhaps this is because landlords in the area tend to be poorer than tenants and are often female-headed households. This may imply that their bargaining power is relatively weak. Also, we could not find any strong evidence of contract choices depending on tenants' bargaining power.

H2. Resource-Pooling Hypothesis

This is the altruistic or more equal sharing of benefits model. The results in Table 8-5 provide some support for the resource-pooling hypothesis on the landlord side (M_2 and M_4) because tenants covered more of the input costs for landlords with high dependency ratios (cw-ratio was significant and positive at 1% level in both models). On the tenant side (M_1 and M_3) we also found support for the resource-pooling hypothesis since tenants with more male work force (significant at 1% level in both models) and larger farm size (significant at 1 and 10% levels) were correlated with tenants covering more of the input costs. This may also be in line with landlords screening and selecting tenants with more resources.

H3. Land Quality-Adjusted Contracts Hypothesis

According to this hypothesis, tenants who rent in good quality land cover more of the input costs. The results for soil depth were highly significant in both models for landlords as well as for tenants (significant at 1% levels for both soil depth variables in all but one model, where the level of significance was at 5% level) and with positive signs related to the deeper and more fertile soils. This is very strong support for the hypothesis that contracts are adjusted for land quality such that tenants have to pay more of the inputs when they rent in better quality land. The results for the other land quality-related variable, plot exposure to erosion, were more mixed.

H4. Duration-Adjusted Contracts Hypothesis

The results in Table 8-5 revealed that tenants were having contracts that required them to cover significantly (at 1 and 5% levels) more of the input costs when contracts were for more than one year and up to three years (durdum2), but the results were mixed when the contracts were for more than three years (durdum3), as compared to annual contracts. It appears that fixed-rent contracts were associated with contracts longer than three years as the durdum3 variable switched from positive and significant (at 5% level) to negative and significant (at 1% level) when the fixed rent contracts were taken out.

H5. Distress Rental Hypothesis

The analytical results in Table 8-5 provide little evidence of this hypothesis as access to credit was insignificant in both landlord models.

H6. Spatial Variation Hypothesis

Consistent with this hypothesis arguing that land rental markets are fragmented, we found that several peasant association (PA) dummy variables were significant in the alternative specification of tenant and landlord models in Table 8-5 and three of the PA dummies were significant in the landlord model. These results, however, were not robust when we compared models with and without fixed-rent contracts, pointing at spatial discrepancies in the role of fixed-rent contracts. These are signs of poor integration of land rental markets but this could possibly also be explained by other unobserved village characteristics. As a further visualization of the variation of contracts across PAs, we present the distribution of contracts for each of the six PAs in Table 8-6.

The results from the overall χ^2 tests showed significant contract variations within PAs vs. across PAs, except in PA5 (due to few observations there). This result also supports our hypothesis (i.e., land rental markets in the study area are fragmented).

H7. Reputation and Trust Hypothesis

We found that young tenants were significantly more likely to have contracts in which they had to cover more of the input costs (significant at 10 and 5%

TABLE 8-6 Distribution of contracts across the six peasant associations

Contract Choices	*Peasant Associations (PAs)*						Total
	PA1	*PA2*	*PA3*	*PA4*	*PA5*	*PA6*	
Y1	0	0	12	6	0	0	18
Y2	14	19	11	3	0	6	53
Y3	126	118	109	35	14	28	430
Y4	91	58	44	11	6	15	225
Y5	0	24	3	8	0	12	47
Y6	19	60	19	8	0	6	112
Total	250	279	198	71	20	67	885
$\chi^2_{(5)}$	49.46***	49.03***	32.44***	24.31***	7.42	25.94***	

Note: As can be seen from Table 8-6, except in PA5 land rental contracts vary within and across Peasant Associations at 1% level of significance. Plot level observations are reported for each contract type across PAs.

levels). So if young age is a sign of limited reputation, and trust has to be earned over time, this result may be in support of this hypothesis. We also had information from landlords and tenants on whether they paid attention to reputation and/or trust when they selected their contract partners. However, when we tried to use this information in the analysis we found no robust evidence in support of the hypothesis. Some of the variables included in the analysis require more cautious interpretation. These include the age of head of household in the landlord model, access to alternative partners (accaltpart), and a variable capturing the landlords' perceived freedom to rent out land (iftensec1) on cash basis without informing the local administration. Only the first of these three was significant and consistent (at 1 and 5% levels). Old age of landlords was associated with tenants covering more of the input costs. This could be in line with our resource-pooling hypothesis as old age may be a sign of lower capacity to cover cash input costs. Therefore the cost may be covered by the (younger) tenants who also may be relatives. It could, however, also be in line with the bargaining power hypothesis if old age is associated with more bargaining power.

Conclusions

In this chapter we have examined diverse forms of input cost-sharing land rental contracts in the southern highlands of Ethiopia and alternative hypotheses that could possibly explain this contract diversity. First, we found that tenants were richer than landlords in terms of income per capita, the endowment of male labor force, and land per capita, while landlords had significantly larger family size and female work force and with an older head of household. We tested the alternative hypotheses regarding the variations in land rental contracts after controlling for household level unobserved heterogeneity. This was done by using household random intercept models when analyzing farm plot panel data with multiple plots per household covering two rainy seasons.

The analysis gave little support to the bargaining power hypothesis on the landlord side. On the other hand, more wealthy tenants were found to be associated with contracts in which they covered more of the cash inputs, in line with the resource-pooling hypothesis. This may also imply that more resource-poor tenants are rationed out of the market like that found by Holden and Ghebru (2006) and Chapter 4 in northern Ethiopia. Landlord households with higher dependency ratio and with older household head also had tenants to cover more of the cash input costs, also in line with the resource-pooling hypothesis.

Furthermore, we found evidence on land quality adjustment in contracts as tenants covered significantly more of the cash inputs if they rented land of better land quality in terms of deeper soils. We also found some evidence of tenants covering more inputs when contracts were for more than one year.

There was also significant spatial variation in the composition of contracts indicating possibly the fragmented nature and poor geographical integration of land rental markets.

We may therefore conclude that the diversity of land rental contracts largely reflects rational adjustments in the environment of rural economies characterized by market imperfections in land, labor, credit and insurance markets, variation in household resource endowments, and land quality. This implies that in general a variety of land rental contracts observed in the southern highlands of Ethiopia are induced to facilitate efficient use and allocation of land and nonland resources. Furthermore, they are found to be conducive to equity. Thus, land tenure policies that favor tenure security and longer-term contracts should be encouraged, as they are likely to reduce the transaction costs and enhance the efficiency of the land rental market. Allowing more long-term land rental contracts may facilitate a more dynamic off-farm sector as an alternative source of employment and income.

CHAPTER 9

Gender and Land Productivity on Rented Land in Ethiopia

STEIN T. HOLDEN AND MINTEWAB BEZABIH

There have been many studies on women's land rights in Africa (see, e.g., Lastarria-Cornhiel 1997; Whitehead and Tsikata 2003). However, few studies have analyzed the market and productivity implications of the typically weaker position of women that is found in most African settings when it comes to their land rights and land-related decision-making positions. Chapter 6 in this book investigated whether gender-related inheritance and residence systems affected land rental market participation. In this chapter we go deeper into assessing whether such partly gender-associated market participation also has land productivity implications in our study in the Amhara region in Ethiopia, and we are not aware of any earlier studies on this. As was found in Chapters 3 and 4, female-headed households in Ethiopia are frequently renting out their land.

When looking at the agricultural production both within and across households some interesting gender-specific differences have been identified. Udry et al. (1995) found that land productivity on plots controlled by women was lower than that on plots controlled by men in Ghana. Their study showed that men were more able to mobilize labor and other factors of production for their plots than women were within the family. Another study by Udry (1996) found 30% lower productivity on female plots than that of male plots within households in Burkina Faso. Goldstein and Udry (2005)attributed the productivity differential between men's and women's plots in Ghana to women's higher level of tenure insecurity, making them less able to invest in the land in form of fallowing because they could then risk losing the land (Goldstein and Udry 2005).

In Ethiopia, Holden et al. (2001) found systematic lower productivity on owner-operated land of female-headed households than that of male-headed households. Bezabih and Holden (2006) found lower land productivity on land of female landlords than that of male landlords. We build on and expand from the analysis of Bezabih and Holden. They used a double moral hazard model with tenure insecurity and transaction costs to explain the productivity differential. Female landlord households were assumed to be more tenure insecure and were therefore considered less able and less likely to use threat of eviction and contract renewal as an instrument to enhance productivity on rented-out land. Alternatively, female landlords faced higher transaction costs in the land rental markets and therefore had larger difficulties in evicting inefficient tenants and in searching for and finding more efficient tenants.

In this study we test a number of alternative models and hypotheses that possibly may explain the productivity differentials between female- and male-headed landlord households using the same data as in Bezabih and Holden (2006). First, we test a simple transaction cost model for participation in the land rental market and its possible consequences for land productivity differentials. Second, we test a model with endogenous contract choice to see whether Marshallian inefficiency due to more limited monitoring capacity can explain the productivity differential. Third, we test whether the productivity differential may be explained by screening and/or adverse selection of tenants possibly due to female-headed households' lower ability to screen and select good tenants and evict bad tenants. This could also be due to the tenure insecurity of female landlords, making them less able to use threats of eviction (see Chapter 10; Kassie and Holden 2006; Bezabih and Holden 2006). Finally, we test whether kinship contracts, involving blood-related tenants or in-laws of male and female landlords, are less efficient, like that found by Kassie and Holden (2006). This may be another reason for female landlords being unable to use threat of eviction as a device to enhance efficiency. Such a finding may give reason to doubt that partner choice in the rental market is voluntary. Coercion may play an important part and possibly explain the productivity differential.

The Setting and Basic Household Characteristics

This study uses information from the 230 landlord households in the sample of approximately 2,000 households in two districts, East Gojjam and South Wollo, in the Amhara National Regional State in the northern and central highlands of Ethiopia. East Gojam is a fertile plateau receiving good average rainfall, whereas South Wollo is characterized by degraded farm plots receiving lower and more erratic rainfall. An overall sample of 130 male and 100 female landlords is included as a result.

As has been noted in the conceptual framework of the book, households may or may not engage in the land rental market, by virtue of which they are categorized as "autarkic," "landlords," or "tenants." For those who engage in the land rental market as landlords, they might do so partially or fully (i.e., by renting out all or part of the plots that belong to them). If they rent out all their land, we call them "pure landlords."

Basic characteristics of landlord, nonparticipant, and tenant households are presented in Table 9-1. The table shows that about 40% of the landlord households are female-headed, whereas only about 3% of the tenant households are female-headed. Landlord households also have significantly less male and female labor, oxen, and other livestock, but there is no significant difference in area owned between landlord and tenant households. This demonstrates the "reverse-tenancy" nature of the land rental market in Ethiopia, which also has been demonstrated in other studies (see Chapter 3; Holden and Ghebru 2006; Chapter 10; Kassie and Holden 2006), and in other African countries, like Madagascar and Eritrea (Bellemare 2006; Tikabo et al. 2008).

Thus, for our purpose, our analysis focuses on male and female landlords who are among the poorest of the poor in the area of study (northern-central highlands of Ethiopia). While there are some households who rent out their land fully, a considerable proportion of them are owner-cum landlords who cultivate a part of their land themselves. This allows us also to compare land productivity of rented out land and owner-operated land of the same landlords. A set of land quality variables is included to control for land quality. Table 9-2 provides data on the endowments of male and female landlord households. We see that male landlords have significantly more male labor (*madultha*) and livestock (*livestha*), larger farm size (*farmarea*), and more education (*heduc1*), whereas female landlords have significantly more female labor (*fadultha*).

TABLE 9-1 Comparison of the characteristics of landlord, non-participant and tenant households

		By Land Market Participation Status		
Variable Household Characteristics	*Total*	*Landlords*	*Non-participants*	*Tenants*
Female head dummy (%)	15.37	39.91***	11.82	3.29***
Head literate (%)	41.31	29.13***	41.24	51.14***
Age of head in years	48.38	51.61***	48.71	45.11***
Owned area in ha	1.08	1.04	1.11	1.06
Number of adult males	1.49	1.04***	1.57	1.70***
Number of adult females	1.46	1.33***	1.51	1.46*
No. of dependents (< 15 or > 60)	2.72	2.26***	2.74	3.05***
Net land rented in	0.01	−0.68***	0.00	0.60***
Owned cultivated area in ha	0.94	0.36***	1.10	1.05
Total cultivated area in ha	1.09	0.36***	1.11	1.65***
Owns bulls or oxen (%)	67.3	23.7***	72.4	91.2***
No. of bulls/oxen (for owners)	1.19	0.39***	1.21	1.78***
Value of other livestock in ('000) Birr	1.20	0.57***	1.20	1.69***

Signiificant differences from non-participants at 10% (*), 5% (**), and 1% (***).

TABLE 9-2 Comparison of the characteristics of male and female landlord households

Sex of Head	*Madultha*	*Fadultha*	*Hage1*	*Heduc1*	*Oxha*	*Livestha*	*Farmarea*
Male	2.001	2.039	56.231	1.777	1.023	3.805	1.321
	.135	.176	1.628	.082	.157	.633	.072
	129	129	130	130	129	129	130
Female	.912	2.937	53.563	1.240	.822	2.545	.840
	.129	.183	1.737	.066	.190	.433	.057
	96	96	96	96	96	96	96
Total	1.536	2.422	55.097	1.549	.937	3.268	1.117
	.101	.131	1.193	.0575	.121	.409	.050
	225	225	226	226	225	225	226

Note: The table presents the mean, the standard error of the mean and the number of households for each household group.

Theoretical Models

We refer to the transaction cost models for land rental markets in Chapter 2 of this book. Can the simple transaction cost model explain gender productivity differentials? This model does not include sharecropping, which is the dominating contract form in our study area. We therefore also introduce a model with sharecropping and transaction costs, which opens for testing of a richer set of hypotheses to explain gender productivity differentials on rented land.

Simple Transaction Costs Model

How can the simple transaction cost model presented in Chapter 2 explain gendered productivity differentials? If female-headed households possess on average less nonland resources relative to land resources than male-headed households, they are more likely to rent out land than are male-headed households. They are therefore more likely to choose one of the first two land strategies specified in Table 2-1 and Figure 2-3. The transaction costs in the land rental market cause land productivity to be higher on rented in land (managed by tenants) than on the owner-operated land of landlords. With a high proportion of the female-headed households in the category of landlord households (land strategy 2) and a low proportion of female-headed households in the tenant category of households (land strategy 4), land productivity of female-headed households should on average be lower than that of male-headed households. In this case the productivity on the land rented out by female-headed households should not be different from that on owner-operated land of tenant households. And land productivity of land rented in from female landlords should not be different from productivity on land rented in from male landlords. For pure owner-operators (land strategy 3) average land productivity should fall between the average land productivity on owner-operated land of landlords and that of tenant households (rented in as well as owner-operated land).

Furthermore, if female-headed households face higher transaction costs in the land rental market, average land productivity on owner-operated land of female landlords should be lower than that of male landlords. This should also lead to female landlord households operating with lower nonland versus land factor ratios on their owner-operated land.

Model with Sharecropping, Transaction Costs, and Tenant Characteristics

A landlord household has land endowment $\overline{L}$ and rents out L^s of this land with $L^s \leq \overline{L}$. The characteristics (γ^l) of the landlord household include nonland resources (N^l) and sex of household head (g). The household expects to produce output q at price p using land and nonland resources in production. It is also assumed that sex of household head affects production in form of farm skill (nontradable nonland resource) in a society in which males traditionally are responsible for farming activities. Production is increasing in land and nonland resources use at a decreasing rate. Land and nonland resources are assumed to be complementary. Land is rented out in a sharecropping arrangement in which the landlord gets α share of the expected output. This share is, for simplicity, assumed to be fixed (exogenous) in the short run. Expected output on rented-out land is a function of land rented out and the expected use of nonland resources by the tenant (N^{t0}). The amount of nonland resources used by the tenant is expected to depend on his or her characteristics (γ^t) and the sex of the head of the landlord household ($g = 1$ if female). We expect that $\partial N^{t0}/\partial g < 0$ if female landlords have lower monitoring and enforcement capacity and there is no difference in endowments of tenants of male and female landlords' tenants. We hypothesize that there is a higher risk of losing the land if it is rented out to the same tenant for more than one cropping season. This tenure insecurity (ϕ) is an increasing function of the share of the land that is rented out $(L^s/\overline{L})^1$, and increasing with the number of years the land has been rented out to the same tenant (C_{-t}), and is higher for female landlords than for male landlords. In case the landlord decides to evict the tenant that was contracted last year, she or he will face an eviction cost and a search problem related to finding a new tenant and an uncertainty about the characteristics of the new tenant and his or her productivity because tenants' characteristics are only partially observable. We assume that search efficiency ($S(g)$), $\partial S/\partial g < 0$, $\partial \gamma^{t1}/\partial S > 0$, is lower for female than for male landlords and that this affects the quality of tenants they are able to find. If female landlords face a higher eviction cost, $c^E = c^E(g)$, $\partial c^E/\partial g > 0$, they may be stuck with inefficient tenants. This may be in-laws or blood-related kin who take control over their land. The landlord's choice problem is stated in equation (9.1) below, assuming uniform land:

$$V^{l}\left(y^{l};\gamma^{l}\right)=\text{org. max}\left\{\begin{bmatrix}V^{l}\left(pq\left(\overline{L}-L^{s},N^{l}(g)\right)+\alpha pq\left(L^{s},N^{t0}\left(\gamma^{t0}(g),g\right)\right)-\phi\left(\frac{L^{s}}{\overline{L}},C_{-t},g\right)L^{s};\gamma^{l}\right),\\ V^{l}\left(pq\left(\overline{L}-L^{s},N^{l}(g)\right)+\alpha pq\left(L^{s},N^{t1}\left(\gamma^{t1}\left(S(g)\right),g\right)\right);\gamma^{l}\right)-c^{E}(g)-c^{S}(g)\end{bmatrix}\right\} \tag{9.1}$$

The landlord will not renew the contract if the first term (above) inside the big parenthesis is larger than the second term (below) inside the big parenthesis.

We are particularly interested in studying the implications of the model for the difference between male and female landlords in terms of their expected land productivity on own and rented-out land.

From the model we derive the following hypotheses, given that the productivity differential between male and female landlords persists after controlling for differences in plot quality, sample selection in relation to plots that are rented out, and other landlord characteristics:

1. Female landlords have less monitoring and enforcement capacity. They are therefore likely to experience more Marshallian inefficiency than male landlords. Contract choice therefore explains the productivity differential.
2. Screening and search cost: Female landlords are less mobile due to more commitments at home and due to cultural restrictions. They are therefore less able to search and find good tenants. The productivity differential can therefore be explained by the difference in tenant characteristics of female and male landlords.
3. Eviction cost: Female landlords are less able to evict inefficient tenants, particularly if these are blood-related kin or in-law tenants. The productivity differential is explained by the fact that female landlords have more blood-related kin and in-law tenants that are less efficient.
4. Duration of partnership: Longer duration is a sign of tenure security and is associated with higher productivity for male landlords, whereas it may be associated with less efficient kin contracts for female landlords.
5. Pure landlords, renting out all their land, are less able to monitor and enforce contracts. Pure landlords are more likely to be female landlords and this explains the productivity differential.
6. Tenure secure landlords may use threat of eviction to enhance the efficiency of tenants. Contract renewal conditional on performance is more used by male landlords that are more tenure secure than female landlords and this explains the productivity differential.

Hypotheses 3 and 4 can be contradictory. High eviction costs may be a sign of tenure insecurity and can lead to longer duration contracts.

Methods

Empirical testing of gendered productivity differentials may reveal which of these propositions are true. We do this in several steps. The structure of the data is such that we have multiple plots per household. We cannot be sure that plot quality is the same on land for male and female landlords or that land quality and productivity are the same on owner-operated and rented-out land. We use matching models and selection models to control for observable and unobservable plot characteristics. We compared nonparametrically the land productivity on owner-operated and rented-out land of male and female landlords. We used propensity score plot matching to do this comparison. The propensity score was constructed on the basis of observable plot characteristics. We tested that the common support requirement was satisfied with our data and used kernel matching and nearest neighbour matching methods (Becker and Ichino 2002; Dehejia and Wahba 2002).

To control for selection bias related to unobservable plot characteristics we have applied selection models in which the selection equation related to whether plots were rented out or not (trade status T_{hp} for household h and plot p).

$$T_{hp} = \begin{Bmatrix} 1 \text{ if rented out} \\ 0 \text{ if owner-operated} \end{Bmatrix} = T_{hp}\left(S_{hp}, Z_h\right) \tag{9.2}$$

For identification we used observable plot characteristics (S_{hp}), including soil type, distance to plot for landlords, plot size, and landlord household resource endowments (Z_h) in the form of male and female labor per unit owned land, oxen, and other livestock endowments per unit of own land. Many of these variables were highly significant. We considered also using household-fixed effects for the selection equation but the fact that a considerable share of the landlord households rented out all their land kept us from doing it. Following Deaton (1997) we used a polynomial of the predicted probability of a plot being traded in the selection equation to control for selection bias in the second stage productivity equations on rented-out land. We were unable to use household fixed effects also in the second stage, unlike Shaban (1987), since we are interested in comparing land productivity of male and female landlord households, and this gender variable is a household level variable. We used bootstrapping to get corrected standard errors in the second stage generalized least squares (GLS) models.

Our analytical strategy for the parametric models was to introduce new variables for the testing of alternative hypotheses in a stepwise approach and to carefully examine the stability of the earlier results when new variables (that were potentially endogenous) were introduced. The sequence of addition of new variables in the second stage GLS selection models of land productivity on rented-out plots of male and female landlords was as follows:

1. Plot characteristics (used in the propensity score matching), to control for plot quality variation, and gender dummy for sex of head of household, to see whether differences may be explained by land quality or whether the gender variable has a separate effect.
2. Contract choice in terms of pure sharecropping as this type of contract is likely to have the largest disincentive effect (Marshallian inefficiency) as compared to fixed rent and cost-sharing contracts. Contract choice in terms of pure sharecropping is endogenous and is predicted. An interaction variable for pure sharecropping and gender was also included to assess whether this problem is more severe for female landlords.
3. Nonland/land factor ratios and other characteristics of tenants, to see whether tenants' characteristics alone can explain the productivity differential (meaning that male landlords are able to get better tenants).[2]
4. Duration of partnership with the same tenant, which may not be the same as contract length because it may be a result of multiple renewals of short-term contracts. (Test whether significance of gender dummy disappears when this variable is added, which would indicate that land productivity on land rented out by male landlords is higher because they use more long-term contracts [perhaps because they are more tenure secure].)
5. Interaction of landlord gender dummy with duration of partnership. This allows us to assess whether contract renewal and duration of partnership has a different effect on land productivity for female than for male landlords.
6. Dummy for pure landlords (these are likely to be the weakest of the landlords, and are more likely to be female). They are more likely to have problems with monitoring and enforcing land use efficiency on their rented-out land. This variable was used to replace the share of land rented out variable in the theoretical model. Different specifications were tried to test the hypothesis, including the predicted probability of being a pure landlord as this variable is endogenous.

We included the alternative variables in a stepwise fashion to test for their relevance and partly assess the potential endogeneity problems.

Results and Discussion

We compared the land productivity of male and female landlords on all plots and on owner-operated and rented-out plots separately. We may then examine this in the light of hypotheses of the theoretical models and this may help us to reject some of the hypotheses. The results of this comparison are shown in Table 9-3. The table reveals that average land productivity was significantly lower (1% level) for female landlords than for male landlords when we compare all their plots. When we compare land productivity separately for owner-

TABLE 9-3 Propensity score matching results of gender of household head on land productivity on owner-operated and rented-out plots of landlord households[a]

	All Plots		*Owner-Operated Plots*		*Rented-Out Plots*	
Variable	*Kernel Matching*	*Nearest Neighbor*	*Kernel Matching*	*Nearest Neighbor*	*Kernel Matching*	*Nearest Neighbor*
Land productivity						
Female landlords	1366.02	1366.02	1430.85	1430.85	1345.05	1345.05
Male landlords	1806.08	1952.30	1799.02	2266.90	1815.49	1753.60
Difference	–440.06	–586.28	–368.16	–836.04	–470.44	–408.55
Bootstrapped st. error	126.21	177.76	255.65	414.59	156.26	210.83
t-statistic	–3.487***	–3.298***	–1.440	–2.017**	–3.011***	–1.938**
Number of observations						
Female landlords	439	439	124	124	315	315
Male landlords	820	477	329	153	484	255

[a]The comparison is based on propensity score kernel and nearest neighbor matching (Dehejia and Wahba 2002, Becker and Ichino 2003).

Significance levels (after correction for bias): *:10% level, **:5% level, ***:1% level.

operated plots, we see that this result is less robust, as a significant difference is retained only when we used nearest neighbor matching. However, we find a significant difference with both matching methods on rented-out plots.

So where does this leave us with respect to the explanatory power of the theoretical models? This latter result that land productivity is lower on land rented out by female landlords than on land rented out by male landlords clearly cannot be explained by our first theoretical transaction-cost model. This model predicted that land productivity should be higher on rented-out plots than on owner-operated plots if we have transaction costs in the land rental market. We do not find this to be true in our empirical results, however. The other proposition of the first model was that land productivity should be lower on owner-operated plots of female landlords if they face higher transaction costs in the land rental market than male landlords. We find evidence on this but the results are not very robust, as the productivity differential was significant only with the nearest neighbor matching method. This may also be due to the relatively small number of owner-operated plots that were owned by female landlords.

We now focus our analysis on explaining the productivity differential on rented-out land. Does this differential mean that female landlords have more problems with Marshallian inefficiency on the rented-out land while this is a less significant problem for male landlords? Table 9-3 does not reveal a larger gap in land productivity between owner-operated plots and rented-out plots of female landlords than of male landlords that would have been consistent with higher Marshallian inefficiency. The gap is too small in both cases to be significant but the variation in characteristics across households and plots may mask important differences.

In order to test more of our hypotheses from the second model we ran a number of parametric models on the rented-out land. We refer to Table 9-4 for an explanation of variable names in the regression tables that follow. We corrected for plot selection bias with a plot trade selection model that is presented in Table 9-5 together with the results from the three first parametric productivity models. We see from the trade selection equation in Table 9-5 that female landlords were more likely to rent out.

The first model (Landlord model 1) to explain productivity differentials on rented-out plots is a village fixed effects (GLS) model with sample selection. This model does not take contract choice into account. While controlling for plot, village, and other household characteristics, land productivity was still significantly lower on their rented-out land. One may wonder why they rent out more land when they seem less able to enforce efficiency on their rented-out land and if they are more tenure insecure and face high risk losing the land they rent out. We therefore wanted to test whether female landlords used pure sharecropping contracts more than male landlords and whether that could explain the productivity differential. Contract choice is endogenous and we predicted the probability that landlords chose pure sharecropping contracts over fixed-rent, cost-sharing, and mixed contracts. The predicted variable (*puresharp*) was included in Landlord Model 2. We see from Table 9-5 that the variable was significant at the 5% level and with a negative sign, while the gender variable (*hsex*) still remained significant. It appears, then, that there is more to the productivity differential than Marshallian inefficiency since there is a significant lower land productivity on land rented out by female landlords after we have controlled and tested for Marshallian inefficiency. In order to test for a gender-specific difference in Marshallian inefficiency we created an interaction variable between the predicted pure sharecropping contract variable and the gender variable (*hsex* = 1 for female landlords, zero otherwise). This new variable was added in the Landlord Model 3. We see from Table 9-5 that this variable also was insignificant, although it had a negative sign. Thus, it appears that female landlords are not significantly less able to enforce efficiency in relation to their pure sharecropping contracts. We see that the gender variable (*hsex*) became insignificant in this last model. The insignificance of these variables could be due to multicollinearity because their standard errors increased.

To test our hypotheses on screening and adverse selection of tenants as an explanation for the productivity differential we included the following tenant

TABLE 9-4 Variable definitions and descriptions

Variable Name	*Description*
LANDLORD	CHARACTERISTICS
Heduc1	Head's formal education (1 = read and write; 2 = read only; 3 = none)
Hage1	Age of household head
Hsex	Gender of the household head
Madultha	The number of male working-age family member of the landlord per ha
Fadultha	The number of female working-age family member of the landlord per ha
Livestockha	The number of livestock per ha
Oxha	The number of oxen per ha
Extreme riskav.	Extreme risk aversion (elicited from experiment)
Severe riskav.	Severe risk aversion
Intermediate riskav.	Intermediate risk aversion
Security	Whether the landlord expects increase, no change or decrease in the land size in the coming five years (1 = decrease; 2 = no change; 3 = increase)
Conflict	Whether the landlord has experienced any conflict regarding the land (1 = conflict, 0 = no conflict)
Changeland	Whether the landlord has experienced change in the landownership in the last five years (1 = change, 0 = no change)
TENANT	*CHARACTERISTICS*
Tage	Tenant's age
Toxcd	The number of oxen owned by the tenant
Btenant	Dummy for blood-related tenant
Stenant	Dummy for in-law related tenant
Clength1	Duration of partnership, years
OTHER VARIABLES	
Tradep, tradep2	Predicted probability that a plot is rented out, square of the same variable
Puresharp	Predicted variable for plot being a pure sharecropping plot with only output sharing but no input/cost-sharing

TABLE 9-5 Generalized least squares selection models for land productivity differentials on rented-out plots: With household random effects and bootstrapped standard errors

	Landlord Model 1 b/(se)	*Landlord Model 2 b/(se)*	*Landlord Model 3 b/(se)*	*Trade Selection Equation b/(se)*
Selection variables				
tradep	–2.912 (2.19)	–3.427 (2.11)	–3.611* (2.04)	
tradep2	2.745 (1.68)	3.326** (1.66)	3.457** (1.59)	
Landlord household characteristics				
Madultha				0.106*** (0.04)
Fadultha				0.157** (0.07)
Fadultha2				–0.011 (0.01)
Totalplot				0.027**** (0.01)
Extreme riskav.				0.079 (0.36)
Severe riskav.				0.289 (0.18)
Intermed. riskav.				–0.001 (0.36)
Security				–0.152*** (0.05)
Conflict				0.242** (0.11)
Changeland				0.111 (0.13)
Hage1				–0.002 (0.00)
Heduc1				0.077 (0.06)
Plotdist				0.009**** (0.00)
Oxha				–0.084** (0.04)
Livestockha				0.012 (0.01)
Hsex	–0.282** (0.14)	–0.286** (0.14)	–0.001 (0.28)	0.526**** (0.12)

TABLE 9-5 *(Cont.)*

	Landlord Model 1 b/(se)	*Landlord Model 2 b/(se)*	*Landlord Model 3 b/(se)*	*Trade Selection Equation b/(se)*
Predicted contract choice				
Puresharp		−2.093** (0.84)	−1.718* (0.91)	
Hsex*puresharp			−0.724 (0.60)	
Constant	8.869**** (0.80)	9.463**** (0.80)	9.377**** (0.76)	0.071 (0.52)
Prob > chi^2	0.000	0.000	0.000	0.000
Number of obs.	605	591	591	1124

Significance levels: * 10%, ** 5%, *** 1%, **** 0.1%. Village dummies and plot land characteristics dropped in presentation.

TABLE 9-6 Tenant characteristics of male and female landlord households

Variable	*Female Landlords Mean Value*	*Male Landlords Mean Value*	*Bonferroni Test of Significant Difference* F	*Prob > F*
Tenant age (Tage)[a]	2.368	2.232	4.73	0.0090
Oxen owned by tenant (Toxcd)	1.940	2.075	2.99	0.0041
Blod-related tenant (Btenant)	.466	.365	12.81	0.0004
In-law related tenant (Stenant)	.184	.102	17.58	0.0000
Duration of partnership (Clength1)	4.825	2.637	9.17	0.0000

[a]The age of tenants was grouped in three age categories: young, medium, and old.

characteristics related to each of the contracts: age of tenant, oxen ownership of tenant, whether the tenant was blood-related, and whether the tenant was in-law related to the landlord. The mean values of these variables and a test for significance of differences between male and female landlord households are presented in Table 9-6. The table reveals that female landlords had tenants that on average were significantly older and had fewer oxen than male landlords. This may be in line with our screening and adverse selection hypothesis, at least for the oxen variable, although it is less obvious that the age of tenants has a negative effect on land productivity. We also see from the table that female landlords were more likely to have blood-related and in-law related tenants than male landlords. However, we cannot be sure whether this implies a negative productivity effect due to screening/adverse selection of tenants until we have tested the effects on land productivity. We therefore tested this in Table 9-7 by assessing the association between tenant characteristics and land productivity.

The empirical evidence on how kinship contracts affect land contract efficiency are mixed (Chapter 10; Kassie and Holden 2006; Sadoulet et al. 1997). Sadoulet et al. (1997) found that kinship contracts were associated with higher efficiency, whereas Kassie and Holden (2006) and Chapter 10 found kinship contracts to be associated with lower efficiency. Since the study of Kassie and Holden was carried out in Ethiopia (like this study), it is more relevant than that of Sadoulet et al. (a study in the Phillipines) for our study area. They proposed that threat of eviction may be used by landlords as an instrument to enhance land productivity on sharecropped-out land while this instrument may not be as efficient in kin contracts as in non-kin contracts.

We have included duration of partnership (*clength1*) in our analysis to test whether duration of partnership with the tenant can be associated with land productivity. We refer to Bezabih and Holden (2006) for more details on the issue of contract length and contract renewal. We see from Table 9-6 that average contract duration was longer for female landlords than for male landlords even though land productivity was lower for female landlords. This may imply that female landlords were less able to use threat of eviction as a device to induce higher productivity and evict less efficient tenants.

We now go ahead and test for adverse selection effects on the productivity differential using parametric selection models. The results are presented in Table 9-7. We see from Table 9-7 in the Landlord-Tenant Model 1 that the gender variable (*hsex*) remained significant after we included the tenant age (*tage*) and tenant oxen ownership (*toxcd*) variables while these variables were insignificant. The tenant variables are clearly endogenous so we have to interpret them with care. We could only compare models with and without these variables as we had no instruments to predict them. When we added the variables for blood-related tenants (*btenant*) and in-law related tenants (*stenant*) in Landlord-Tenant Model 2, these variables were also insignificant while the gender dummy variable remained significant. The size of the parameter for the gender dummy variable was reduced somewhat, indicating that there may be a grain of truth in the adverse selection hypothesis, given the significant difference in these tenant characteristics between male and female landlords.

To test whether kin contracts could be associated with lower productivity for female than for male landlords, we included interaction variables between the blood-related tenant and in-law tenant variables and the gender dummy variable. We present the results in Landlord-Tenant Model 3. We see that the interaction variable for blood-related tenants was insignificant but it was significant at the 5% level and negative for in-law tenants. The parameter for the

TABLE 9-7 Generalized least squares selection models for rented-out land including tenant characteristics: with household random effects and bootstrapped standard errors

	Landlord-Tenant Model 1 b/se	*Landlord-Tenant Model 2 b/se*	*Landlord-Tenant Model 3 b/se*	*Landlord-Tenant Model 4 b/se*
Selection variables				
Tradep	–3.571*	–3.632	–3.648*	–3.956
	(2.05)	(2.24)	(2.15)	(2.57)
Tradep2	3.435**	3.475**	3.463**	3.645*
	(1.56)	(1.75)	(1.67)	(1.99)
Predicted contract choice				
Puresharp	–2.101***	–2.253***	–2.253***	–2.434***
	(0.81)	(0.87)	(0.85)	(0.86)
Tenant characteristics				
Tage	0.009	0.015	–0.012	–0.107
	(0.11)	(0.11)	(0.11)	(0.13)
Toxcd	0.018	0.024	0.043	0.057
	(0.08)	(0.07)	(0.08)	(0.09)
Btenant		0.080		
		(0.12)		
Stenant		–0.297		
		(0.25)		
Gender and interaction variables				
Hsex	–0.266*	–0.236*	–0.087	–0.004
	(0.14)	(0.14)	(0.15)	(0.24)
Hsex*btenant			–0.092	–0.041
			(0.19)	(0.24)
Hsex*stenant			–0.740**	–0.926**
			(0.37)	(0.42)
Clength1				0.006
				(0.03)
Hsex*clength				–0.027
				(0.03)
Constant	9.469****	9.537****	9.486****	9.860****
	(0.84)	(0.84)	(0.85)	(0.96)
Prob > chi2	0.000	0.000	0.000	0.000
Number of obs.	577	577	577	485

Significance levels: * 10%, ** 5%, *** 1%, **** 0.1%. Village dummies and plot land characteristics dropped in presentation.

gender dummy variable became insignificant and the absolute value was much reduced. This indicates that in-law contracts are associated with the lower land productivity of female landlords' rented out plots. In all the models the predicted contract choice (= pure sharecropping) remained significant at the 1% level, showing the disincentive effect of sharecropping.

In order to assess the importance of duration of partnership (*clength1*) the interaction variable for contract length and gender was included. We see from the Landlord-Tenant Model 4 that this variable also was insignificant and the inclusion did not affect any of the other variables of interest. It appears, therefore, that in-law tenants are less efficient than other tenants and contracts with in-law tenants appear to be of longer duration even though they are less efficient (see Table 9-6). This is in line with anecdotal evidence that in-laws take control over land of female landlords. The female landlords may in such cases face high eviction costs and may therefore fail to evict these tenants and cannot search for more efficient tenants. This appears to be the main form of tenure insecurity that female landlords face.

Finally, we wanted to assess whether pure landlords that are renting out all their land were facing higher problems with inefficiency on their rented-out land, possibly because they may be more resource-poor or absentee and therefore less able to monitor and enforce contracts, or because they may be more tenure insecure and therefore less able to use efficiency-enhancing devices. To test for this we included the dummy variable for pure landlords, the predicted probability of being a pure landlord and an interaction variable for the predicted pure landlord variable and female dummy. We included these variables one at a time and jointly because of possible problems with multicollinearity. None of the specifications made these variables significant. The signs were even positive in case of the interaction variable. Multicollinearity may therefore not be blamed for these variables not being significant and with a negative sign. We therefore reject the hypothesis that lower productivity on rented out land from female landlords is due to their higher probability of being pure landlords.

We did some further tests to see whether the difference could be associated with crop choice (results not included here). We tested whether the results were the same if we did the analysis for plots with cereals only. We found that this was the case, although levels of significance were reduced due to the smaller sample size (about 63% of the cropped plots were cropped with cereals). Alternatively we also ran the regressions with a cereal dummy. Also this did not change the key results while the cereal dummy was significant and positive in most models (with about 20% higher output value than other plots). We therefore found that the main findings were robust to these alternative specifications.

Our theoretical model captured the issue of eviction of tenants with the eviction cost, which is like a transaction cost that may lock these female landlords into a continued contract relationship. This may also be seen as involuntary contract participation and is related to the weak property rights and weak farmer status of female-headed households. We think that our second model capturing transaction costs related to eviction and screening and search for tenants provides an adequate theoretical framework for the problem.

Conclusion

We have tested alternative models and theories to explain the productivity differential on rented-out plots of female versus male landlord households in the Ethiopian highlands. A simple transaction cost model could not explain the differential. The hypothesis of endogenous contract choice and higher Marshallian inefficiency in the case of female landlords also could not explain the whole productivity differential, although the share contract variable was highly significant. We found a significantly higher level of inefficiency linked to contracts of female landlords with in-law tenants. This may be due to the high eviction costs of tenure-insecure female landlords who typically live in the village of their in-laws and who therefore are less able to freely screen and select better tenants.

An important policy implication of our analysis is that strengthening women's land rights may not only be good for equity but also for efficiency of land use. The new land proclamations in Ethiopia push in this direction. The recently introduced land certification may also strengthen the land rights of female-headed households and this may strengthen their bargaining power in the land rental market. A recent study in Tigray region (Dokken 2007) did not find a similar productivity differential when analyzing similar data collected eight years after land certification had taken place. However, more research is needed to investigate whether this also will be the case in our study areas. Joint certification of husbands and wives, like implemented in the Oromiya and SNNP regions of Ethiopia recently, could further strengthen women's land rights and bargaining power (Holden and Tefera 2008). However, there is a long tradition of male dominance in household and farm decision making so it may take time and additional effort in terms of information dissemination and sensitization to reduce the gender biases that are deeply rooted in the culture. Local administrations, including local courts as well as NGOs, have to play important roles in this process. Donor support may be important to provide sufficient resources for this.

NOTES

1. In the recent Ethiopian land proclamations it is stated that one may rent out up to half of the land, making it illegal to rent out more than that. Previously all land

renting was prohibited and inability to cultivate one's land was an indication of too much land and this could increase the probability of losing the land in the next land redistribution.

2. We also did separate tests of whether there were significant differences in the tenant characteristics of male and female landlords (Table 9-6). Tenant characteristics are obviously endogenous so the results have to be interpreted with care.

CHAPTER 10

Kinship, Tenure Insecurity, Input Use, and Land Productivity

The Case of Sharecropping in Ethiopia

Menale Kassie and Stein T. Holden

There have been many empirical studies of efficiency of sharecropping, especially in Asia (Otsuka and Hayami 1988). Most empirical studies compare the efficiency of sharecroppers with owner-operators and fixed-rent tenants, and have failed to control well for tenant and plot characteristics. Most studies have also failed to explain why specific sharecroppers are efficient or not. Exceptions include Sadoulet et al. (1997) who found sharecroppers with a kinship relationship with the landlord not to be affected by the Marshallian disincentive effect, whereas other sharecroppers were. We follow up on this and assess the importance of kinship for sharecropping efficiency in our study in the Ethiopian highlands.

In countries where policy distortions have undermined the functioning of land rental markets, it is important to understand whether land rental markets reallocate land to more efficient producers and therefore contribute to more efficient land use, or whether sharecropping undermines this effect and causes tenants to undersupply inputs on their rented land. In Ethiopia, there have been restrictions on land renting, and the "land-to-the-tiller" philosophy was strong in the 1970s and contributed to the radical land reform in 1975, with prohibition of land renting and hiring of labor. Land redistributions were practiced to sustain an egalitarian land distribution among households in the period up to 1991 when the Derg regime was overthrown. Although the new regime was more market friendly, all land remained state owned, land sales continued to be prohibited, and only short-term renting was allowed. This policy regime may also have caused Marshallian inefficiency to occur in Ethiopia but there have not been many studies that have investigated this with a rigorous methodology. In this chapter, we assess the output levels and fertilizer input levels of kin and non-kin sharecropped tenants' plots, using the sharecropping tenants' owner-operated land as counterfactual. Unlike most other studies of sharecropping efficiency (e.g., Chapter 9 by Holden and Bezabih) we were able to differentiate sharecropping into kin and non-kin sharecropping tenants and control for self-selection due to unobservable household characteristics (by using household fixed effects) and a range of plot level characteristics. This is a more robust approach than that used in Chapter 9 for testing of Marshallian inefficiency. We also used stochastic dominance assessment in the analysis of sharecropping efficiency.

We refer to Kassie and Holden (2007) for a complete representation of a theoretical model with kinship affecting land productivity directly and indirectly through its effect on probability of contract renewal. In this chapter we include new results on fertilizer input use on kin and non-kin sharecropped plots versus owner-operated plots. We found significantly higher output levels on sharecropped plots than on sharecropping tenants' own plots. We also found significantly higher fertilizer input levels on non-kin sharecropped plots than on kin sharecropped plots and than on sharecroppers' owner-operated plots. Both non-kin and kin sharecropped plots stochastically dominated sharecroppers' owner-operated plots in terms of output value and non-kin sharecropped plots stochastically dominated kin sharecropped plots.

Analytical Framework and Hypotheses of the Study

We base our hypotheses on literature in which contracts between landlords and tenants are seen as repeated games. Radner (1981, 1985) used infinitely repeated principal-agent games to show that cooperative solutions may be possible. He also indicated that repeated games provide the principal with the opportunity to punish the agent for inappropriate behavior. The efficiency of such punishment will depend on the degree to which the agent discounts future utility. Bardhan (1984) and Banerjee et al. (2002) developed two-period principal-agent models to show how a threat of eviction on unsatisfactory performance increased the incentive of the agent to work hard in the first period. The effectiveness of eviction threats depends on the tenant's concern about his or her future utility on rented plots, earnings from outside farming activities, and availability of farmlands in the village. Contract nonrenewal would have a negative effect on the reputation of the tenant such that it becomes very difficult for the tenant to obtain a new contract due to loss of reputation (Hayami and Otsuka, 1993b). When the opportunity cost of alternative employment is higher than the benefits that will be obtained from rented land and land is not scarce, the tenant would not care about the contract insecurity, since he can easily move to other alternative income generation activities. If there is rationing of land such that it is not easy for a tenant who has lost his contract to find another landlord from which to obtain land, the threat of eviction is real and has a cost.

To secure future utility from rented in land and to qualify for renewal of contract, the tenant has to cultivate the rented in land with greater care and intensity. It is assumed that the probability of contract renewal (π) in period 2 depends on the amount of output produced (Q_{r1}) during the period 1 of the contract. In addition we assume that kinship relationship between landlord and tenants measured by h, in which h is social distance, affects the probability of contract renewal so that

$$\pi = \pi(Q_{r1}, h), \text{ and } \partial\pi/\partial Q_{r1} > 0, \partial\pi/\partial h < 0, \partial^2\pi/\partial Q_{r1}\partial h > 0 \qquad (10.1)$$

We assume that the probability of renewal (inverse of threat of eviction) is higher among kin and that good performance (leading to high productivity) is more important to reduce the threat of nonrenewal when the social distance is large. We assume it may be harder to impose eviction threats to kin. Kinship relations may cause other things than efficiency to be of highest priority in land tenancy arrangements. Kinship relations may therefore affect the performance (ability and incentives) of tenants in rental arrangements (Sadoulet et al., 1997).

With threat of eviction it is possible to find that the larger the social distance is, the more the tenant will apply nonland resources to rented land. Kinship may therefore reduce the contract insecurity effect on the use of nonland resources on rented land. Land productivity on sharecropped land may therefore be higher for non-kin sharecrop plots than for kin sharecrop plots, and the difference in productivity and input use between own land and rented in land will be smaller for kin than for non-kin tenants. Alternatively, if enforcement costs are high and kinship is essential for enforcement, we get the standard Marshallian inefficiency in non-kin contracts and less so in relation to kin contracts. Kinship may thus be a double-edged sword.

Based on the above theoretical discussion and Kassie and Holden (2007) we developed the following hypotheses. We have focused on the key variables of interest.

H1. *Marshallian inefficiency hypothesis.* Sharing of the output reduces incentives to apply inputs on sharecropped plots (kin and non-kin sharecropped plots) and this causes output on sharecropped land to be lower than on tenants' own plots. Testable implication: Output level and input levels are lower on kin and non-kin sharecropped plots than on share tenants' own plots.

H2: *Threat of eviction hypothesis.* Uncertainty about contract renewal creates incentives to increase output (and therefore input use) on sharecropped plots in order to qualify for contract renewal. Testable implications: Output and input levels are higher on kin and non-kin sharecropped plots than on share tenants' own plots. This is a conservative test when Marshallian inefficiency exists, as it requires that the threat of eviction effect is stronger than the Marshallian inefficiency effect.

H3. *Kinship eliminates/reduces Marshallian inefficiency.* Kinship ties increase the incentive of tenants to use more inputs on kin sharecropped plots. Testable implication: Output on kin sharecropped plots is not lower than on share tenants' own plots, but it is lower for non-kin sharecropped plots.

H4. *Kinship eliminates/reduces threat of eviction.* Kinship ties reduce contract insecurity of kin tenants and the freedom of landlords to evict kin tenants when performance is poor. Testable implication: Output on kin sharecropped plots is not higher than on share tenants' own plots, but it is so for non-kin sharecropped plots.

Sharecropping Efficiency Estimation Methods

Based on the theoretical discussion and the empirical literature on land rental markets, the input and output value per hectare can be expressed as follows:

$$x_{ph} = f(p_x, p_y, \sigma_h^2, \psi_h, h_{ph}, \rho_h, \beta, \gamma_p, z_p, z_h) \quad (10.2)$$

where σ_h^2 and ψ_h is the risk and risk aversion variables, respectively. In the empirical analysis the parameters β, γ, and h will be replaced by tenancy dummy variables (kin and non-kin sharecrop plots, cost share plots, and fixed rental plots using tenants' own plots as a counterfactual). All household level variables that were plot invariant (σ_h^2, $\psi_{h,}$ z_h and ρ_h) were dropped in the empirical estimation due to our estimation procedure using household fixed effects. In cross-sectional data price variation is not often observed across households and could therefore also not be included in our analysis. Households were using the same input and output markets in the area.

We used parametric and nonparametric techniques for the empirical analysis. The nonparametric method was stochastic dominance (efficiency) analysis (SD) to compare yield distribution obtained under different tenancy arrangements. In SD we compared output value on share tenants' own plots with output values on kin and non-kin sharecrop plots based on cumulative distribution functions (CDF) of output values. Similarly, output value distributions for kin and non-kin sharecrop plots were compared.

The advantages of parametric methods are that they may allow us to control for observable and, under certain conditions, unobservable characteristics that otherwise would bias (selection bias, endogeneity bias) the results and possibly lead to wrong conclusions. By applying household fixed effects to our multiple plots per household type of panel data, we have been able to eliminate plot invariant household and partially plot specific effects that may otherwise have created such biases.

The general regression model to be estimated is

$$y_{hp} = x_{hp}\beta + c_h + e_{hp} \quad (10.3)$$

where y_{hp} is output value per hectare or input use per hectare for household h on plot p, c_h is the unobserved household heterogeneity that captures unreported household characteristics such as farm management ability (tenants' quality), average land fertility, unobserved dimensions of wealth, and tenants' connections with landlords, households' risk preferences, time preferences, and so on, that affect input use and productivity. e_{hp} is the random variable that summarizes the plot-specific component other than the ones reported in the survey such as unobserved variation in plot quality and plot-specific production shocks (e.g., plot level variation in rainfall, hail, frost, floods, weeds, pests, and diseases infestations), x_{hp} includes both plot-invariant and variant observed explanatory variables, and β is a vector of parameters to be estimated.

Data and Structure of Land Rental Market in the Study Area

Household and plot level data were collected in 2001 for the 2000 production season from a random sample of 148 households operating 1,290 plots in one of the villages in Dembecha district, West Gojjam zone of northwestern Ethiopian highlands. Table 10-1 presents means and standard deviations of key variables for the entire sample and for the sub-samples of plots by tenancy type. The 1,290 plots are categorized into kin sharecrop plots (5.2%), non-kin sharecrop plots (1.6%)[1], cost share plots (4.1%), fixed-rental plots (4.3%), landlord-owned cultivated plots (6.3%), share tenants' own plots (25.8%), cost share tenants' own plots (10.9%), fixed-rental tenants' own plots (10.9%), and owner-operators' own plots (31.0%).

One has to be very cautious when comparing average output and input levels across contracts and plot types in Table 10-1. We therefore leave this for the econometric analysis. Still, we may make some comparisons. There were no statistically significant wealth differences between kin tenants versus non-kin tenants as well as kin landlords versus non-kin landlords (Tables 10-2 and 10-3). The mean number of years that the same tenant had cultivated the plot was 4.3 and 3.7 for kin and non-kin sharecrop plots, respectively. Of the total sample of plots, 49% received fertilizer. About 37% of kin and non-kin sharecrop plots were fertilized. Rented in plots were more distant from the residence compared to tenants' own plots. About 35% of the rented in plots were accessed outside the village due to shortage of land supply within tenants' residence. It took 3 to 7 hours (single trip) to reach these plots.[2] The average farm size in the area was 0.35 hectare per adult equivalent.

The survey also contains a wide range of information on plot characteristics, the household's economic conditions, and contract information (output and input sharing rules, kinship relations). The most prevalent output and input sharing formula was a 50–50% split of the output between the landlord and the tenant. About 89.7, 7.4, and 2.9% of the sharecropped plots were

TABLE 10-1 Descriptive statistics for key variables by tenancy types

Variables	*Owner-Operated Plots*	*Cost Sharing Tenants Own Plots*	*Cost Shared Rented in Plots*
Plots received fertilizer (dummy)	0.48(0.50)	0.53(0.50)	0.89(0.32)
Output value per ha (continuous)	783(757)	887(723)	1047(609)
Fertilizer value per ha (continuous)	102(165)	108(136)	199(179)
Seed value per ha (continuous)	120(128)	131(117)	105(91)
Oxen-days per ha (continuous)	17(30)	16(13)	14(12)
Man-days per ha (continuous)	67(71)	65(48)	65(43)
Threshing labor per ha	20(15)	20(16)	16(13)
Weeding labor per ha	18(35)	18(24)	22(20)
Harvesting labor per ha	12(10)	11(8)	13(9)
Ploughing labor per ha	17(30)	16(13)	14(12)
Good fertile soil (dummy)	0.21(0.41)	0.11(0.31)	0.19(0.39)
Medium fertile soil (dummy)	0.52(0.50)	0.53(0.50)	0.53(0.50)
Poor fertile soil (dummy)	0.28(0.48)	0.36(0.48)	0.28(0.45)
Shallow soil depth (dummy)	0.21(0.40)	0.33(0.47)	0.26(0.45)
Medium soil depth (dummy)	0.37(0.48)	0.29(0.45)	0.25(0.43)
Good soil depth (dummy)	0.43(0.49)	0.39(0.48)	0.49(0.50)
Plot size in ha (continuous)	0.24(0.14)	0.25(0.15)	0.32(0.14)
Plot distance (continuous: minutes)	13(15)	11(14)	50(74)
Plot slope in degree (continuous)	17(10)	17(8)	16(4)
Conservation intensity (continuous, ha)	0.01(0.01)	0.01(0.01)	0.004(0.01)
Homestead plots (dummy)	0.25(043)	0.34(0.47)	0
Intercropped plots (dummy)	0.27(0.45)	0.26(0.44)	0.21(0.41)
Pulses and oil crops (dummy)	0.15(0.35)	0.14(0.34)	0.06(0.23)
Teff crop (dummy)	0.20(0.40)	0.16(0.37)	0.19(0.39)
Barley crop (dummy)	0.32(0.47)	0.35(0.48)	0.09(0.30)
Potato crop (dummy)	0.06(0.24)	0.09(0.28)	0.02(0.14)
Wheat crop (dummy)	0.05(0.21)	0.07(0.26)	0.11(0.32)
Improved maize crop (dummy)	0.10(0.30)	0.09(0.28)	0.47(0.50)
Local maize crop (dummy)	0.13(0.33)	0.11(0.31)	0.06(0.23)
Household head age in years (continuous)	40(12)	38(12)	
Extension visit (dummy)	0.76(0.43)	0.59(0.50)	
Oxen power per ha (continuous)	0.77(0.47)	0.69(0.42)	
Other livestock in TLU per ha (continuous)	1.30(0.88)	1.32(0.72)	
Off farm activities participation (dummy)	0.12(0.33)	0.14(0.35)	
Farm size per capita (continuous)	0.37(0.10)	0.34(0.12)	
Duration of rented in plots (continuous)			2.9
Number of observations	400(48)	140(22)	53

Note: Standard deviations in parentheses except in the number of observations row where figures in parentheses are number of households

Share Tenants Own Plots	Kin Sharecrop Plots	Non-Kin Sharecrop Plots	Fixed Rental Tenants Own Plots	Fixed Rental Plots	Landlords' Own Plots**	All Samples
0.45(0.50)	0.37(0.49)	0.37(0.50)	0.50(0.50)	0.67(0.47)	0.31(0.46)	0.49(0.50)
791(668)	893(650)	1033(818)	912(747)	889(807)	659(422)	827(708)
100(251)	70(179)	121(199)	114(156)	133(128)	53(105)	104(186)
128(157)	124(137)	112(141)	148(164)	123(109)	87(83)	124(136)
17(17))	18(16)	11(7)	19(19)	17(18)	20(34)	17(23)
71(65)	60(56)	42(28)	77(69)	59(52)	67(52)	68(63)
20(18)	14(5)	19(23)	23(16)	19(29)	19(14)	20(17)
22(35)	13(16)	10(17)	23(38)	15(16)	15(17)	19(31)
12(11)	9(4)	12(14)	12(9)	10(11)	13(10)	12(10)
17(17)	18(16)	11(7)	19(19)	17(18)	20(34)	17(23)
0.20(0.40)	0.22(0.42)	0.16(0.37)	0.19(0.40)	0.22(0.42)	0.19(0.39)	0.19(0.39)
0.51(0.50)	0.50(0.50)	0.42(0.51)	0.48(0.50)	0.47(0.50)	0.46(0.50)	0.50(0.50)
0.29(0.46)	0.28(0.45)	0.42(0.51)	0.33(0.47)	0.31(0.47)	0.36(0.48)	0.30(0.46)
0.22(0.41)	0.41(0.50)	0.26(0.45)	0.27(0.42)	0.25(0.44)	0.28(0.45)	0.25(0.43)
0.35(0.48)	0.22(0.42)	0.37(0.50)	0.23(0.42)	0.31(0.47)	0.31(0.46)	0.32(0.47)
0.43(0.50)	0.37(0.49)	0.37(0.50)	0.50(0.50)	0.44(0.50)	0.41(0.49)	0.43(0.49)
0.24 (0.14)	0.27(0.18)	0.31(0.18)	0.22(0.12)	0.25(0.12)	0.27(0.13)	0.25(0.14)
13(14)	40(52)	40(39)	12(15)	38(42)	14(19)	17(28)
17(9)	16(5)	19(10)	19(12)	16(6)	19(11)	17(9)
0.004(0.01)	0.003(0.031	0.01(0.01)	0.004(0.01)	0.002(0.01)	0.01(0.02)	0.02(0.04)
0.27(0.45)	0.01(0.12)	0	0.23(0.42)	0.09(0.29)	0.25(0.43)	0.23(0.42)
0.26(0.44)	0.04(0.21)	0.05(0.23)	0.27(0.45)	0.07(0.26)	0.22(0.42)	0.24(0.43)
0.12(0.33)	0.15(0.36)	0.16(0.37)	0.14(0.34)	0.13(0.34)	0.15(0.36)	0.13(0.34)
0.19(0.39)	0.35(0.48)	0.32(0.48)	0.18(0.39)	0.20(0.40)	0.20(0.40)	0.20(0.40)
0.32(0.47)	0.37(0.49)	0.32(0.48)	0.33(0.47)	0.20(0.40)	0.30(0.46)	0.31(0.46)
0.08(0.27)	0.03(0.17)	0	0.04(0.21)	0.13(0.34)	0.04(0.19)	0.06(0.24)
0.04(0.21)	0.04(0.21)	0.16(0.37)	0.09(0.28)	0.09(0.29)	0.06(0.24)	0.06(0.24)
0.09(0.28)	0.03(0.17)	0	0.09(0.28)	0.22(0.42)	0.07(0.26)	0.106(0.31)
0.16(0.37)	0.03(0.17)	0.05(0.23)	0.13(0.34)	0.04(0.19)	0.19(0.39)	0.12(0.33)
39(13)	38(11)	41(0.14)	37(8)		43(14)	39(12)
0.54(0.50)	0.57(0.50)	0.42(0.51)	0.39(0.50)		0.45(0.52)	0.59(0.49)
0.68(0.31)	0.68(0.32)	0.67(0.27)	0.79(0.26)		0.65(0.45)	0.72(0.39)
1.20(0.75)	1.16(0.74)	1.29(0.79)	1.35(0.87)		1.11(0.63)	1.26(0.79)
0.17(0.38)	0.19(0.40)	0.08(0.29)	0.33(0.49)		0.27(0.47)	0.18(0.38)
0.33(0.13)	0.32(0.12)	0.38(0.18)	0.32(0.09)		0.38(0.06)	0.35(0.0.11)
	4.3	3.7		2.3		
333(49)	68(37)	19 (12)	141(18)	55	81(11)	1290(148)

No statistically significant difference between kin versus non-kin tenants' household level variables using t-test

** indicates landlords residing outside the village and pure landlords are not included.

TABLE 10-2 Tenants and landlords: Resources comparison

	Tenants		*Landlords*			*Tenants*		*Landlords*	
Resources	*Mean*	*Std. Err.*	*Mean*	*Std. Err.*	*Resources*	*Mean*	*Std. Err.*	*Mean*	*Std. Err.*
Oxen holding (number)	1.95***	0.077	0.57	0.105	Beehives (number)	0.63***	0.164	0	0
Cow holding (number)	1.65***	0.122	0.73	0.118	Labor in standard unit	2.96***	0.120	2.30	0.185
Small ruminant (number)	3.32***	0.341	0.68	0.153	Family size in standard unit	4.67***	0.183	3.09	0.204
Chicken (number)	1.70	0.185	1.89	0.280	Household head age	38***	1.236	50	2.342
Pack animals (number)	0.82***	0.104	0.05	0.030	Farm size in ha	1.42***	0.056	1.18	0.291
Number of observations	88		56						

*** significant at 1%. We did not ask livestock age for those landlords residing outside the study area and thus livestock holding is not converted into standard units.

TABLE 10-3 Kin versus non-kin landlords' resource endowments comparison[a]

Resources	*Non-Kin Landlord*		*Kin-Landlord*		*Resources*	*Non-Kin Landlord*		*Kin-Landlord*	
	Mean	*Std. Err.*	*Mean*	*Std. Err.*		*Mean*	*Std. Err.*	*Mean*	*Std. Err.*
Oxen holding (number)	0.55	0.12	0.61	0.20	Labor in standard unit	2.51	0.24	1.86	0.39
Cow holding (number)	0.68	0.14	0.83	0.23	Family size in standard unit	3.04	0.20	3.19	0.40
Small ruminant (number)	0.55	0.17	0.94	0.31	Household head age	51	2.89	47	4.03
Chicken (number)	1.71	0.31	2.27	0.58	Farm size in ha	1.14	0.27	1.28	0.70
Number of observations	18		38		Number of observations	18		38	

[a]No statistically significant differences in means were found. Landlords residing outside the village and pure landlords are included.

under 50–50, 33–67, and 25–75% share arrangements (in which the first share goes to the landlord). The landlord and the tenant were sharing the cost of fertilizer and seed equally in cost-sharing contracts. The landlord did not share draft power and labor inputs.

Empirical Evidence of the Study

Regression results for output value per hectare are presented in Table 10-4. The parametric regression results suggested that conditional plot characteristics, crop types, inputs and household fixed effects, kin and non-kin sharecropped plots were more productive than share tenants' own plots. The stochastic dominance analysis supported this result in which the yield from kin and non-kin sharecropped plots unambiguously dominated the yield from share tenants' own plots (Figure 10-1). This is consistent with the threat of eviction hypothesis that tenants work harder to qualify for contract renewal in the second period (hypothesis H2). Non-kin sharecropped plots were more productive than kin sharecropped plots in each specification. The non-kin sharecropped plots' yield distribution unambiguously dominated the kin sharecropped plots' yield distribution. This result is consistent with the hypothesis H4 that threat of eviction is stronger among non-kin than among kin partners. First-order stochastic dominance findings indicated the dominance of the threat of eviction effect over the Marshallian inefficiency effect and kinship reducing the threat of eviction effect, implying that it is more difficult to evict kin tenants than non-kin tenants.

The results showed that inequality in plot quality might not be the culprit in the observed productivity gap between kin and non-kin sharecrop and own plots. Controlling for plot quality did not eliminate the observed significant productivity difference between kin and non-kin sharecrop and share tenants' own plots (see Table 10-4, columns A and B). The yield difference between share tenants' own plots, and kin and non-kin sharecropped plots was not a consequence of access to different crop types and technologies. The coeffi-

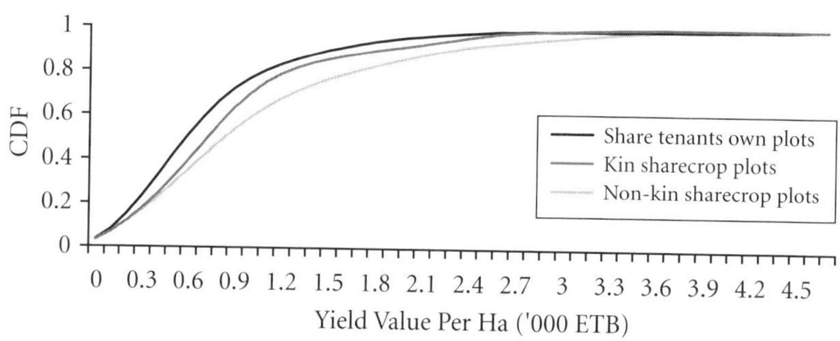

FIGURE 10-1 CDF for impacts of share tenancy on yield: First order stochastic dominance analysis

TABLE 10-4 Linear household fixed effects estimates of the determinants of plot level output value (Dependent variable: log of output value per hectare)

Explanatory Variables	*Without Plot Quality* *A*	*With Plot Quality* *B*[a]	*With Plot Quality and Crop Types* *C*[b]	*With Plot Quality, Crop Types and Inputs* *D*[c]
Kin sharecrop plots	0.168(0.145)	0.301(0.152)*	0.295(0.130)**	0.205(0.116)*
Non-kin sharecrop plots	0.310(0.167)*	0.439(0.162)**	0.453(0.150)***	0.334(0.124)***
Joint F test for plot quality indicator variables*		5.30(0.000)***	3.86(0.001)***	1.34(0.239)
Joint F test for crop dummy variables*			7.86(0.000)***	6.64(0.000)***
Joint F test for inputs*				25.67(0.000)***
Constant	6.379(0.025)***	6.319(0.335)***	6.157(0.359)***	3.836(0.315)***
Observations	420	420	420	420
R-squared	0.15	0.23	0.36	0.62
Model test	F(2, 47) = 2.42(0.10)	F(12,47) = 5.77(0.000)***	F(18,47) = 7.90(0.000)***	F(24,47) = 32.33(0.000)***

Robust standard errors adjusted for clustering effects in parentheses

* significant at 10%; ** significant at 5%; *** significant at 1%.

[a]Plot quality indicator variables include good fertile soil, medium fertile soil, shallow soil depth, medium soil depth, ln (plot distance from residence), ln (plot slope), homestead plot, conservation intensity, and intercropped plot.

[b]Crop dummy variables include pulses and oil crops plot, teff plot, barley plot, potato plot, wheat plot, and improved maize plot.

[c]Factor input include ln (weeding labor per ha), ln (plowing labor per ha), ln (threshing labor per ha), ln (harvesting labor per ha) ln (value of seed per ha), and ln (fertilizer value per ha).

TABLE 10-5 Linear household fixed effects estimates of the determinants of fertilizer intensity using full sample data (using all tenants' own plots as reference group)

Independent Variables	*Without Plot Quality* A	*With Plot Quality* B	*Without Potential Endogenous Variables* C
Kin sharecrop plots	−0.159(0.275)	−0.215(0.243)	−0.215(0.251)
Non-kin sharecrop plots	0.689(0.175)***	0.880(0.221)***	0.552(0.199)***
Fixed rental plots	−0.073(0.132)	−0.138(0.132)	−0.008(0.132)
Cost shared plots	0.270(0.162)*	0.262(0.144)*	0.572(0.159)***
Plot quality indicator variables^	No	2.04(0.059)*	2.49(0.001)***
Crop dummy variables^	1.44(0.212)	2.37(0.013)**	No
Constant	5.050(0.136)***	6.013(0.380)***	5.973(0.385)***
Observations	629	629	629
R-squared	0.38	0.48	0.39
Robust standard errors in parentheses			

Note: These variables are the same as those in Table 10-4.

* significant at 10%; ** significant at 5%; *** significant at 1%.

cients of interests were still statistically significant with and without crop types and inputs (Table 10-4, columns B, C, and D).

As indicated in Tables 10-2 and 10-3, kin versus non-kin tenants as well as kin versus non-kin landlords appear to have similar production capacity since there were no statistically significant differences in resource endowments between them. The duration of contracts was relatively shorter for non-kin than for kin sharecropped plots. This may indicate more frequent occurrence of eviction for tenants that are socially distant from landlords.

The parameter estimates for the lognormal regression model for fertilizer intensity estimation are presented in Table 10-5. Fertilizer use is consistently higher on non-kin sharecropped plots than on tenants' own plots and on kin sharecropped plots in each specification. There is no significant fertilizer use difference between kin sharecropped plots and tenants' own plots. The larger portion of the variation in fertilizer use between share tenants' own plots and non-kin sharecropped plots is not explained by omitted plot quality. The same argument as in the productivity estimation may explain this result.

Conclusions and Implications

With the more robust approach on household fixed effects we did not find significant signs of Marshallian inefficiency. Efficiency on rented land was higher than that on owner-operated land of the same tenant households; tenants used more fertilizer on their rented in plots than on their own plots as this would enhance their probability of contract renewal. Specifically, the results indicated that the threat of eviction effect is dominating over the Marshallian inefficiency effect and that kinship reduced the threat of eviction effect. This may be because it is more difficult to evict kin tenants than non-kin tenants. This implies that non-kin tenants feel less secure about their tenure and therefore use more fertilizer and are more productive on their sharecropped plots to increase the probability of contract renewal.

Although the quantitative results derived in this study are somewhat specific to a single community, the policy recommendations that follow may apply equally to any rural areas of the country that have similar settings of our study area.

The results have the following important policy implications. First, the results of the study suggested that share tenancy contracts are not inefficient in all circumstances and may not be an impediment to growth. Second, tenure insecurity may be higher in the Amhara region where this study and the study in Chapter 9 were carried out because the last land redistribution was carried out as late as in 1997. The land reform relevant to land registration and certification and the new land laws that were implemented from 2003 and onward, some two years after our fieldwork, is likely to contribute to strengthening the land rights and should reduce tenure insecurity. This may also explain why there appeared to be higher transaction costs in the land rental markets in Amhara region (Chapter 3 by Deininger, Ali, and Alemu) before the reform than there were in Tigray (Chapter 4 by Ghebru and Holden) five years after the land registration and certification reform. Finally, we should note that higher land productivity on rented in land than on owner-operated land of tenants may also involve inefficiency because of overuse of fertilizer on rented in land. It is possible that total output could have been improved if some of this fertilizer had been reallocated to the tenants' own plots. If the following land registration and certification contributes by enhancing the allocative efficiency of the land rental market, as found by Holden et al. (2007a) in the Tigray region, this could also enhance the access to rented land by tenant households. This reform could then possibly also contribute to reducing the overuse of fertilizer on rented land in our study area, but further empirical studies would be required to test this.

NOTES

1. Some landlords reside outside the study village and some are pure landlords in that they entirely rent out their lands.
2. Tenants (migrant tenants) went to some of these plots one day before the cultivation day and stayed there with their relatives since it was difficult to start work the same day after long hours walking together with oxen and farm implements.

PART IV

Land Markets, Land Tenure, and Land Management

CHAPTER 11

Land Rental Markets and Land Management in the Highlands of Ethiopia

SAMUEL BENIN AND JOHN PENDER

Land is a crucial resource in agrarian economies and, thus, efficient land markets (including sales and rentals) are important for sustainable land management and agricultural development. Land markets allow land to be used by farmers who are more capable to earn the highest return from it, particularly where markets for other scarce factors of production do not function perfectly because of asymmetric information, transaction costs, and incentive problems (Binswanger and Rosenzweig 1986; Chapters 1 and 2). Unfortunately, land sales markets are also usually poorly developed in developing countries due to asymmetric information about land quality, lack of land titles, undeveloped credit markets, inability of poorer farmers to pay for the collateral value of the land, and/or various policy distortions (Binswanger and Rosenzweig 1986; Binswanger et al. 1993). Where land sales markets do not function well, or are prohibited—as in Ethiopia—the effective functioning of land rental markets is particularly critical for agricultural development.

Land rental markets generally promote improved short-term agricultural efficiency (relative to the absence of such markets) (Binswanger and Rosenzweig 1986; Otsuka et al. 1992), although they do not always lead to fully efficient outcomes (Bell 1977; Shaban 1987; Otsuka and Hayami 1988; Sadoulet et al. 1994; Laffont and Matoussi 1995; Chunrong et al. 1996; Ahmed et al. 2002). However, rental contracts are often very short term in nature due to tenure insecurity of landowners, especially where "land to the tiller" laws or other redistribution schemes may penalize landowners who rent out land (Pender and Kerr 1999). In northern Ethiopia, restrictions on the duration of land leases (limited to no more than 2 years unless tenants use "modern technology") also contribute to short-term leases, which are usually no more than 1 year in duration (Benin 2006; Pender and Gebremedhin 2004). As a result, the incentives of tenants to invest in sustainable land management practices may be very limited in such circumstances.

Although there is a large amount of literature investigating the short-term efficiency impacts of land lease contracts (especially sharecropping contracts), the impacts of such contracts on land management practices have been less well studied, especially in the African context. The objective of this chapter is to investigate these relationships in the context of the northern Ethiopian highlands, where there are very serious concerns about the sustainability of land management, and where it is possible that the short-term nature of land rental contracts undermines incentives to invest in and manage rented land in a sustainable way. Using community, household, and plot level data from the highland areas of Tigray and Amhara regions, we examine the impacts of land rental on land management practices. Earlier chapters have shown that land rental markets in Ethiopia are very active although allocative efficiency is low. Rented land is transferred to more efficient producers, but whether they have sufficiently strong incentives to conserve the land they rent in is the remaining issue to be resolved empirically in this chapter.

Tenure Security and Land Management

A conceptual model of the link between land markets, investment in land improvements, agricultural productivity, and land quality was given in Chapter 2. Of relevance to this chapter is the *tenure security effect*, which hypothesizes that farmers who purchase land will have more incentive to title their land and privatize it away from local customary tenure arrangements that may impinge on investment decision making. On the contrary, rented plots are typically of limited duration and carry with them significant limitations in terms of investments that can be made by the tenant. Since there are no land sales in Ethiopia, only the second part of the hypothesis can be tested, with rented plots being hypothesized to have lower physical investments or long-term land management practices.

Empirical Approach

We assess hypotheses about the impacts of land tenancy on farmers' land investments and land management practices in the highlands of Tigray and Amhara using descriptive and econometric analysis of the community, household, and plot survey data, which are presented later on. In the descriptive analysis, we seek to uncover potential relationships between land tenancy and land management and possible explanations for these relationships. Then we test the significance of these relationships after accounting for the influence of other factors using econometric analysis.

The specification of the econometric models is as follows[1]:

$$LI^{pr}{}_{hp(t1-t0)} = LI(T_{hpt0}, NC_{hpt0}, LI_{hpt0}, LI^{mm}{}_{hp(t1-t0)}, LU_{hpt0}, PC_{ht0}, HC_{ht0}, SC_{ht0}, IS_{ht0}, X_{hv}, u_{LIhpt01}) \qquad (11.1)$$

$$LM_{hpt1} = LM(T_{hpt1}, NC_{hpt1}, LI_{hpt1}, LU_{hpt1}, PC_{ht1}, HC_{ht1}, SC_{ht1}, IS_{ht1}, X_{hv}, u_{LMhpt1}) \qquad (11.2)$$

where $LI^{pr}{}_{hp(t1-t0)}$ refers to private land investment by household h on plot p during period $t1$–$t0$ (where $t0$ is 1997 and $t1$ is 1999 for the Tigray analysis),[2] $LI^{mm}{}_{hp(t1-t0)}$ is mass mobilization land investment on the plot during $t1$–$t0$, LM_{hpt1} is use of annual land management practices by household h on plot p during year $t1$, T_{hpt0} is the tenure status (whether rented in or owner operated) of the plot in year $t0$, NC_{hpt0} are the "natural capital" characteristics of the plot (area, elevation, slope, position on the slope, soil depth, color, texture, perceived fertility, plot quality ranking, presence of land degradation problems such as gullies and water logging, distance of the plot to the farmers' residence), LI_{hpt0} is the initial stock of land investments on the plot at the beginning of $t0$, LU_{hpt0} is the land use of the plot in $t0$ (i.e., whether the plot is a homestead plot, cultivated or irrigated outfield, or a fallow plot), PC_{ht0} is the "physical capital" or asset endowments of the household in $t0$ (land, oxen, other cattle, other livestock, radio), HC_{ht0} is the "human capital" of the household in $t0$ (education, age and gender of the household head, the size of household, the share of dependents, and the female share of the labor supply), SC_{ht0} is the "social capital" of the household in $t0$ (membership in local organizations, which include farmers' associations, women's associations, youth associations, marketing cooperatives and agricultural cadres, and leadership in the local community[3] or village council), IS_{ht0} is the income strategy of the household in $t0$ (reflected by the secondary income source of the household, since cereal production is the primary income source of almost all households), X_{hv} are relatively fixed household or community level factors affecting the comparative advantage of different activities and costs, returns and risks of land investments and management practices (average annual rainfall, population density, distance to the nearest town, road, and bus service), and $u_{LIhpt01}$ and u_{LMhpt1} are unobserved error terms, assumed to be normally distributed.

The dependent variables are measured as binary (yes/no) variables. This fact, together with the assumption of normally distributed error terms and linear relationships between the explanatory variables and the index functions determining the level of the dependent variables, results in use of the probit model to estimate equations (11.1) and (11.2) (Maddala 1983).

One concern in estimating equations (11.1) and (11.2) is the endogeneity of farmers' decisions about land leasing, which could result in endogeneity bias. We address this by estimating the rental decision and the investment or land management decision jointly, using a bivariate probit model. We use the same explanatory variables as in equations (11.1) and (11.2) as determinants of land rental, excluding the tenure variable (T_{hpt0} or T_{hpt1}) and the mass mobilization land investment variable ($LI^{mm}{}_{hp(t1-t0)}$). In the bivariate version of the land investment regression, we dropped several explanatory variables for investment that were jointly statistically insignificant in the univariate model in order to achieve convergence of the model. This was not necessary in the bivariate versions of the land management regressions, as these models were identified without such exclusion restrictions.[4] In the bivariate probit models, the error terms in the two equations are assumed to be jointly normally distributed with a correlation of rho. A test of whether rho is significantly different from zero is a test of whether there is endogeneity bias in the univariate model.

In addition to testing for endogeneity, we also tested for multicollinearity among the explanatory variables. The maximum variance inflation factor was 4.8 and 11.8 in Tigray and Amhara regions, respectively, indicating that multicollinearity was not a major concern (Mukherjee et al. 1998).

Study Sites and Data

We use survey data from community, household, and plot level surveys that were conducted in the highland areas (above 1,500 meters above sea level [masl]) of the Amhara and Tigray regions between 1999 and 2001. A stratified random sample of 99 peasant associations (PAs, usually consisting of three to five villages), two villages (198 in total) randomly selected from each PA, and four or five households (934 in total) randomly selected from each village were selected from the highland areas of the two regions.[5] In addition, all plots operated by the selected households were surveyed, and the stratification was based on indicators of agricultural potential, market access, and population density (Benin, 2006; Pender and Gebremedhin, 2006). Information was collected through structured surveys about household structure and endowments, the household's access to infrastructure and services, plot characteristics (mode of acquisition, size, slope, quality, crops cultivated, etc.), land investments, land management practices, inputs, and agricultural production in 1999 and 2000 cropping seasons. Recall methods were also used to obtain information prior to 1999–2000 (specifically for 1997 and 1998 in Tigray and for 1991 in Amhara).[6] Data on altitude were collected using a global positioning system (GPS). The primary data were supplemented by secondary information on rainfall and population.

Past Land Policies and Their Implications for This Research

Prior to 1975, land tenure in Ethiopia was based on a feudal system in which land was owned by the emperor, who granted tenure rights subject to feudal obligations. Land tenure forms known as *rist* and *gult* were then practiced (Rahmato, 1984). The rist was a communal form of land ownership in which an individual was entitled to usufruct rights in a commune if he or she could prove some blood relation to the founding patriarch of the commune. Usufruct rights were lifelong and extended over many generations. The gult was linked with the legal and political institutions of the crown and those involved, *gultegnas*, were responsible for levying and collecting tax from *ristegnas*, who tilled the land. After the overthrow of the monarch, lands were nationalized and declared as the collective property of the people in the land reform proclamation of March 1975, and households were given use rights only, with occasional redistribution of farmland to accommodate landless households. Since 1991, under the current government of the Ethiopian Peoples Revolutionary Democratic Front (EPRDF), households have been given the right to use the land indefinitely, lease it out temporarily to other farmers, and transfer it only to their children. However, they cannot sell or mortgage the land. Land redistribution has been phased out, except where major infrastructure investments such as irrigation necessitate redistribution.

Leading up to the time of the data collection, however, there were significant differences across regions with respect to the implementation of land redistribution. For example, in the Tigray region, land redistribution was stopped in 1991, and the policy of no future redistribution was made official by a land use and tenure policy in 1997. In the Oromiya region, too, there had not been a land redistribution since 1992 (Tefera et al. 2000). In the Amhara region, on the other hand, land redistribution was common, with the most recent major land redistribution occurring in 1997–1998. There was at least one land redistribution since 1991 in about 73% of the villages surveyed in Amhara, with three being the average number. One of the villages interviewed had experienced as many as fourteen land redistributions since 1975. Compared to other countries, land is relatively equally distributed in the Ethiopian highlands because of the radical land reform program begun in 1975 (Rahmato 1984; Bruce et al. 1994; Abate 1995; Amare 1995) and the continued prohibition of land sales and mortgages, a policy enshrined in the new Ethiopian constitution.[7]

Land Management in Tigray Region

In the highlands of Tigray, about 12% of operated plots in 1997 were rented in, almost exclusively by sharecropping (Table 11-1).[8] Of these lease contracts, about 94% were for one year or less, and only 2% were for more than two years. The lease agreement was verbal in almost all cases; in only 5% of cases was a written contract used. The dominant sharecropping arrangement was with the landowner receiving 50% of the harvested crop, although in a few cases the landowner received a one-third or one-fourth share. In 80% of the cases, the landowner received no share of the crop residue, while in the remainder, the landowner received either half or all of the crop residues. In some cases, landowners share input costs (labor, oxen, fertilizer, or seeds), although this is relatively uncommon. (The landowner shares the cost of seeds in less than 20% of cases, labor in about 10% of cases, and oxen or fertilizer in even fewer cases.)

The most common form of land investment occurring in the region is investment in stone terraces. Stone terrace investments were made on about 15% of the plots surveyed during 1997 to 1999. Most of these investments

TABLE 11-1 Investments in stone terraces and land management practices in Tigray region, by land tenure (standard errors in parentheses)

Variable	*Sharecropped or Rented*		*"Owned" (Allocated by Tabia, Inherited, Received as Gift)*	
Number of plots in 1997[a]	252[b]		1,628[c]	
% of total plots in 1997	11.86	(1.12)	88.14	(1.12)
% of plots with investment in stone terraces in 1997–99				
Private investment	5.48	(1.91)	5.74	(0.77)
Labor mass mobilization	13.58	(2.87)	9.09	(1.02)
Food or cash for work	0.65	(0.64)	0.60	(0.25)
% of private investments in stone terraces				
By "owner"	19.46	(16.93)	96.51	(2.03)
By tenant	74.30	(15.92)	0.00	(0.00)
By others	6.24	(5.48)	11.63	(3.07)
% of plots where operator expects to operate the plot 10 years in the future	52.35	(7.22)	96.48	(0.70)
% of plots with land management practices in 1999				
Fertilizer	22.36	(3.26)	30.86	(1.79)
Manure	11.91	(2.83)	24.76	(1.38)
Contour plowing	92.86	(1.90)	85.79	(1.32)
Reduced tillage	7.04	(1.65)	13.49	(1.53)

Source: IFPRI/ILRI/Mekelle University household survey.

[a]Excludes 4 borrowed plots.

[b]Includes 246 sharecropped plots and 6 rented plots.

[c]Includes 1,398 plots allocated by the *tabia*, 31 inherited plots, 33 gift plots, and 166 plots with means of acquisition unspecified.

were made through community mass mobilization of labor, according to the regional policy that each adult was required to contribute at least 20 days of labor for soil and water conservation and other community activities, while a few investments (affecting less than 1% of plots) were promoted by food for work or cash for work programs. Privately initiated investments in stone terraces were made on about 6% of plots.

These private investments were made on a similar share of rented in plots as on owner-operated plots. Interestingly, the tenant made the investment on rented-in plots in three-fourths of the cases, despite the short duration of lease contracts. This may be because such tenants expect these lease contracts to be renewed in the future, with their investment effort possibly perceived as increasing the likelihood of renewal. Consistent with this, we find that slightly more than half of the tenants expected to operate the same rented in plots 10 years later. Although this is much less than the proportion of owner-operated plots that the owners expect to continue operating in the long term (almost 100%), it is apparently a large enough proportion to enable many tenants to invest in longer-term land improvements. Of the 12 cases in which a tenant invested in stone terraces on a plot during 1997 to 1999, the tenant expected to operate the same plot for at least 10 years in 7 cases. More puzzling are the five cases in which a tenant invested, but did not expect to operate the same plot for even five years. Further research is necessary to try to understand why a tenant would make investments in such cases.

The most common annual land management practices used by farmers in the Tigray highlands are contour plowing, use of fertilizer and manure, and reduced tillage. Use of fertilizer, manure, and reduced tillage are all more common on owner-operated plots than on rented in plots, although contour ploughing is somewhat more common on rented in plots.

Land Management in Amhara Region

The average land holding per household is about 1.3 hectares in the Amhara region, and significantly higher (about 75% more) among male-headed households than that of their female counterparts. Up to 89% of the plots were cultivated by their owners (i.e., by those receiving the land directly from the government [peasant association] or through gift, inheritance, or permanent exchange) (Table 11-2). The remaining 11% of the plots were obtained through temporary farmer-to-farmer exchanges in the form of rental, mostly sharecropping. Although both male- and female-headed households participate equally in the land rental market, the males tended to rent in more while females tended to rent out more. Contracts for rented plots were very short (one season or one year on average) and *equl* sharecropping (one-half of crop output to landowner) was the common practice. For fixed leases, rents were about $30 to $65 per hectare, depending on the quality of the land.[9]

TABLE 11-2 Descriptive statistics of plot acquisition and land management practices by type of plot acquisition in Amhara region (% of plots)

		Tenure	
	All Plots	*Owner Cultivated*[a]	*Rented*[b]
How plot acquired		88.1	11.9
Type of practice			
Reduced tillage	24.5	25.2	19.2
Contour plowing	69.7	69.7	69.9
Crop residues	63.1	64.1	56.4
Crop rotation	61.8	63.8	47.6
Chemical fertilizers	35.4	32.1	59.7
Number of plots	1187	1057	130

Source: IFPRI/ILRI/Amhara National Bureau of Agriculture household survey.

Note: Sample means are adjusted for stratification, weighting, and clustering of sample.

[a]Owner cultivated includes, PA-allocated (74.9%), gifted (7.3%) and inherited, or exchanged-in permanently, etc (5.9%).

[b]Rented is mainly sharecropped.

Land rights associated with exchange, transfer, and making long-term investments were exclusive to owner-cultivated plots. There seems, however, to be a high level of restriction on rented plots even for simple activities such as crop choice and grazing animals. Tenants could not choose what type of crops to plant on 50% of the rented plots, while they could not graze their animals on 28% of the rented plots. Expectations to operate the plot in the next 5 or 10 years or to bequeath the plot were almost perfect on owner-cultivated plots, but were between 20 and 28% on rented plots.[10] The main reasons for not expecting to operate the plot in the future were fear of land redistribution on owner-cultivated plots and termination of rental contract or uncertainty of renewal of contract on rented plots.

The most common types of investments, in terms of proportion of plots having them, were drainage ditches (39%) followed by stone terraces (22%) and constructed fences (16%). Although the incidence of stone terraces, live fences, and trees was significantly higher on owner-cultivated plots compared to rented plots, the incidence of drainage ditches was lower. In fact, there were no trees on rented plots. These results reflect the relatively long-term versus short-term return to different types of investments, in which those with longer-term benefits such as trees are preferred on more tenure-secure plots, while those with immediate or shorter-term benefits such as drainage structures are preferred on rented plots. Generally, there are other factors that influence the decision to invest in land improvement. However, due to lack of adequate data on investments in earlier years, econometric analysis could not be carried out to assess the influence of different factors on making the investments.

Compared to the types of land investments discussed here, households tended to engage more in various short-term land management practices, with the most common practices including contour plowing, plowing in crop residues, and crop rotation (Table 11-2). Notable differences across different types of plot were found for crop rotation, in which the incidence was higher on owner-cultivated than on rented plots, while use of chemical fertilizers was higher on rented plots. It is likely that renters have more resources, including access to credit to finance the purchase and use of chemical fertilizers, which is consistent with the *distribution effect* hypothesis presented in Chapter 2. Also, since sharing the cost of purchased inputs between the tenant and landowner is a common feature associated with sharecropping arrangements, the cost (and risk) of applying chemical fertilizers and improved seeds is reduced for each party, which increases the likelihood of using chemical fertilizers and improved seeds on sharecropped plots.

Next we investigate whether the differences in land investments in Tigray (stone terraces) and land management practices in both regions on leased versus owned plots are found after accounting for the effects of other factors such as differences in household endowments and land quality, using econometric analysis.

Econometric Results

Determinants of Investment in Stone Terraces in Tigray Region

Econometric analysis of the determinants of investment in stone terraces shows no statistically significant difference in the probability of investment on rented in plots versus owner-operated plots (Table 11-3). This and most other results are consistent between the univariate and bivariate probit models, indicating that endogeneity of the lease variable is not seriously affecting the results of the univariate probit model.[11] The lack of significance of land tenure as a factor in explaining differences in stone terrace investment is consistent with the descriptive results discussed earlier, and suggests that tenants' expectations of tenure security due to renewable short-term tenure contracts is sufficient to promote land investments.

Variables that do have a statistically significant (at 10% level or less) association with investment in stone terraces in both the univariate and bivariate models (except where noted below) include investments in stone terraces on the same plot using labor mass mobilization (negative effect (–)); literacy training (+); secondary income source of the household (+ for cereals or perennial crops production, cattle, or food for work as the secondary income, compared to households with no secondary income after cereals production);

TABLE 11-3 Determinants of private investment in stone terraces and land rental in Tigray region in 1997–99

Variable	Univariate Probit Models				Bivariate Probit Model			
	Stone Terrace		Sharecropped/Rented		Stone Terrace		Sharecropped/Rented	
	Coeff.	*Sig.*	*Coeff.*	*Sig.*	*Coeff.*	*Sig.*	*Coeff.*	*Sig.*
Plot tenure (cf., owned)								
Sharecropped/rented	0.148				−1.067			
Labor mass mobilization investment in stone terraces on plot	−0.403	**			−0.390	**		
Household characteristics								
Share of dependents	−0.922	*	0.676		−0.537		0.618	
Education of household head (cf., none)								
1–2 years	0.335		0.505	**	0.398	*	0.479	**
3 or more years	−0.617	*	0.692	***	−0.399		0.662	***
Literacy training	0.492	**	−0.189		0.394	*	−0.168	
Secondary income source (cf., none)								
Cereals	1.097	***	−0.197		0.972	*	0.064	
Perennial crops	0.561	*	0.827	**	0.664	**	0.862	**
Cattle	0.365	*	0.413	*	0.338	*	0.444	*
Small ruminants	−0.317		NE		−0.644		−7.174	**
Beekeeping	−0.009		1.392	***	0.200		1.406	***
Food for Work	0.943	**	−0.248		0.772	*	−0.164	
Salary employment	0.371		1.538	***	0.677		1.537	***
Farm employment	NE		0.778	**	−7.727	***	0.763	***
Trading	0.483	*	−0.338		0.253		−0.201	
Food aid/other assistance	NE		−0.192		−6.545	***	−0.156	
Membership in organizations								
Farmers' association	0.155		0.612	***	0.235		0.566	***

Women's association	–0.081		–0.466	**	–0.188		–0.513	**
Youth association	0.485	***	0.232		0.463	***	0.276	*
Tabia (PA) council	–0.871	*	0.407		–0.523		0.344	*
Village council	1.246	***	–0.822**	**	0.949	*	–0.782	
Agricultural cadre	0.322		–1.852***	***	0.062		–2.269	***
Assets owned								
Land (ha)	–0.050		–0.034		–0.065	*	–0.038	
Other cattle (number)	–0.059	***	0.024	*	–0.045	**	0.029	*
Pack animals (number)	0.195	**	0.172	**	0.212	***	0.157	*
Walking time to roads and markets (hours)								
Seasonal road	0.209	**	–0.098		0.179	*	–0.088	
Woreda (district) town	–0.078	*	–0.088	**	–0.080	*	–0.086	*
Residence	–0.395	**	–0.420	***	–0.479	**	–0.439	***
Average annual rainfall (mm)	–0.002	**	0.002	**	–0.001		0.001	**
Population density (persons/km^2)	–0.001		–0.003	**	–0.002	*	–0.003	**
Altitude (m.a.s.l.)	–0.001	**	0.000		–0.001	**	0.000	
Slope of plot (cf., flat)								
Gentle	0.453	**	0.006		0.475	**	0.007	
Steep	0.665	**	–0.371		0.553		–0.452	*
Position on slope (cf., top)								
Bottom	0.632	***	–0.494	**	0.437		–0.506	**
Plot quality rank (1 = highest quality)	–1.115	***	1.271	***	–0.535		1.272	***
Area of plot (ha.)	0.166		0.718	**	0.337		0.679	*
Soil color (cf., black)								
Grey	0.013		0.510	*	D		0.516	**
Waterlogging problem on plot	0.330	*	–0.048		0.245		–0.036	
Gullies on plot	0.532	*	–0.421		0.339		–0.475	

Continued

TABLE 11-3 *(Cont.)*

Variable	*Univariate Probit Models*				*Bivariate Probit Model*			
	Stone Terrace		*Sharecropped/Rented*		*Stone Terrace*		*Sharecropped/Rented*	
	Coeff.	*Sig.*	*Coeff.*	*Sig.*	*Coeff.*	*Sig.*	*Coeff.*	*Sig.*
Land use (cf., homestead plot)								
Rainfed cultivated	–0.291	*	1.253	***	D		1.192	***
Irrigated cultivated	–0.397		0.767	**	D		0.720	**
Initial investments on plot in 1997								
Stone terrace	0.110		–0.293	**	D		–0.283	**
Constructed fence	0.378		–0.877	*	D		–1.095	**
Intercept	0.621		–4.315	***	0.210		–4.105	***
Rho					0.736			
Number of positive observations/ Number of observations	121/1395		201/1420		121/1452		201/1452	
Pseudo R^2	0.2395		0.3036					

Note: Variables with insignificant coefficients have been omitted to reduce the size of the table. These include: household characteristics (gender and age of head, gender share in labor supply, size of household), some secondary income sources (perishable crops, other non-farm activities), membership in marketing cooperative, ownership of some livestock (oxen, small ruminants, camels), and radio, position of slope (middle, elsewhere), soil color (brown, red), access to certain infrastructure/services (all-weather road, bus), perception of soil depth, texture, and fertility, land use (fallow), and presence of some land investments (soil bund, trees, live fence).

*, **, *** mean coefficient is statistically significant at 10%, 5%, and 1% levels, respectively.

D = Variables dropped from bivariate probit model to enable identification. A Wald test of the significance of these variables in the univariate model for stone terrace investment indicated these were jointly insignificant (P = 0.1361).

NE = Coefficient not estimable because no investment occurred for positive values of this explanatory variable. These observations and this variable were dropped from the estimation.

membership in a youth association or village council (+); ownership of cattle other than oxen (–) and pack animals (+); walking distance to a seasonal road (+), district town (–) or to the farmer's residence from the plot (–); elevation (–); slope of the plot (greater on gentle or steep slopes than flat plots, but significant only in the univariate model); slope position (greater on bottom than top of slope); and plot quality (greater on higher quality plots, but significant only in the univariate model).

These results are generally as one might expect. The negative coefficient for investment via labor mass mobilization indicates that such investments displace private investment, although the magnitude of this coefficient (less than one in absolute value) indicates that the crowding out effect is not complete. The positive effect of literacy training and membership in some organizations may be because education and social capital increase farmers' awareness of the benefits of soil and water conservation measures (Ervin and Ervin 1982; Pender and Kerr 1998). The positive effect of various secondary income sources may be because such income sources enable farmers to overcome financial constraints and high discount rates that may undermine the propensity to invest (Pender 1996; Holden et al. 1998; Clay et al. 1998). The negative association of cattle ownership with investment may be because farmers pursuing greater cattle production are less interested in investing in improving crop production and/or because cattle grazing may be damaging to terrace investments. The association of investments with better access to towns and the farmer's residence suggests that better economic activity and convenience of access to the plot promote investment by increasing the net benefits. However, it is not clear why terrace investments are less common closer to a seasonal road but more common closer to a town. Stone terrace investments are less common at the top of a slope, at higher elevations, and on flat terrain probably because they serve less purpose and provide less benefit in such circumstances. Investments are greater on farmers' higher-quality plots probably because farmers expect greater returns from investing on such plots.

The factors significantly associated with sharecropped or rented plots, compared to owner-operated plots, include education of the household head (+); secondary income source (+ for households with perennial crops, cattle, beekeeping, salary employment or farm employment as a secondary income source, negative for households with small ruminants as a secondary source); membership in a farmers' association (+), women's association (–), or agricultural cadre (–); ownership of cattle other than oxen (+) and pack animals (+); distance to a district town or to the farmer's residence (–), average rainfall (+); population density (–); position on slope (less common on bottom than top); plot quality rank (+); prior land use (more common for cultivated outfields than homestead plots); and the presence of initial investments on the plot (less common where stone terraces or a fence already on the plot). Again, most of these results are as one would expect. Households with more education, income sources, and assets, and with better access to markets, and better agroecological conditions are better able to rent in land. The plots that they rent in tend to be close to their homestead, but households are not likely to rent in their homestead plots or plots with substantial investments already on them. The plots that are rented in tend to be of lower quality than owner-operated plots, consistent with our asymmetric information hypothesis. This last finding suggests that farmers may tend to invest less on rented in plots because of plot quality differences, separately from any effect of tenure security. However, our descriptive results suggest that any such effects are not quantitatively very large, given the finding of little difference in the propensity of private investments to occur on rented in plots versus owner-operated plots.

Land Management Practices in Tigray Region

As for investments in stone terraces, we find statistically insignificant associations between land tenure and use of different land management practices, including use of fertilizer, manure, contour plowing, and reduced tillage (Table 11-4). Also as for stone terraces, our results are very similar between the univariate and bivariate probit models.

Several household characteristics influence several land management practices. For example, households with a larger female share of the household labor supply are less likely to apply fertilizer or use contour plowing, probably due to both labor and capital constraints, as well as a cultural taboo in Ethiopia against women using oxen to plow. Households for whom cereals are a secondary income source are less likely to apply manure and more likely to use reduced tillage, possibly because they have higher labor opportunity costs due to other primary economic activities. Households with farm employment as a secondary income source are more likely to use fertilizer, manure, and contour plowing, probably because they have relatively low labor opportunity costs and need to intensify their food production. Members of a farmers' association are less likely to apply manure but more likely to use contour plowing; the reasons for this are not clear. Members of a marketing cooperative are more likely to use contour plowing and reduced tillage; apparently they are more oriented toward conservation than other farmers. Members of an agricultural cadre are more likely to use fertilizer and less likely to use manure; such cadres are supposed to be "progressive farmers" and hence tend to substitute modern methods for more traditional means of soil fertility management. Farmers with more land are less likely to apply fertilizer or use contour plowing, but more likely to use reduced tillage. Such farmers generally farm each plot less intensively than farmers with less land. By contrast, farmers with more oxen are more likely to apply manure and less likely to use reduced tillage, since they have the means to use more intensive methods. Similarly, farmers with more cattle besides oxen are also more likely to apply manure and use contour plowing.

TABLE 11-4 Determinants of land management practices in Tigray region in 1999

	Fertilizer				*Manure*				*Contour Plowing*				*Reduced Tillage*			
	Probit		*Biv. Probit*		*Probit*		*Biv. Probit*		*Probit*		*Biv. Probit*		*Probit*		*Biv. Probit*	
Variable	*Coeff.*	*Sig.*	*Coeff.*	*Sig.*	*Coeff.*	*Sig.*	*Coeff.*	*Sig.*	*Coeff.*	*Sig.*	*Coeff.*	*Sig.*	*Coeff.*	*Sig.*	*Coeff.*	*Sig.*
Plot tenure (cf., owned)																
Sharecropped/rented	–0.170		0.506		0.030		–0.801		–0.061		–1.362		–0.215		–0.373	
Household characteristics																
Female head	0.054		0.087		–0.486	**	–0.512	**	0.042		0.012		0.244		0.234	
Female share of labor supply	–0.566	**	–0.548	*	–0.003		0.014		–0.795	**	–0.749	**	0.238		0.238	
Household size	–0.024		–0.013		0.024		0.007		0.002		–0.012		–0.124	***	–0.126	***
Education of hh head (cf., none)																
1–2 years	–0.195		–0.244		–0.337	*	–0.256		0.325		0.393	*	0.194		0.206	
Secondary income source (cf., none)																
Cereals	0.402		0.437		–1.177	*	–1.252	**	–0.206		–0.216		1.356	***	1.345	***
Perishable annual crops	–0.256		–0.233		–0.235		–0.236		–0.292		–0.272		0.624	*	0.623	*
Cattle	–0.068		–0.098		0.047		0.085		0.165		0.230		0.362	*	0.369	*
Small ruminants	–0.708		–0.667		–0.406		–0.454		NE		7.321	***	0.187		0.178	
Beekeeping	–0.400		–0.656		–0.390		–0.041		–1.187	**	–0.514		0.574		0.618	
Food for Work	0.312		0.355		0.043		–0.016		–0.090		–0.113		1.017	***	1.002	***
Farm employment	1.419	***	1.310	***	1.766	***	1.819	***	NE		6.146	***	0.318		0.331	
Trading	0.220		0.268		0.158		0.085		–0.382		–0.360		0.824	***	0.811	***
Food aid/other assistance	–0.403		–0.381		–0.061		–0.085		0.016		–0.055		1.060	***	1.051	***
Membership in organizations																
Farmers' association	0.187		0.146		–0.673	***	–0.589	***	0.389	*	0.409	**	0.152		0.161	
Tabia (PA) council	–0.143		–0.254		–0.465		–0.375		NE		6.052	***	–0.042		–0.018	
Village council	0.030		0.046		0.708	**	0.676	*	0.851		0.794		–0.110		–0.113	
Marketing cooperative	–0.086		–0.144		0.088		0.160		0.881	***	0.854	***	0.517	***	0.530	***
Agricultural cadre	0.859	**	0.817	**	–1.042	**	–1.014	**	0.860		0.725		0.006		0.004	
Assets owned																

Continued

TABLE 11-4 *(Cont.)*

Variable	Fertilizer				Manure				Contour Plowing				Reduced Tillage			
	Probit		Biv. Probit		Probit		Biv. Probit		Probit		Biv. Probit		Probit		Biv. Probit	
	Coeff.	Sig.	Coeff.	Sig.	Coeff.	Sig.	Coeff.	Sig.	Coeff.	Sig.	Coeff.	Sig.	Coeff.	Sig.	Coeff.	Sig.
Land (ha)	−0.058	**	−0.051	*	−0.017		−0.021		−0.052	*	−0.055	*	0.093	***	0.092	***
Oxen (number)	0.021		0.018		0.210	***	0.205	***	−0.151		−0.136		−0.197	**	−0.196	**
Other cattle (number)	0.026		0.020		0.062	***	0.069	***	0.074	**	0.079	***	−0.009		−0.007	
Small ruminants (number)	−0.007		−0.008		0.007		0.008		0.029	***	0.029	***	0.021	**	0.022	***
Camels (number)	−0.015		−0.027		−0.105		−0.084		0.471	*	0.403	*	−0.029		−0.026	
Radio	0.061		0.068		−0.029		−0.045		−0.435	**	−0.437	**	0.191		0.189	
Walking time to roads and markets (hours)																
All-weather road	−0.073		−0.062		0.013		0.005		0.044		0.028		0.230	***	0.227	***
Seasonal road	−0.029		−0.034		−0.159		−0.141		−0.020		−0.010		−0.243	**	−0.242	**
Residence	−0.445	***	−0.403	***	−1.764	***	−1.755	***	−0.329	**	−0.342	**	0.655	***	0.649	***
Average annual rainfall (mm)	−0.002	***	−0.002	***	−0.004	***	−0.003	***	0.001		0.001		0.000		0.000	
Population density (#/km^2)	0.003	***	0.003	***	0.002	*	0.001		0.001		0.001		0.000		0.000	
Altitude (m.a.s.l.)	−0.001	***	−0.001	***	0.000		0.000		0.001	**	0.000	*	0.000		0.000	
Slope of plot (cf., flat)																
Gentle	0.162		0.114		−0.376	**	−0.314	*	0.164		0.201		0.003		0.011	
Position on slope (cf., top)																
Middle	0.263		0.235		−0.004		0.014		−0.106		−0.087		−0.527	**	−0.522	**
Bottom	0.156		0.168		−0.015		−0.036		0.040		−0.021		−0.409	**	−0.414	**
Not on slope	0.298		0.276		−0.730	***	−0.690	***	0.043		0.038		−0.311		−0.308	
Plot quality rank	−0.476	***	−0.602	***	−1.147	***	−0.929	**	0.217		0.398		0.009		0.040	
Area of plot (ha.)	0.598	***	0.589	***	0.275		0.277		0.416		0.416		0.096		0.094	
Soil depth (cf., deep)																
Medium	0.204		0.189		−0.394	**	−0.370	**	0.054		0.045		0.174		0.176	
Shallow	0.594	***	0.579	***	−0.077		−0.071		0.117		0.103		0.614	***	0.614	***
Soil color (cf., black)																

Brown	0.039	0.017	0.449 *	0.436 *	−0.951 ***	−0.903 ***	−0.045	−0.038
Grey	−0.321 *	−0.332 *	0.453 *	0.441 *	−0.679 ***	−0.659 ***	0.183	0.191
Red	−0.142	−0.137	−0.088	−0.112	−0.652 ***	−0.668 ***	−0.021	−0.017
Soil texture (cf., clay)								
Loam	0.219	0.213	−0.291	−0.257	0.214	0.227	0.545 **	0.542 **
Sand	0.404 **	0.391 **	0.008	0.043	0.152	0.190	−0.146	−0.149
Silt	0.738 ***	0.705 ***	−0.271	−0.213	0.212	0.262	−0.068	−0.064
Soil fertility (cf., high)								
Infertile	−0.591 ***	−0.610 ***	0.425 *	0.483 **	−0.263	−0.216	−0.367	−0.355
Land use (cf., homestead plot)								
Rainfed cultivated	0.219 *	0.119	−1.745 ***	−1.565 ***	−0.324	−0.252	−0.277 *	−0.256
Irrigated cultivated	0.677 ***	0.628 ***	−1.250 ***	−1.161 ***	−0.178	−0.155	0.436	0.442
Initial investments on plot in 1999								
Stone terrace	0.268 ***	0.282 ***	0.228	0.193	0.004	−0.012	−0.129	−0.133
Soil bund	0.361 **	0.370 **	0.345 **	0.304 *	−0.045	−0.080	−0.332	−0.336
Trees planted	0.394 *	0.415 *	−0.304	−0.328	0.057	−0.047	0.794 ***	0.788 ***
Constructed fence	−0.589 **	−0.560 **	0.953 ***	0.891 ***	0.134	0.085	−0.019	−0.023
Intercept	1.863 **	1.821 **	4.129 ***	4.000 ***	0.064	0.178	−1.760	−1.752
Rho		−0.405		0.525		0.718		0.097
Number of positive observations/ Number of observations	523/1460	523/1460	380/1457	380/1457	1299/1392	1364/1457	195/1457	195/1457
Pseudo R^2	0.1813		0.4888		0.1794		0.2964	

Note: Variables with insignificant coefficients in any of the investment regressions have been omitted to reduce the size of the table. These include: composition of household (age of head, share of dependents), education of head (3 or more years, literacy training), some secondary activities (perennial annual crops, salary employment, other non-farm), membership in women's and youth associations, ownership of pack animals, access to services (distance to bus service and district town), slope of plot (steep), perception of moderate soil fertility, problems associated with waterlogging and gullies, and presence of live fence.

*, **, *** mean coefficient is statistically significant at 10%, 5%, and 1% levels, respectively.

NE = Coefficient not estimable because contour plowing used for all positive values of this explanatory variable. These observations and this variable were dropped from the estimation.

Plot level factors also influence land management. On plots that are closer to the farmer's residence, the farmer is more likely to use fertilizer, manure, and contour plowing, and less likely to use reduced tillage. Generally, more intensive methods are used on more accessible plots, as one would expect. Both fertilizer and manure use are less common in higher rainfall areas. Since greater rainfall generally increases the return and can reduce the risk of using inorganic fertilizer, the negative impact on fertilizer use is surprising. The negative effect is less surprising for manure use, since manure can help conserve soil moisture in dryer areas. Fertilizer use is less common but contour plowing is more common at higher elevations, probably because erosion risk is greater on higher and usually steeper slopes. Both fertilizer and manure are used less on lower-quality plots, again probably because the returns to such measures are greater on better-quality plots. Fertilizer use and reduced tillage are more common on plots with shallower soils, probably because such methods are more necessary on shallow soils. Fertilizer is less likely to be used and manure more likely on plots with soils judged to be infertile, possibly because the near term response to inorganic fertilizer is low on infertile soils, while manure may be more effective as a means of rehabilitating fertility over a longer term. Fertilizer use is more common and manure use less common on irrigated plots, probably because the response to fertilizer is quite favorable on such plots. Use of fertilizer is more common on plots with stone terraces, and both fertilizer and manure are more common on plots with soil bunds, suggesting complementarity between such land investments and soil fertility management practices because these help to conserve soil moisture as well as reduce losses of soil nutrients through runoff. Fertilizer and reduced tillage are more likely to be used on plots where trees have been planted; the latter association may be because trees make tillage operations more difficult. Fertilizer is less likely and manure more likely to be used on plots with a fence; this may be because animals are kept on such plots during certain times of the year, increasing the availability of manure and reducing the need for inorganic fertilizer. There are additional associations between particular explanatory variables and particular land management practices that we will not discuss here.

Overall, these results demonstrate that many factors influence farmers' land investments and annual land management practices in Tigray, particularly those that affect the costs, returns, and risks of such measures or farmers' capacities to undertake them. Land tenure, by contrast, has no measurable impact on these practices.

Land Management in Amhara Region

Contrary to the Tigray region, the results for the Amhara region show that not accounting for endogeneity of land rental leads to falsely concluding that land tenure has no effect on the land management practice decisions analyzed here (i.e., use of reduced tillage, contour plowing, crop residues, and crop rotation) (Table 11-5). Regarding use of fertilizer, however, land tenure has no significant effect and the endogeneity of rental decision does not matter. After accounting for endogeneity of the rental decision, the incidence of use of contour plowing, crop residues, and crop rotation was significantly higher on owner-cultivated plots than on rented plots, which is consistent with the greater likelihood of using reduced tillage practices on rented plots.

The results presented in Table 11-5 also show that there are several other factors that affect land management practice decisions, and the impacts are not substantially affected by controlling for endogeneity of the rental decision. That is, with the exception of very few variables, the magnitudes and signs of the coefficients of the variables are very similar in the both univariate and bivariate probit models. Several of the findings here are consistent with those in the case of Tigray, although some are contradictory also. We discuss some of the key results only.

Larger plots were associated with greater likelihood of using crop rotation and chemical fertilizers and improved seeds, but lower likelihood of using contour plowing. More sloped plots, on the other hand, were associated with a lower likelihood of being tilled or using chemical fertilizers and a greater likelihood of being contour plowed when tilled, which are expected. Irrigated plots were surprisingly associated with a lower likelihood of using chemical fertilizers, but, as expected, the presence of stone terraces was associated with a greater likelihood of using chemical fertilizers, which is similar to the finding in Tigray region and suggests complementarity between such soil and water conservation investments and soil fertility management practices.

Consistent with the finding in the Tigray region, female-headed households were more likely to use reduced tillage and less likely to plow in crop residues or use chemical fertilizers on their farm plots, which is expected given the customary prohibition of women in undertaking oxen plowing in many places in the highlands of Ethiopia. More educated households were less likely to plow in crop residues but more likely to till their plots many times and use chemical fertilizers. The latter finding is perhaps because education enhances the ability of individuals to utilize technical information associated with use of such modern inputs. Ownership of livestock or where livestock is a secondary source of income was associated with greater tillage, which is expected, but is also associated with a lower likelihood of using chemical fertilizers, perhaps because of the potential of applying manure obtainable from the livestock. Compared to others, households with nonfarm employment as the secondary income source are less likely to practice crop rotation but more likely to use reduced tillage or apply chemical fertilizers. Several of these findings are consistent with those found in the Tigray region.

The impacts of membership in various organizations and access to different public services were mixed, suggesting differences in the underlying incentives to farmers of those factors in promoting or hindering use of different land management practices. For example, being farther away from an

TABLE 11-5 Determinants of land management practices in Amhara region in 1999

	Reduced Tillage				*Contour Plowing*				*Crop Residues*				*Crop Rotation*				*Fertilizers*			
	Probit		*Bivariate Probit*		*Probit*		*Bivariate Probit*		*Probit*		*Bivariate Probit*		*Probit*		*Bivariate Probit*		*Probit*		*Bivariate Probit*	
Explanatory variable	*Coeff.*	*Sig.*	*Coeff.*	*Sig.*	*Coeff.*	*Sig.*	*Coeff.*	*Sig.*	*Coeff.*	*Sig.*	*Coeff.*	*Sig.*	*Coeff.*	*Sig.*	*Coeff.*	*Sig.*	*Coeff.*	*Sig.*	*Coeff.*	*Sig.*
Plot characteristics																				
Tenure (owned vs. rented)	−0.332		−1.638	***	0.064		1.980	***	−0.057		1.502	***	0.259		1.750	***	0.516		−0.090	
Expectation to operate	0.260		0.554		−0.209		−0.562	**	−0.100		−0.335		0.446		0.187		−0.756		−0.683	
Walking time to residence	0.001		0.000		0.003		0.006	*	0.002		0.004	*	−0.007	***	−0.004		−0.004		−0.005	
Ln Size	0.028		0.050		−0.184	**	−0.217	***	0.139	*	0.088		0.277	***	0.214	***	0.323	***	0.338	***
Ln Slope	0.029	***	0.030	***	0.078	***	0.072	***	0.008		0.007		0.006		0.004		−0.049	***	−0.047	***
Soil depth (cf. deep)																				
Medium	0.629	***	0.548	**	−0.014		0.152		0.111		0.221		−0.102		0.014		−0.005		−0.060	
Shallow	0.671	***	0.538	*	0.196		0.361	*	0.310		0.418	**	−0.123		0.024		0.041		−0.033	
Soil color (cf. black)																				
Brown	−0.253		−0.265		0.368	*	0.379	**	0.235		0.234		−0.059		−0.022		−0.043		−0.059	
Gray	0.109		0.028		0.341		0.497	*	0.172		0.272		−0.213		−0.045		0.109		0.025	
Red	0.393	*	0.402	**	0.362	*	0.280		0.185		0.172		0.231		0.193		0.389	*	0.387	*
Presence of gullies	0.326		0.313		−0.124		−0.174		−0.336		−0.334		0.092		0.044		−0.557	*	−0.507	
Waterlogging problem	0.070		0.082		0.292		0.200		0.512	***	0.430	**	−0.209		−0.254		−0.045		−0.012	
Presence of investments																				
Irrigation	−0.206		−0.180		−0.132		−0.189		−0.161		−0.231		0.656	*	0.527		−1.509	***	−1.440	***
Stone terrace	−0.117		−0.065		1.088	***	0.980	***	−0.106		−0.106		0.069		0.031		0.680	***	0.677	***
Drainage ditch	0.021		−0.050		0.343	**	0.442	***	0.253	*	0.323	***	−0.309	**	−0.150		0.284	*	0.234	
Live fence	−0.892	***	−0.794	***	0.074		−0.063		−0.386	*	−0.575	***	0.023		−0.140		−0.041		0.043	
Fence	−0.173		−0.191		0.843	***	0.802	***	−0.075		−0.010		0.142		0.178		−0.144		−0.162	
Household characteristics and assets																				
Male head	−1.357	***	−1.287	***	0.087		0.063		0.966	***	0.780	***	−0.166		−0.141		0.957	**	0.944	*

Ln Age of head	−0.773 **	−0.692 *	−0.483	−0.556 *	−0.554 **	−0.612 **	−0.220	−0.269	−1.134 ***	−1.075 **
Average education	−0.097 **	−0.110 ***	−0.008	0.025	−0.086 ***	−0.063 *	−0.036	−0.008	0.087 **	0.077
Ln Household size	0.507 *	0.515 *	0.134	0.206	0.309	0.255	0.289	0.242	0.341	0.337
Dependency ratio	−0.201	−0.204	−1.796 ***	−1.789 ***	−0.175	−0.086	−0.487	−0.490	−1.057 *	−1.014 *
Ln Size of farmland	0.286 *	0.253 *	0.152	0.150	−0.076	−0.035	0.013	0.063	0.006	−0.007
Number of oxen	0.212 *	0.152	0.019	0.121	−0.055	0.027	−0.117	−0.045	0.338 ***	0.303
Tropical Livestock Units	−0.103 **	−0.083 *	−0.033	−0.065	0.036	0.013	0.071	0.044	−0.035	−0.025
Secondary income source (cf. non-farm)										
Cereals	−1.098 *	−1.245 **	−0.251	−0.055	−1.861 ***	−1.523 ***	1.783 ***	1.876 ***	−0.164	−0.242
Cash crops (annuals)	−0.927 *	−1.049 **	0.431	0.451	1.688 ***	1.696 ***	0.928 ***	0.978 ***	−1.114 **	−1.135 **
Cattle	−1.278 ***	−1.425 ***	0.310	0.474	−0.584	−0.347	0.775 ***	0.901 ***	−0.827 **	−0.906 *
Small ruminants	−1.320 ***	−1.454 ***	0.149	0.332	−0.622	−0.414	0.498	0.611 *	−1.164 ***	−1.211 ***
Membership in organizations/co-ops										
Input	1.593 ***	1.557 ***	−1.067 ***	−0.906 ***	2.742 ***	2.438 ***	−0.021	−0.084	1.867 ***	1.865 ***
Marketing	−0.587	−0.488	0.699 *	0.440	−1.938 ***	−1.782 ***	1.344 ***	1.265 ***	−1.437 ***	−1.415 ***
Women/youth	−1.700 ***	−1.449 ***	0.592	0.365	−3.524 ***	−3.272 ***	−0.131	−0.299	−1.138 ***	−1.072 **
Church	0.153	0.171	−1.329 ***	−1.316 ***	−1.187 ***	−1.124 ***	−0.390	−0.357	0.298	0.273
Water users	0.634 ***	0.548 ***	0.023	0.098	−0.555 ***	−0.397 ***	0.118	0.221	−0.315 *	−0.359
Village level and other fixed factors										
Access (km) to services										
Ln All-weather road	0.468 ***	0.446 ***	0.197 ***	0.194 ***	−0.138 *	−0.127 *	−0.175 **	−0.135 *	0.111	0.107
Ln Development agent	−0.004	−0.005	−0.133 *	−0.127 *	0.077	0.066	−0.085	−0.079	−0.362 ***	−0.354 ***
Ln Bus service	−0.471 ***	−0.463 ***	−0.023	−0.025	0.209 **	0.223 **	0.219 **	0.217 **	−0.020	−0.031
Ln District town	−0.179	−0.101	−0.270 **	−0.321 ***	0.800 ***	0.598 ***	0.014	−0.101	0.111	0.151
Land redistribution	−0.057	0.002	−0.030	−0.050	0.190	0.080	−0.640 ***	−0.684 ***	0.153	0.208
Ln Household density	−1.250 ***	−1.176 ***	−0.034	−0.029	0.773 ***	0.688 ***	0.372 ***	0.362 ***	0.139	0.129
Ln Altitude	1.346 ***	1.536 ***	1.749 ***	1.460 ***	1.146 **	0.954 **	−0.577	−0.758 *	2.183 ***	2.266 ***

Continued

TABLE 11-5 *(Cont.)*

	Reduced Tillage				*Contour Plowing*				*Crop Residues*				*Crop Rotation*				*Fertilizers*			
	Probit		*Bivariate Probit*		*Probit*		*Bivariate Probit*		*Probit*		*Bivariate Probit*		*Probit*		*Bivariate Probit*		*Probit*		*Bivariate Probit*	
Explanatory variable	*Coeff.*	*Sig.*	*Coeff.*	*Sig.*	*Coeff.*	*Sig.*	*Coeff.*	*Sig.*	*Coeff.*	*Sig.*	*Coeff.*	*Sig.*	*Coeff.*	*Sig.*	*Coeff.*	*Sig.*	*Coeff.*	*Sig.*	*Coeff.*	*Sig.*
Ln Rainfall	1.520	***	1.174	**	1.277	***	1.367	***	−1.314	***	−0.834	***	0.013		0.325		0.192		0.066	
Sub-regional location (cf. Southern)																				
Northern	−0.485	**	−0.523	**	0.053		0.143		−1.776	***	−1.512	***	0.143		0.199		0.869	**	0.836	*
Western	−4.483	***	−4.203	***	1.086	***	0.951	***	−3.333	***	−2.908	***	0.638	**	0.664	**	0.879	**	0.841	*
Eastern	1.286	***	1.176	***	0.764	***	0.811	***	−2.287	***	−1.898	***	−0.770	***	−0.594	***	1.121	***	1.059	**
Intercept	−13.627	***	−12.388	***	−19.629	***	−19.246	***	1.530		−1.345		2.173		0.325		−17.670	***	−17.002	**
Pseudo-R^2	0.439				0.307				0.405				0.200				0.493			
Rho			0.620	**			−0.945	**			−0.820	***			−0.838	***			0.351	
Number of positive observations/Number of observations	190/1187				878/1187				711/1187				712/1187				351/1187			

Note: Variables with insignificant coefficients in any of the land management regressions have been omitted to reduce the size of the table. These include: distance to the homestead plot, gender share of labor supply, and secondary income associated with perennial cash crops.

Ln means variable is transformed by natural logarithm.

Sub-regional location of household: southern = North Shewa zone; northern = North/South Gondar zones; western = Awi/East/West Gojjam zones; and eastern = North/South Wollo zones.

*, **, *** mean coefficient is statistically significant at the 10%, 5% and 1% level, respectively.

all-weather road was associated with greater likelihood of using reduced tillage and a lower likelihood of plowing in crop residues or practicing crop rotation, whereas being farther away from a bus service or the district town had an opposite effect. However, being closer to a development agent was associated with greater likelihood of using chemical fertilizers, as expected. More densely populated villages were associated with a lower likelihood of using reduced tillage and a greater likelihood of plowing in crop residues or practicing crop rotation, with the high labor to land ratios driving intensification of food production. Households located in the southern part of the Amhara region (i.e., North Shewa zone) had greater likelihood of using reduced tillage or plowing in crop residues but a lower likelihood of using contour plowing or chemical fertilizers.

Overall, these results demonstrate that land tenure is one of several factors that influence farmers' land management practices in the Amhara region. Other relevant factors affect the costs, returns, and risks of undertaking such practices or farmers' capacities in undertaking them.

Conclusions and Implications

Using community, household, and plot level survey data from the highland areas of Ethiopia, this chapter examined the impacts of land rental on land management practices. With the existing ban on land sales and restrictions on the duration of land leases in this region, only short-term rentals, mainly sharecropping arrangements, were used by farmers. In the Tigray region we find statistically insignificant differences between rented and owner-operated plots in farmers' private investments in stone terraces and in use of annual land management practices. In the Amhara region, by contrast, we find less use of various labor-intensive land management practices—contour plowing, incorporation of crop residues, and crop rotation—and more use of reduced tillage on rented plots.

These results highlight the dependence of impacts of land tenancy arrangements on the local context and policies, and suggest that the rental market functions better in the Tigray region than in the Amhara region in terms of promoting use of sustainable land management practices. In the Tigray region, most tenants feel sufficiently confident about their long-term ability to renew their lease contracts that they have incentive to make investments in land improvements and to adopt more sustainable land management practices, despite the short-term nature of the contracts. In the Amhara region, by contrast, tenants appear to have a shorter-term perspective. This may be because the relatively recent (at the time of the surveys) experience of land redistributions in the Amhara region has caused landowners to be reluctant to renew leases with the same tenants over an extended period. Thus, even though the current policy of the regional government of the Amhara region is not to have further land redistributions (and this is being supported by a program of land certification), at the time of the surveys analyzed in this chapter the 1997–1998 land redistribution was still fresh in the minds of landowners, and they feared further redistributions.

These findings imply that land rental markets need not inhibit even longer-term investments in sustainable land management, if the policy environment is sufficiently favorable, as in the Tigray region. Where the tenure insecurity of landlords has been undermined by land redistributions, as in the Amhara region, the ability to assure sustainable land management on rented land is more likely to be compromised. Avoiding such policies in the future can thus help to assure more sustainable land management. The land registration and certification that also started in the Amhara region (after this survey) may have positive impacts on land management, as it is likely to enhance tenure security. The most recent land laws in Ethiopia also emphasize monitoring and enforcement of more sustainable land management through establishment of Land Administration Committees at the community level. Future research should investigate the outcomes of these new policy initiatives.

NOTES

1. These specifications are similar to the specifications used in Nkonya et al. (2004), Jansen et al. (2006), Pender and Gebremedhin (2006), Pender and Gebremedhin (2007), and Benin (2006). The derivation of the specification is discussed in detail in those references. Some explanatory variables that were used in Pender and Gebremedhin (2006, 2007) were dropped in this analysis because they were found to be statistically insignificant in those studies, and present potential concerns about endogeneity (e.g., participation in the agricultural extension program, use of credit).
2. The analysis of stone terrace investment was not done for the Amhara region due to lack of adequate recall data on the explanatory variables and the initial stock of investment in a prior year for the Amhara region.
3. The local community or peasant administration (PA), which is called *kebele* and *tabia* in the Amhara and Tigray regions, respectively, is the lowest unit of government.
4. Exclusion restrictions are not strictly necessary in the first stage of a recursive bivariate probit model due to the nonlinearity of the model (Wilde 2000). Nevertheless, such restrictions can improve the identification of the model.
5. PAs that are predominantly below 1,500 masl in elevation or have more than 50% of their total area below 1,500 masl were excluded from the sample frame.
6. As noted above, we judged the 1991 recall data for the Amhara region to be inadequate for use in the econometric analysis.
7. As a result of the predominant *rist* land tenure system that existed in northern Ethiopia prior to the 1975 land reform (which also involved periodic land redistribution and prohibited land sales and mortgages), land ownership was not greatly unequal in most places even before 1975 (Bruce et al. 1994).

8. Only 6 of 252 rented in plots in the Tigray survey were rented in for cash rent; the rest were sharecropped.
9. At the time of the surveys, U.S. $1 ≈ Ethiopian Birr 8.50.
10. On rented plots, "bequeath" refers to the expectation of transferring the rental contract to an heir in the event that the current contract holder is unable to continue farming.
11. Also consistent with this is the finding shown in Table 11-3 that the estimated correlation between the error terms in the two equations in the bivariate model is statistically insignificant, although the estimated correlation is positive in magnitude.

CHAPTER 12

Land Rental Markets and Land Management

Evidence from Uganda

EPHRAIM NKONYA, JOHN PENDER, SAMUEL BENIN, AND EDWARD KATO

Uganda has implemented innovative economic and agricultural reforms, making the country one of the showcase studies in Africa (Dijkstra and van Donge 2001). In the case of land markets, Uganda's Land Act of 1998 is one of the few land laws enacted in Africa with an objective of giving land users full ownership and rights in order to promote a more secure and free land market. This is contrary to the land policies in many African countries where the land is owned by the government and the land operators are given only user rights (e.g., Burkina Faso, Ethiopia, Malawi, Rwanda, Tanzania, and Zambia). In Chapter 7 we learned that both land sales markets and land rental markets are active in Uganda. Chapter 11 investigated how the more restricted land rights in Ethiopia with prohibition of land sales and restrictions on duration of land rental contracts may affect management of rented land. In this chapter we assess land management on rented land under the less restrictive Ugandan land policy regime.

There are four major land tenure types in Uganda (customary, *mailo*, freehold, and leasehold) and each type has its own associated land rights and obligations (Republic of Uganda 1998) and sense of tenure security. Customary land tenure, as the name implies, is regulated by customary rules and is the most common land tenure system in Uganda and often involves limitations on the holder's right to sell or mortgage land, though usufruct and bequest rights are usually secure. Transfer of rights under this system is mainly through inheritance, and landholders require approval of clan chiefs regarding any land transaction. However, under the 1998 Land Act (Republic of Uganda 1998), a customary landholder may apply for a certificate of ownership from the District Land Board and, once such a certificate is issued, the landholder may lease, mortgage, sell, give away, or bequeath the land by will. Under the freehold land tenure system, the landholder is allowed to own the land for an unlimited time and can use it for any lawful purposes, including selling or leasing, or as collateral to obtain a loan. This form of tenure provides holders with the most complete rights and the fewest obligations or restrictions on use. The *mailo* land tenure system is similar to the freehold system in the sense that the landholder is allowed to own the land for an unlimited time. However, in most cases, *mailo* land has been occupied by long-term tenants for a considerable period of time. Leasehold tenure also has some similarities to freehold, except that the rights are specified for a fixed (usually relatively long) period. The Land Act gives leasehold landholders the freedom to use the land in any lawful way he or she deems fit during the lease period, including sub-leasing the land as long as the sub-lease does not exceed the duration of the leasehold contract.

Despite this market-friendly legal framework, Uganda still faces many constraints related to land transactions, equitable access to land for vulnerable groups and the impact of land transactions on sustainable land management. One of the objectives of the 1998 Land Act is to promote more sustainable land management by formalizing the land market (Republic of Uganda 1998; Rugadya 2003). Hence, we conducted this study to determine the impact of land markets on land management and to investigate the determinants of participation in the land markets.

This study uses community, household, and plot level data to analyze the determinants of the decision to participate in the land rental market and the impact of the land rental market on land management in Uganda. The study contributes to literature since it uses plot level data that have better representation of land quality—a factor that is important in making land transaction decisions. Most past studies on land markets in Uganda and Africa have used household level data (e.g., Deininger and Mpuga in Chapter 7; Baland et al. 2007).

Analytical Methods

We build on the conceptual framework presented in Chapter 2. Two types of empirical analyses are used to assess the impact of land rental markets on land management. First, descriptive analysis of means and statistical tests are used to examine differences in various indicators of long-term land investments (e.g., soil and water conservation [SWC] measures, trees, irrigation); practices that have impacts over an intermediate term (e.g., fallowing, use of manure, incorporation of crop residues, and crop rotation); and practices that have mostly short-term effects (e.g., fertilizer application, use of slash and burn) across different land acquisition methods. Second, we use econometric analysis to examine the impact of the land rental market on selected land management practices. We control for several other factors that may affect land management practices, including household natural, physical, and human resource capital endowments, land tenure of plot, and village level factors that affect the comparative advantage of agriculture and other economic activities.

Since there are many land management practices, we selected only those that were reported by a large number of respondents. Crop rotation, slash and burn, fallow, incorporation of crop residues, and application of animal manure and inorganic fertilizer are the most common land management practices reported. Hence, we will analyze only these six land management practices. In the econometric analysis, we focus only on determinants of such land management practices that are decided each year and do not investigate determinants of longer-term land investments, since we lack information on when investments were made and on the explanatory variables prior to such investments. However, we do explore relationships between the stock of investments and land markets using descriptive statistics.

We analyzed the determinants of participation in the land rental market and land management practices assuming that the error terms of these two equations have a nonzero covariance. To address the nonzero covariance of the two models, we use the maximum likelihood bivariate probit model (Greene 2003), which is specified as follows:

$$\begin{aligned} y_1{}^* &= \beta_1 x_1 + e_1 \\ y_2{}^* &= \beta_2 x_2 + \tau y_1 + e_2 \end{aligned} \tag{12.1}$$

where $y_1{}^*$ is a latent variable determining participation in the land rental market; $y_2{}^*$ is a latent variable determining use of land management practices, $y_1 = 1$ if $y_1{}^* > 0$, $y_2 = 1$ if $y_2{}^* > 0$; $x_i =$ vectors of explanatory variables for the participation in land rental market and land management practices equations, $i = 1, 2$; β_i, and τ are coefficients associated with x_i, y_2, and y_1, respectively, $i = 1, 2$; e_1 and e_2 are error terms for the participation in land rental market and land management practices equations, which are assumed to be independently and identically distributed as bivariate normal (IIDN)—that is:

$$\begin{pmatrix} e_1 \\ e_2 \end{pmatrix} \sim IIDN\left[\begin{pmatrix} 0 \\ 0 \end{pmatrix}, \begin{pmatrix} 1 & \rho \\ \rho & 1 \end{pmatrix}\right] \tag{12.2}$$

where $\rho = cov(e_1, e_2)$.

We estimated the land management practices using specifications described in Nkonya et al. (2004) and Nkonya et al. (2005b). We also used the same specification for the decision to participate in the land rental market. The explanatory variables included in the empirical models are specified in equation (12.3):

$$x_1 = (NC_{hp}, T_{hp}, PC_h, HC_h, X_h, SVY) \tag{12.3}$$

Description of the independent variables included in x_1 is given in Table 12-1. To improve identification of the model, some of the variables included in x_1 were excluded from x_2 based on Wald tests showing some of these variables to be jointly insignificant in the land management regressions (to be discussed).

We expect that a farmer who rents a plot will be more likely to invest in short-term land management practices than longer-term ones due to the short-term nature of the lease contract. However, a farmer renting a plot may also use long-term investments to increase tenure security of the rented plot if this is allowed by the owner (Place et al. 2001). This suggests that land rental markets can affect land management practices in complex ways. We do not include land management variables (y_2) in the land rental regression since land management decisions are taken after a farmer rents a plot. The empirical model is thus recursive in structure, with land management decisions depending on land rental but not vice versa.

TABLE 12-1 Description of the independent variables used for the land management and decision to participate in the land rental market regressions

Variable	*Description*
NC_{hp}	Natural capital (size and quality land) of household *h* on plot *p*. This includes the plot slope, sand content of the soil, and land improvement investments existing on the plot, including planted trees, soil and water conservation (SWC) structures, irrigation structures, other natural resource management (NRM) investments (fish ponds, fences, paddocks, and pasture management)
PC_h	The household's endowment of physical capital, which includes the value of farm equipment and the amount of livestock in tropical livestock units (TLU)[a]
T_{hp}	The tenure category of the plot, which includes customary, *mailo*, leasehold and freehold tenure, and tenure security which is measured by the respondent's expectations of access to plot for the next 10 years.
HC_h	Human capital, which includes level of education of female and male members of the household, proportion of area owned by female members of the household, and gender and primary activity of the household head.
X_v	Village and higher-level factors influencing comparative advantage (agro-climatic potential, access to markets and roads, population density, and village level wage rate
SVY_i	Year the survey was conducted. SVY1 = 1999/00, SVY2 = 2000/01, SVY3 = 2002/03 (see details below)
e_i	Random error term

[a]A TLU is equivalent to a standard animal with live weight of 250 kg (Defoer et al. 2000). Average TLU for common livestock in Uganda area: Cow = 0.9, oxen = 1.5, sheep or goat = 0.20, and calf = 0.25.

To improve the identification of the model, we dropped some exogenous variables that were jointly insignificant using a Wald test. The value of farm equipment and the potential market integration (PMI)[1] variables were insignificant in the land management regressions and were dropped from the land management bivariate probit equations. The year the survey was conducted was dropped from the land rental equation for the same reason. These restrictions, although not strictly necessary for identification (due to the nonlinear nature of the bivariate probit model), improved the identification of the model, as observed by Wilde (2000).

Since the bivariate model specified above is recursive, it addresses the endogeneity problem of the participation in the land rental market regressor in the land management equation (Greene 2003). We tested the statistical significance of the covariance of the error terms of the land management and participation in the land rental market equations and observed it was significantly different from zero (at $p = 0.01$) for the slash and burn and animal manure equations. For inorganic fertilizer, fallow, crop rotation, and incorporation of crop residues, $cov(e_1, e_2)$ was not significantly different from zero (at $p = 0.10$). Hence, we used univariate probit regressions for these equations, treating land rental as statistically exogenous in these models.

We tested for multicollinearity of the explanatory variables and observed a maximum inflation factor of 7.75, suggesting that multicollinearity was not a serious problem (Mukherjee et al. 1998). Standard errors are corrected for heteroscedasticity using the Huber/White Sandwich estimator of the covariance matrix (Stata version 9.0).

Data

The data used in this study were obtained from three surveys conducted by the International Food Policy Research Institute in 1999–2000, 2001–2002, and 2003.[2] A stratified random sample of 270 communities (LC1s, the lowest administrative unit, usually consisting of a single village) and 4 to 10 households (1,426 in total) were randomly selected from each village. All plots operated by the selected households were surveyed and the stratification was based on indicators of agricultural potential, market access, population density, and altitude (Nkonya et al. 2004; Nkonya et al. 2005b). The total number of plots surveyed was 5,391. At the household level, information collected includes household composition, endowments of assets, income and expenditures, and adoption of agricultural and land management technologies. At the plot level, information collected includes land tenure, plot quality characteristics, land management practices, and use of inputs and outputs from the plot. The survey information was supplemented by secondary information collected from the 1991 population census and available digitized map information incorporated into a geographic information system. These data were used to determine the agricultural potential of the communities selected.

Land Markets and Land Management: Descriptive Analysis

Methods of Land Acquisition

Land markets in Uganda have been allowed to develop without much intervention from the government. About 34% of the plots surveyed were purchased, 2% rented in (predominantly by cash rental), and 57% inherited (Table 12-2). The share of rented in plots is much smaller than the share (36%) reported by Deininger and Mpuga (2002) but comparable to Baland et al. (2007), who reported that 12% of land parcels equivalent to 4% of total farm size were rented in. In general, land transactions are informal, as they involve little legal recourse (Rugadya 2003). Hence, one of the objectives of the Land Act of 1998 is to formalize the land market, with a proposal to convert customary land to freehold status (Republic of Uganda 1998).

Table 12-2 shows that male-headed households were more likely to buy or rent in land than their female counterparts. This is expected, since female-headed households have lower per capita household income (Nkonya et al. 2005b). Female-headed households were more likely to borrow land than were male-headed households, which also is expected, since poor households tend to rely on social capital to acquire capital and services. With average buying and rental price of land estimated at about $285 and $27 per hectare,[3] respectively, land markets are likely to increase the inequality of land between male- and female-headed households. Surprisingly, the share of female- and male-headed households who acquired land through inheritance was not significantly different (58% and 56% for female- and male-headed households, respectively). Given that most customary institutions encourage household heads to bequeath their land to their sons (Bikaako and Ssenkumba 2003),

TABLE 12-2 Land acquisition methods (percentage of plots)

Tenure/Plot Acquisition Method	*Gender of Household Head*			*Paired T-test (P-value)*[a]
	Female	*Male*	*All*	
Purchased	26.9	31.8	31.2	0.01
Inherited, other permanent acquisition (e.g., exchange)	52.3	50.9	51.0	0.478
Rented	1.0	2.4	2.3	0.025
Other temporary acquisition (e.g., borrowed, squatting)	9.3	5.4	5.8	0.000

Source: Community and household surveys (see text for details).

[a]Compares male-headed and female-headed households.

TABLE 12-3 Duration of lease contracts

Duration (years)[a]	*% of Plots*
0–1	45.2
2–3	29.0
Over 3	26.8
Number of rented in plots = 133	

[a]Years approximated to whole numbers

it was expected that a significantly lower share of female-headed households acquire their land through inheritance.

The duration of leasing in contracts is very short (Table 12-3). About 45% of the rented in plots were rented for one year or less. The short duration of the rented in contracts suggests that land rights under such contracts are so insecure that there may be limited possibilities of improving such plots using long-term investments. We explore this issue further in the next subsection.

Land Markets and Land Investments

The presence of long-term land investments on the plots surveyed was generally low (Table 12-4). Tree planting was the most common type of investment, followed by trenches and bench terraces. All of these investments were more common on purchased or inherited plots than on rental plots, consistent with the expected negative effects of lower tenure security on rental plots. Other investments were fairly rare, although in a few cases (e.g., grass strips, irrigation wells, and soil bunds), investments were more common (though still rarely) on rented than other plots. The higher incidence of grass strips and soil bunds on rental plots may be due to the low cost and short-term nature of these investments.

In general, however, these results reflect the relative long-term versus short-term return to different types of investment, in which those with longer-term benefits such as trees are preferred on more tenure-secure plots, whereas those with immediate or shorter-term benefits such as drainage structures are preferred on rented plots. The mixed results also indicate that there are other factors that influence the decision to invest in land improvement.

Land Management Practices

The land management practices discussed in this section are short-term management practices such as fertilizer application, crop rotation, and slash and burn. We also discuss animal manure and fallow practices that have medium-term benefits. Compared to the types of land investments discussed in the previous section, households tended to engage more in various short-term land management practices.

TABLE 12-4 Land investments by method of plot acquisition

Type of Iinvestment	All Plots	Method of plot acquisition: Owner-Cultivated: Purchased	Owner-Cultivated: Inherited	Borrowed	Rented	t-Test
		% of Plots				
Bench terrace	6.1	6.1	6.3	6.0	1.8	
Fanya juu terrace	0.1	0.2	0.0	0.0	0.0	
Fanya chini terrace	0.3	0.3	0.3	0	0	
Soil bunds	1.0	1.1	0.9	0.6	1.9	a
Stone wall	0.7	0.3	1.0	0	0	e
Trenches	9.5	8.8	10.5	5.7	6.5	f
Grass strips	2.5	2.7	2.5	0	3.7	d,f
Planted trees	14.3	16.9	13.3	8.6	4.6	a,c,d,e
Irrigation well	0.1	0.0	0	0	1.9	a,b,c

Source: Community and household surveys in Uganda (see text for details).

Note: Paired tests for significant difference at the 10% level: [a] is rented vs. purchased; [b] is rented vs. borrowed; [c] is rented vs. inherited; [d] is purchased vs. borrowed; [e] is purchased vs. inherited; and [f] is borrowed vs. inherited.

Application of mulch, animal manure, compost, and household refuse was more common on purchased plots (and in several instances inherited plots as well) than on rented plots, whereas application of chemical fertilizers was higher on rented plots than the others (Table 12-5). Regarding other practices, slash and burn tended to be used more on borrowed and rented plots, and fallowing was used more on inherited and purchased plots.

Again, these results seem to reflect the relative long-term versus short-term return to different types of management practices, where those with longer-term benefits such as manuring, mulching, and composting are preferred on more tenure-secure plots, whereas those with immediate or season-to-season benefits such as chemical fertilizers and slash and burn are preferred on rented plots.

Land Markets and Land Quality

We examine the soil nutrient stocks of the macronutrients—namely, nitrogen (N), Phosphorus (P), and Potassium (K). Soil samples were taken from each plot and analyzed in the lab to determine the nutrient stock and flow.[4] Purchased plots had the greatest nutrient stocks (N, P, and K in kilogram/hectare) of all types of plot, borrowed plots had the lowest stocks, and rented plots had intermediate nutrient stocks (Table 12-6). These results support the hypothesis that plots traded on short-term markets tend to be of inferior quality, due

TABLE 12-5 Land management practices by method of plot acquisition

Type of Practice	*All Plots*	*Method of plot acquisition*				*t-Test*
		Owner-Cultivated		*Borrowed*	*Rented*	
		Purchased	*Inherited*			
	% of plots					
Slash and burn	16.7	14.7	17.4	20.8	20.6	d,e
Fallow	20.9	17.9	24.1	11.7	13.1	c,d,e,f
Crop rotation	28.6	30.3	27.3	25.9	35.5	
Crop residues	17.1	18.6	16.0	15.1	24.3	e
Mulch	3.3	4.3	2.9	2.4	0.9	e
Compost	1.3	1.9	1.2	0.3	0.0	d
Household refuse	4.6	6.3	3.9	0.8	1.8	d,e,f
Manure	8.1	10.8	7.5	0.5	8.1	a,d,e,f
Inorganic fertilizer	2.1	1.9	2.3	0.5	3.6	

Source: Community and household surveys in Uganda (see text for details).

Note: Paired tests for significant difference at the 10% level: [a] is rented vs. purchased; [b] is rented vs. borrowed; [c] is rented vs. inherited; [d] is purchased vs. borrowed; [e] is purchased vs. inherited; and [f] is borrowed vs. inherited.

TABLE 12-6 Land quality by method of plot acquisition

Type of Practice	*Method of Plot Acquisition*				*t-Test*
	Owner-Cultivated		*Borrowed*	*Rented*	
	Purchased	*Inherited*			
Nutrient stock in topsoil					
Nitrogen (kg/ha)	4353.0	3697.0	3663.0	4038.0	d,e
Phosphorus (kg/ha)	2762.0	2437.0	1904.0	2389.0	d,e,f
Potassium (kg/ha)	8223.0	6788.0	6591.0	6528.0	d,e
Nutrient balances					
Nitrogen (kg/ha/yr)	–84.9	–77.3	–65.5	–49.8	d
Phosphorus (kg/ha/yr)	–14.0	–16.7	–10.7	–10.7	f
Potassium (kg/ha/yr)	–90.8	–85.4	–53.5	–55.1	d,f

Source: Community and household surveys in Uganda (see text for details).

Note: Sample means are adjusted for stratification, weighting and clustering of sample.

Paired tests for significant difference at the 10% level: [a] is rented vs. purchased; [b] is rented vs. borrowed; [c] is rented vs. inherited; [d] is purchased vs. borrowed; [e] is purchased vs. inherited; and [f] is borrowed vs. inherited.

to information asymmetry and other market imperfections. However, the N and K balances (in kilogram/hectare/year)[5] for the purchased plots were lower than those measured for plots acquired through other methods, despite the seemingly better land management practices associated with purchased plots as discussed previously. This could be due to the higher crop yield on purchased plots (Nkonya et al. 2004). Crop yield is one of the leading channels of nutrient depletion in Uganda (Nkonya et al. 2005a).

Econometric Results

Our econometric analysis estimated both the determinants of participation in the land rental market and the effects of land rental and other factors on land management practices.

Determinants of Participation in the Land Rental Market

Several factors are associated with the decision to rent a plot (Table 12-7). Controlling for other variables, plots on steeper slopes are less likely to be rented. This could be due to the need for additional investment to control erosion on steep plots. If no SWC structures (e.g., terraces, stone walls, trenches, and soil bunds) are used, plots on steeper slopes are also likely to show visible soil erosion characteristics, which could discourage farmers from renting such plots. Plots with more sandy soils are less likely to be rented, whereas the presence of agroforestry and other NRM investments[6] is positively associated with land rental.

TABLE 12-7 Determinants of decision to participate in land rental market and land management practices (bivariate probit)

	Slash and Burn	*Animal Manure*	*Land Rental Market[b]*
Rental plot	1.678***	–1.882***	—
Land quality			
Ln (plot slope %)	0.046	0.111***	–0.271***
Ln (% sand)	0.119	–0.209**	–0.828***
Land investments			
Agroforestry	0.031	0.125**	0.197
SWC structures	0.225***	0.246***	–0.018
Perennial crop (cf annual crops)	0.281***	0.245***	–0.049
Irrigation	0.080	—	–0.382
Other NRM investments[a]	0.287**	0.277**	1.220***

Continued

TABLE 12-7 *(Cont.)*

	Slash and Burn	*Animal Manure*	*Land Rental Market*[b]
Physical capital			
Ln (value of farm equipment, Ush)	—	—	–0.056***
Ln (TLU)	–0.0.55*	0.109***	–0.072
Ln (plot area ha)	0.075***	0.036	—
Ln (farmsize ha)	–0.029	–0.165***	–0.081
Land tenure (cf customary tenure)			
Freehold	–0.275***	0.042	–0.341*
Mailo	–0.099	0.294**	–0.544*
Leasehold	0.354**	0.068	1.443***
Have access to plot in the next 10 years? (cf no)			
Yes	–0.143*	0.390*	–1.435***
Uncertain	–0.274*	0.273	0.385**
Ln(household size)	–0.080	0.059	–0.011
Share of land area owned by women	0.144	0.106	–73.099
Male household head	0.176	0.153	0.967**
Primary activity of household head (cf crop production)			
Non-farm	–0.213***	–0.117	–0.510***
Livestock production	0.035	–0.317	–6.149
Village level factors			
Ln(distance to plot km)	0.022	–0.201**	–0.134
Ln(distance to all-weather road km)	0.131***	0.014	0.081
Ln(village wage rate USh/day)	0.257***	0.096*	–0.057
Population density	0.016	–0.053*	0.083
Likelihood ratio test (P-value)	0.000	0.000	0.000
Sample size	4204	4204	4204
% of positive observations	15.5	7.6	2.3

* significant at 10%; ** significant at 5%; *** significant at 1%.

[a]This includes fish ponds, fence, paddock, and pasture improvement.

[b]To save space, the following variables are not reported: agroecological zones, level of education of household members and PMI, the dummies for the year the surveys were conducted and the constant. These results are available upon request from authors.

Note: The following variables were dropped from the land management equation after jointly failing the Wald test: value of farm equipment and PMI. The year of survey variable also failed the Wald test in the land rental market equation and was dropped. These variable exclusion restrictions improved identification of the bivariate estimation.

These results indicate that farmers use observable soil quality characteristics and investments when making renting decisions. The positive association between agroforestry and other NRM investments and renting thus is consistent with the capitalization effect hypothesis that land owners invest in visible land improvements in order to attract tenants. It is also consistent with the alternative hypothesis that tenants make investments (such as short-term improved fallows) that yield relatively short-term returns, or invest to increase their tenure security on rented plots. As reported by Place et al. (2001), tenant farmers who invest in long-term improvements on rented in plots may do so with an objective of increasing their stake in owning a rented plot. Further information on who is investing on rented plots (whether the landlord or the tenant) and when (whether before or after the plot is rented) would help to disentangle which of these hypotheses is most salient in this case.

The land tenure system has a strong impact on participation in the land rental market. Rented plots are more likely to be under leasehold tenure than under customary tenure. These results are consistent with expectations that farmers owning plots under customary land tenure are less likely to participate in land rental markets due to the customary institutions that promote land acquisition through inheritance rather than through the market (at least compared to leasehold tenure). However, rented plots are more likely to be under customary land tenure than freehold or *mailo* tenure. This suggests that customary tenure is not a major constraint to land rental markets and that customary land tenure has flexibility in accommodating land transactions using the market rather than inheritance. Baland et al. (2007) also report similar results.

Female education has no significant impact on the decision to rent in a plot, reflecting the weak role that women play in land transaction decisions in Ugandan households. By contrast, households with more educated males are less likely to rent land. It is likely that male farmers with a higher level of education will look for alternative livelihoods in order to meet the household subsistence requirements, rather than seeking to rent in land. Consistent with this, households whose household head has a nonfarm activity or livestock production as the primary activity are less likely to rent a plot than those whose primary activity is crop production. The results imply the potential that education and alternative livelihoods offer to relieve the pressure on land for areas with unsustainably high population density such as the southwestern and eastern highlands.

Male-headed households are more likely to rent in a plot than are female-headed households. This is consistent with the descriptive analysis in which we observed that female-headed households are less likely to buy or rent a plot than are male-headed households. Contrary to our expectation, ownership of farm equipment decreases the probability to rent a plot. This is likely due to the ability of wealthy households to buy instead of leasing in. The correlation between buying a plot and value of farm equipment was positive and significant at p = 0.01. Similarly, the correlation between the value of farm equipment and farm size was significant (at p = 0.01) and positive.

Household size did not have a significant impact on the probability to rent in a plot. Larger families have greater demand for more land to provide their subsistence requirements. At the same time, larger households in Uganda have lower per capita income than smaller rural households (Nkonya et al. 2005b). This suggests their low financial ability to participate in the land rental market. The nonsignificant impact of household size may be due to these opposing effects on the decision to participate in the land rental market. Contrary to the distribution effect of land markets discussed in the conceptual model, population density and farm size did not have significant impacts on participation in the land rental market. It was expected that areas with higher population density and smaller farms would be more likely to participate in the land rental market than areas with lower population density or larger farms.

Security of rented plots is also an important determinant of the decision to rent in plots. As shown in the descriptive analysis, the duration of the land rental contract is shorter than 3 years for 74% of the rented plots. This suggests that a farmer holding land under lease contract is not likely to have access to the plot over the next 10 years. The econometric results are consistent with this, showing a negative relationship between security of tenure and rented plots. These results imply that farmers seeking longer-term access to land are more likely to buy a plot than rent one. For example, a correlation between buying a plot and having a 10-year access to a plot was significant at p = 0.01.

Impacts of Land Rental and Land Tenure on Land Management Practices

Tables 12-7 and 12-8 show that incorporation of crop residues and slash and burn are more common on rented plots than on plots acquired through other methods, and use of animal manure is less common on rented plots. These results are consistent with the descriptive statistics results discussed above. The effects of land rental on slash and burn and manure use are consistent with the land tenure security effect hypotheses, since they show that farmers refrain from using an intensive practice with medium- to long-term benefits (manure) and are more prone to use slash and burn, which provides a short-term boost in yield but contributes to longer-term depletion of the soil. The positive association of land rental with incorporation of crop residues is more surprising, since the near term productivity benefits of this practice are not always evident (Nkonya et al. 2004) and can in some circumstances reduce yields if residues are low in nitrogen content, by immobilizing available nitrogen (Giller et al. 1997). Still, some studies have found significant positive short-term impacts of crop residue incorporation on yields in Uganda (e.g., Nkonya et al. 2005b), suggesting that these impacts are context dependent. Where such positive impacts do occur in the near term, tenants are more likely to be willing to make the effort to incorporate crop residues.

TABLE 12-8 Determinants of land management practices (univariate probit models)

	Fallow[a]	*Residues*[a]	*Fertilizer*[a]	*Rotation*[a]
Rental plot	–0.021	0.664***	–0.591	0.17
Freehold	0.284***	–0.263**	0.012	0.455***
Mailo	–0.234	0.01	0.046	–0.142
Leasehold	–0.176	–0.422	D	–0.153
Have access to plot in the next 10 years? (cf no)				
Yes	0.347**	0.172	–0.670***	–0.004
Uncertain	–0.174	–0.108	–1.037***	–0.008
Pseudo R[a]	0.07	0.136	0.246	0.087
Sample size	4204	4204	3921	4204

* significant at 10%; ** significant at 5%; *** significant at 1%.

[a]Only variables most relevant to land markets are reported. These are: land tenure, perceived tenure security, and method of land acquisition. For full list of variables used in the model, please see Table 12-1. Full results are available upon request from authors.

D = dropped due to predicting failure 100%

Also consistent with the expected effects of tenure security are the impacts of the tenure security variable (expected access to the plot for the next 10 years) in Tables 12-7 and 12-8. On plots that are more secure, fallowing and use of manure are more likely, whereas use of fertilizer and slash and burn are less likely. Again we find that practices with longer-term benefits, such as fallowing and manure use, are favored by longer-term security, while practices with shorter-term benefits, such as fertilizer and slash and burn, are more common on less secure plots.

Conclusion and Implications

Our descriptive statistical analyses show that male-headed households are more likely to buy or rent in a plot than are female-headed households. The econometric results also show that male-headed households are more likely to rent in a plot than are female-headed households. These results suggest that women are not benefiting as much as men from the land markets. This has been observed in other African countries (e.g., Walker 2002). These results suggest the need to find strategies to help women and other poor farmers to participate in the land market.

Our results support the capitalization effect hypothesis by showing that purchased plots had higher nutrient stocks and more planted trees than plots acquired through most other methods. However, purchased plots had lower soil nutrient balances than other plots. This implies the downside of the land markets in which farmers who buy land may use practices that lead to higher soil nutrient depletion as they attempt to recover sunk costs using practices that increase productivity but mine soil fertility. Furthermore, on rental plots, short-term considerations may dominate farmers' decisions due to the short-term nature of most rental contracts. Consistent with this, we find that use of manure is less common and slash and burn is more common on rented plots. Similarly in the Amhara region of Ethiopia, farmers were less likely to use various labor-intensive land management practices—contour plowing, incorporation of crop residues, and crop rotation—and more likely to use reduced tillage on rented plots. These findings suggest that development of land markets may lead to more rapid land degradation, and that promotion of land rental markets should be accompanied by promotion of sustainable land management practices through agricultural extension and natural resource management programs. The results also imply the need to conduct more research in agroforestry and other organic soil fertility management technologies that are low cost and have both short-term and longer-term benefits. As observed in this study, the short-term benefits are appealing to farmers who rent in land using short-term contracts.

The negative impact of nonfarm activities on the land rental market suggests that farmers often use alternative livelihoods to meet their household subsistence requirements. This suggests the potential that nonfarm activities offer to relieve the unsustainable land pressure in high population density areas such as the highlands in Africa. Nonfarm activities could also help farmers buy land, which gives them more tenure security that fosters long-term investments.

As observed in this study, land purchases are likely to lead to long-term land investment and improvement. However, there is concern—at least in the short run—of soil nutrient depletion. Land rental markets are particularly of concern, since they contribute to unsustainable land management practices such as slash and burn and reduce use of more sustainable practices such as applying manure. This implies the need to promote sustainable land management practices and to help farmers negotiate and enforce contracts that restrict renters from using unsustainable land management practices. More fundamentally the question is why land tenancy contracts are short term, which seems to be a root cause for tenure insecurity and unsustainable management practices. Another question is the extent to which the new land law in Uganda is effective in improving tenure security. According to our analysis, the clear effects of the new law have not yet been confirmed.

NOTES

1. PMI is estimated as travel time to the nearest five markets, weighted by their population (Wood et al., 1999).
2. For details of these surveys, see Nkonya et al. (2004), Pender et al. (2004), and Nkonya et al. (2005b).

3. The standard deviations of these estimates are quite large, due to the high land value in the high potential areas in the central region and the low value of land in the northern and northwestern marginal areas.
4. For more details, see Nkonya et al. (2005a) and Nkonya et al. (2004).
5. Soil nutrient balance is the difference between soil nutrient inflow into and outflow out of a plot (or larger land area). The more negative the nutrient balance, the more rapid the rate of depletion of soil fertility.
6. These include fish ponds, fences, paddocks, and pasture management.

CHAPTER 13

Land Conflicts and Soil Management

Evidence from Kenya[1]

TAKASHI YAMANO AND
KLAUS DEININGER

Land is increasingly becoming a source of conflicts in Sub-Saharan Africa, where land access had traditionally been characterized as relatively egalitarian. It has been shown that local land conflicts can erupt into large-scale civil strife and political movements (Andre and Platteau 1998; Fred-Mensah 1999; Daudelin 2002). The most recent example is the postpresidential election violence in Kenya, started in December 2007, where historical land conflicts between ethnic groups have amplified the postelection violence, especially in Rift Valley (Mathenge 2008). Some underlying factors, such as population pressure,[2] agricultural commercialization, and urbanization have contributed to the increasing number of land conflicts, and the current land tenure systems in Africa may not be well equipped to resolve such conflicts (Cotula et al. 2004; van Donge 1999). In many African countries, formal institutions for land administration were often simply superimposed on traditional structures without a clear delineation of responsibilities and competencies, implying that they lack both outreach and social legitimacy (Deininger 2003).

Despite the increasing incidences of land conflicts, previous studies on this topic have been limited to some specific incidences that are related to large-scale civil strife or politically motivated conflicts. A recent study in Uganda, however, shows that rural households experience small-scale land conflicts with relatives, neighbors, landlords, or local governments, and that such small-scale conflicts may have significant impacts on their agricultural productivity (Deininger and Castagnini 2006).

The purpose of this chapter, therefore, is to assess the prevalence of land conflicts, examine who suffers from such conflicts, and measure the impacts of land conflicts on soil management and tree stocks in Kenya. We estimate the determinants of land conflicts at three levels of conflicts (concern about future conflicts, pending current conflicts, and resolved past conflicts) with the Multinomial Logit model at the parcel level.[3] Then, we estimate the impacts of land conflicts on soil management and tree stocks. Because Kenya has one of the most advanced land titling systems in Africa, the findings from this country will provide valuable lessons to other African countries that are in the process of modernizing their land titling systems.

Data and Descriptive Analysis

Data

The data used in this chapter come from 899 households interviewed in a survey conducted in 2004. This survey was conducted as part of the Research on Poverty and Environment and Agricultural Technology (RePEAT) Project.[4] The households were randomly sampled from three surveys conducted by the International Livestock Research Institute (ILRI) in 1996, 1998, and 2000. The three ILRI surveys used a similar sampling method and covered about 3,300 households who resided in central and western regions of Kenya. From the sub-locations in which the ILRI samples are located, we selected 100 sub-locations randomly and 10 households from each of the 100 chosen sub-locations.

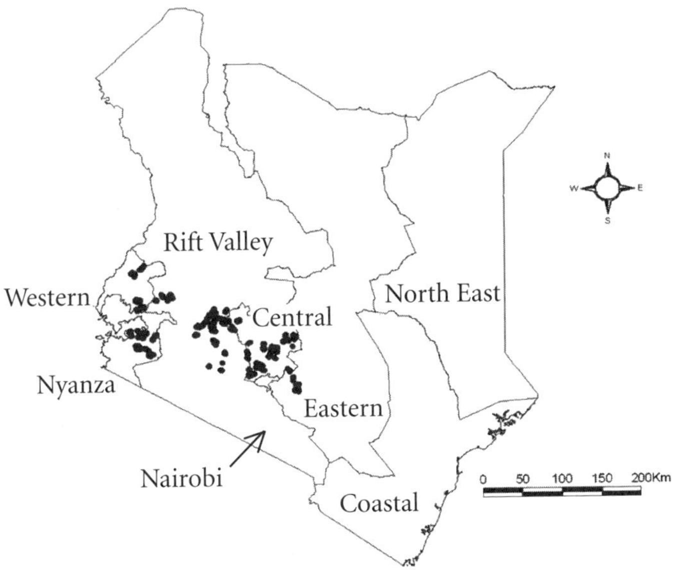

FIGURE 13-1 Distributions of sampled sublocations (map of Kenya with provincial borders)

Note: Black dots show the sampled sublocations. Names and boundaries of provinces are shown in the map.

In Figure 13-1, we present the locations of the 100 sub-locations in the map of Kenya. Two waves of surveys were conducted in 2004 over a period of six months on the selected sample households. The first wave was conducted in February 2004, asking respondents about the previous six months—that is, from August 2003 to January 2004. In October 2004, the second wave took place to cover the following six months, starting from February 2004.

Table 13-1 shows the number of sampled households and parcels across provinces and ethnic groups in Kenya. To focus on owned parcels, we exclude 381 parcels that were rented in for rent payments. We still include, however, parcels that were borrowed from relatives or others for free. We find that the sampled households are worried about future land conflicts on 9.3% of all parcels and have pending conflicts on 4.3% of their parcels. They have resolved conflicts on 8.1% of the parcels in the past. There are some differences in land conflicts across provinces. In Nyanza, land conflicts do not appear to be a prevailing problem, whereas in a nearby province, western province, households have more pending conflicts than in any other province. This difference seems to be driven partly by the small land size, which is just 1.1 hectares per household, in western province.

Unlike western province, the average land size in Rift Valley province is large at 2.6 hectares per household, the largest across the provinces. Yet, households are still worried about future conflicts and have pending conflicts. This could be partly due to the high land productivity and partly to the mix of ethnic groups in Rift Valley province. In Rift Valley province, people from different ethnic groups have purchased land from white farmers since independence.

To observe the associations between land conflicts and ethnicity, we also divide the samples by ethnicity in Table 13-1. Masai households worry about future conflicts on more than 14% of their parcels, and this is probably because they have large grazing land (the average parcel size is 12 hectares), which border on many households. However, they worry only about less than 5% percent of their parcels and have resolved conflicts on 19% of their parcels in the past. This may indicate that they have mechanisms to resolve conflicts and may not have pressing issues that cause them to worry about future conflicts. Other ethnic groups, on the other hand, have fewer pending conflicts currently than the Masai, but worry more about future conflicts. Especially those who have small land size worry about future conflicts. For instance, the Luhya, who reside mainly in western province, worry about future conflicts on more than 9% of their parcels, and the average land size is only 1.1 hectares.

Descriptive Analyses

Relationship between Land Titling and Conflicts. Land scarcity and agricultural commercialization are expected to increase land value and lead to the individualization of land rights, creating opportunities to establish institutions

TABLE 13-1 Sampled households and land conflicts by province

	Number of Households	Number of Own[a] Parcels	Household Land Size in ha	Land Conflicts at the Parcel Level[b]		
				Concerned about Future Conflicts	Pending Conflicts	Resolved
	(A)	(B)	(C)	(D)	(E)	(F)
	Number	Number	ha	%	%	%
All	897	1,167	1.75	9.3	4.3	8.1
Province						
Eastern	71	125	1.70	3.2	4.8	4.8
Central	318	378	1.47	7.7	3.7	8.7
Rift Valley	222	264	2.71	9.5	4.9	8.7
Western	112	135	1.08	14.8	10.4	13.3
Nyanza	174	265	1.47	6.8	2.3	9.1
Ethnic Group						
Maasai	17	21	12.0	4.8	14.3	19.0
Kamba	69	123	1.68	3.3	4.9	4.9
Kikuyu	469	516	1.53	8.3	3.7	7.6
Kalenjin	71	87	2.55	10.3	3.4	4.6
Kisii	106	148	1.12	8.8	2.7	13.5
Luhya	124	151	1.10	13.9	9.3	13.9
Luo	71	121	1.94	4.1	3.3	3.3

[a]381 parcels that were rented in are not included in this table and the following analyses.

[b]These categories are defined as mutually exclusive.

to better define and enforce property rights (Otsuka and Place 2001). In Kenya, however, the formal individualization of land rights has been in place since independence. The 1954 Swynnerton Plan granted secure individual land titles to African farmers, and the Plan was reinforced further by the Native Lands Registration Ordinance 1959, replaced after independence by the Registered Land Act 1963 and the Land Adjudication Act 1968 (Place and Migot-Adholla 1998). Although the registration process might have increased tenure security for many landowners, it has also created new forms of disputes, such as challenges over registered land and conflicts over land sales (Shipton 1988). Moreover, the high cost of registration has discouraged updating the registrations after land transactions, such as inheritance and sales.

In Table 13-2, we find that many sampled households have outdated land registrations. Out of the 1,167 parcels, more than 81% of them are registered, while the rest are unregistered. More than half of the registered parcels are under the name of either the household head or spouse who cultivates the parcels. However, the other half of the registered parcels bears names that are other than the household head or spouse. Instead, they bear the names of living parents, deceased parents, deceased husbands, or others. This situation arises when new landowners neglect to update the registration because of high costs or are prevented from doing so because they have not been able to reach an agreement among relatives over inheritance.

TABLE 13-2 Land title holder and land conflicts

		Land Conflicts at the Parcel Level[a]		
	Number of Parcels (*A*)	*Concerned about Future Conflicts* (*B*)	*Pending Conflicts* (*C*)	*Resolved* (*D*)
	Number (%)	%	%	%
All	1,167 (100)	8.2	4.5	8.1
Not Registered	217 (18.6)	10.1	9.2	8.8
Registered	950 (81.4)	7.8	3.5	8.4
Title Holder of Registered Parcels				
All registered	950 (100)	7.8	3.5	8.4
Head/Spouse	511 (53.8)	6.8	2.3	7.0
Parent	234 (24.6)	7.7	3.8	10.3
Deceased Parent	46 (4.8)	20.0	3.5	8.2
Deceased Husband	85 (8.9)	2.2	8.7	8.7
Other	74 (7.8)	4.1	6.8	12.2

[a]These categories are defined as mutually exclusive.

Turning to land conflicts, we find that the unregistered parcels have pending conflicts on 9.2%, whereas the registered parcels have pending conflicts on only 3.5%. Households are more worried about future conflicts on unregistered parcels than registered parcels. When the land registration has the name of either the household head or spouse, the landowners have pending conflicts on 2.3% of their parcels and are worried about future conflicts on 6.8% of the parcels. When the title has the name of a living parent as the household head, the situation is similar. However, if the title has the name of a deceased parent, then the current landowners are worried about future conflicts on about 20% of their parcels. The high level of concern could be due to the expectation of potential conflicts with siblings or other relatives who might be interested in claiming the ownership of the parcels left by the deceased parents.

The proportion of pending conflicts is very high at 8.7% if the land titles belong to the deceased husbands. This suggests that widows are experiencing pending conflicts with the deceased husbands' relatives. The prevailing practice after the death of a husband in Kenya is for the wife of the deceased husband to hold land in trust for her male children because customary laws rarely allow widows to legally inherit land (Drimie 2002). In some cases, widows are threatened to leave their land, which belongs to their husbands' ancestral land, especially when they have no children or refuse to marry one of their husbands' brothers (Wanyeki 2003).[5] In Uganda, for instance, Deininger and Castagnini (2006) find that households headed by females and widows experienced more land conflicts than male-headed households. We also saw in Chapter 9 by Holden and Bezabih that female-headed households in Ethiopia face similar problems as their in-laws may want to control and use their land.

Causes and Resolution of Land Conflicts. We further stratify the cases of land conflicts into three groups by the starting year: 32 conflicts started from 1965 to 1989, 45 conflicts from 1990 to 1999, and 67 conflicts from 2000 to 2004 (Table 13-3). To obtain information about actual conflicts, we select conflicts that are either pending or resolved in the past (columns C and D in Table 13-2). As the data indicate, we find more cases per year in recent years. Although this is consistent with informal information that land conflicts are more prevalent in recent years, it is not clear if this is due to the increasing cases of land conflicts or if the respondents neglected to mention past conflicts, although they were encouraged to do so by the interviewers. About half of the land conflicts are over boundaries that occur mainly with neighbors or relatives who live close by. The second most common reason for conflicts is over inheritance, which exclusively occurs among relatives. In the past five years, it appears that there are more conflicts related to land sales. As the value of land increases duc to population pressure, agricultural commercialization, and urbanization, it is expected that the land sales market will develop over time. However, if property rights are not clearly defined, there could be more cases of land conflicts related to land sales. The next cause of land conflicts is

TABLE 13-3 Causes and resolutions of pending or resolved conflicts[a]

	All (A)	Starting Year of Conflicts: 1965–1989 (B)	1990–1999 (C)	2000–2004 (D)
Number of incidences	144	32	45	67
Incidences per year	3.7	1.3	4.5	13.4
Causes				
Boundaries	50.0	50.0	44.4	53.7
Inheritance	28.5	25.0	37.8	23.9
Land sales	8.3	6.3	4.4	11.9
Use rights	4.2	3.1	4.4	4.5
Eviction	5.6	12.5	4.4	3.0
Others	3.5	3.1	4.4	3.0
All	100%	100%	100%	100%
Have resorted to				
Informal institutes (%)	85.4	68.8	91.1	89.6
Formal institutes (%)	38.9	53.1	48.9	25.4
Both institutes (%)	31.9	31.3	44.4	23.9
Resolved?				
Resolved (%)	64.6	59.4	62.2	68.7
Years before resolution	2.0	3.6	3.2	0.5
Average pending years	8.5	18.9	8.5	2.0

[a]These correspond to the pending and resolved conflicts presented in columns C and D of Table 13-2.

related to use rights. This type of land conflict occurs between landowners and tenants who rent in land with a fixed rent. Finally, the last common cause of land conflicts is eviction. Eviction mostly takes place on public land between illegal land occupants and public offices.

In the survey, respondents were asked if they had resorted to informal or formal institutions to resolve land conflicts. Informal institutions include community elders or committees, while formal institutions include land tribunals or other governmental institutions. Over 91% of the land conflicts that occurred from 1990 to 1999 have been brought to informal institutions, and about 49% of them have been brought to formal institutions (Table 13-3). Note that 44% of the cases have been brought to both informal and formal institutions. In a related question, respondents were asked why they did not resort to informal or formal institutions. The dominant answer to this question about not resorting to formal institutions suggests that they had resolved the conflicts informally before turning to the formal institutions. The second most common answer to this question is that it is too expensive to bring a case to a formal institution. Therefore, it seems that most cases are brought to informal institutions first, and if the informal institutions are unable to resolve the conflicts and if the complainants have sufficient resources, then the cases are brought to formal institutions.

Regarding the more recent conflicts that started from 2000 to 2004, we find that over 89% of the cases have been brought to informal institutions but only about 25% of them have been brought to formal institutions. According to the preceding argument, it seems reasonable to expect that many cases will be brought to formal institutions in the future when informal institutions are unable to resolve the conflicts.

Impacts of Land Conflicts on Land Management. There exists a rich literature on land tenure security and farm production and investment (Otsuka and Place 2001). Many of the studies implicitly assume that weak tenure security is associated with land conflicts but do not actually examine land conflicts directly. However, a recent study by Deininger and Castagnini (2006) suggests a 5 to 11% productivity loss due to land conflicts. In Kenya, especially in the western regions near Lake Victoria, the HIV/AIDS epidemic has greatly increased the number of widows. Yamano and Jayne (2004), for instance, show that the death of a working-age male household head reduces the land allocated to high-value crops and results in a large reduction in per capita household crop value production. Although various factors affect crop production after the death of a working-age male head, land conflicts might be a contributing factor to the reduction in crop production.

To investigate this issue, we present the land management and tree stock variables in Table 13-4, stratified by the land conflict status. We find that the average manure application is significantly lower on parcels with pending conflicts: The average manure application is 1,475 kilograms per hectare on pending conflicts in column B, while it is 3,563 kilograms per hectare on parcels with no conflicts in column A. The difference between the two is statistically significant at the 5% level. Because the organic fertilizer application is a long-term investment in soil fertility, farmers may be reluctant to apply organic fertilizer on plots that are under pending conflicts.

Although the chemical fertilizer application is also lower on parcels with pending conflicts, it is not statistically different from the average chemical fertilizer application on parcels with no conflicts: The average chemical fertilizer application is 85.3 kilograms per hectare on pending conflicts, while it is 93.1 kilograms per hectare on parcels with no conflicts. Because chemical fertilizer is only effective for one or two production seasons, farmers may not have a reason to reduce chemical fertilizer application unless they fear losing the parcels during a cropping season due to land conflicts.

TABLE 13-4 Land conflicts and input application at the plot level[a]

		Land Conflicts Level[b]		
	No Conflicts at All	*Pending Conflicts*	*Resolved*	*Concerned about Future Conflicts*
	(A)	*(B)*	*(C)*	*(D)*
	Mean (number)	Mean (number)	Mean (number)	Mean (number)
Soil management				
Manure application (kgs/ha)	3,563	1,475**	3,259	3,481
Chemical fertilizer (kgs/ha)	93.1	85.3	99.0	87.0
Mulching (ratio)	0.20	0.24	0.22	0.16*
Number of plots	(4,281)	(442)	(520)	(260)
Tree stocks				
Timber	125	170	100	119
Coffee	99	99	65	85
Tea	295	301	256	348
Fodder	10.9	4.3	5.2	9.1
Number of parcels	(1,157)	(56)	(145)	(111)

* and ** indicate that the mean is smaller than the mean in column A at the 10 and 5%, respectively.

[a]A plot is defined as a portion of a parcel devoted to one crop or a group of intercrops. Thus, a parcel could be divided into many plots.

[a]These categories are defined as mutually exclusive.

In addition, we find that the mulching is low on parcels in which farmers are concerned about future conflicts. It is not clear, however, why the current conflicts do not have any impacts on mulching. Although we also have tree stock variables in Table 13-4, we do not find any significant differences in tree stocks between plots that are under different land conflict levels. To investigate these issues further, we need to control for the effects of other factors by means of regression analysis.

Estimation Strategies and Variables

Estimation Strategies

Determinants of Land Conflicts. To examine the characteristics of households that are experiencing different levels of land conflicts, we divide parcels into four groups, as we did in Table 13-2, and estimate the following Multinomial Logit (MNL) model at the parcel level:

$$\text{Prob}\ (c_i) = f(\boldsymbol{T}_i, \boldsymbol{P}_i, \boldsymbol{X}_i) \tag{13.1}$$

where $c_i = 1$ if the household is worried about the future conflicts over parcel i, $c_i = 2$ if parcel i has a pending conflict, $c_i = 3$ if parcel i had a conflict that has been resolved, and $c_i = 0$ otherwise. $\boldsymbol{T}_i$ is a set of land titleholder variables of parcel i; $\boldsymbol{P}_i$ is a set of parcel characteristics; and $\boldsymbol{X}_i$ is a set of household and community characteristics. To capture the access to informal institutions for resolution at the community level, we include the number of elder groups per 1,000 households in the community. We also include 14 district dummies to control for regional characteristics. Since the land conflict variables do not change much over cropping seasons within a year, we use the data from the first cropping season of 2004. To make interpretations of estimated coefficients meaningful, we calculate the marginal changes in the probability for each outcome category according to Wooldridge (2002). Estimated coefficients on a continuous variable are the marginal changes in the probability for each outcome category measured at the mean values, and estimated coefficients on dummy variables are changes in the probability for each outcome category when the value of the dummy variables changes from zero to one.

Impacts of Land Disputes on Input Application. Next, we estimate the impacts of land conflicts on farm input application at the plot level:

$$ln(Y_{hi}) = f(c_i, \boldsymbol{T}_i, \boldsymbol{P}_i, \boldsymbol{X}_i), \tag{13.2}$$

where Y_{hi} is the amount of input applied to plot h in parcel i, and the other variables are defined as before. We estimate this equation at the plot level, rather than at the parcel level, because inputs are applied at the plot level. In Kenya, households often have only one or two parcels. To diversify their crop production, they divide one parcel into small plots and plant crops separately. Thus, we can obtain more accurate information about the determinants of input applications if we estimate the model at the plot level. We focus on three important inputs in Kenya: organic fertilizer, chemical fertilizer, and mulching. Theoretically, we need to include all input and output prices that may affect the input application, but we do not have all the price information. Instead, we include 95 sub-location dummies, assuming that all the prices are uniform within the community. Because there are two cropping seasons in most parts of Kenya, we estimate the model with a pooled data of two cropping seasons. Finally, since inputs are not applied on all parcels, we estimate the equation (13.2) with Tobit.

In addition, we estimate the determinants of tree stocks of fodder trees, coffee bushes, tea bushes, and timber trees. Ideally, we would like to estimate the model on tree planting because the current and future land conflicts would have impacts on the current tree planting. However, we only have the current tree stocks (number of trees per hectare) in our data. These variables represent the past behavior. Nonetheless, we investigate the current tree stockings with the current land conflicts levels to obtain any insights about the relationship between the tree planting and land conflicts.

A concern arises about the correlation between some of the explanatory variables and the error term, which includes unobserved parcel and household characteristics. One example of such unobserved characteristics is the quality of the parcel. If land conflicts are more likely to occur because of the high quality of the parcel, then the coefficient of the land conflict variables would be biased positively (or negatively) if the input application is positively (or negatively) correlated with the quality of the parcel. To overcome this problem partly, we include the rent of the parcel for one cropping season in the regression. The rent is estimated by respondents for each parcel under a hypothetical question of "How much rent would you get on this parcel if you were going to rent it for one season?" Although, the subjective nature of the variables raises some concerns about the reliability, we think it is still a useful variable since there is no alternative.

The concern about the omitted variables problem, however, still remains. One way to reduce the remaining concern is to use instrumental variables that are correlated with the land conflicts but not with the input application. Unfortunately, we do not find plausible instrumental variables that satisfy such a condition for this study.[6] Another way to reduce the omitted variables problem is to use the parcel level fixed effects model, which will control for time invariant fixed effects that might be correlated with the land conflict variables. However, since the land conflict variables do not change much over a season, we do not have sufficient variation in the land conflict variables in our data. Because of the lack of plausible instrumental variables and the lack of within parcel variations in the land conflict variables over a season, we simply present the Tobit results of equation (13.2) without controlling for the potential correlation between the land conflict variables and the error term. Thus, the results from the input application should be interpreted carefully. Despite such potential drawbacks, however, the results presented in the later section seem to have some important policy implications.

Variables

As already described in Table 13-2 and other places, we define land conflicts in four stages: concerned about future conflicts, pending conflicts, resolved conflicts, and otherwise. For estimations, we create a dummy variable for each conflict stage. These variables represent c_i in the previous sub-section. On the land titleholder information, $\boldsymbol{T}_i$, we also create dummy variables for different land titleholders as specified in Table 13-2—namely, head/spouse, parents, deceased parents, deceased husbands, others, and no land title. We treat the parcels whose titleholder is the household head or spouse as the base group since this is the largest group among the various land titleholders.

Parcel characteristics, $\boldsymbol{P}_i$, include parcel size in hectares, walking time to the parcel from the homestead in minutes, years since the acquisition of the parcel, a dummy variable for purchased parcels, and the rent for one cropping season. A large parcel is expected to be a source of a land conflict since it has a longer boundary and tends to invite demands for a fraction of it from relatives. A parcel that is farther away from the homestead could be at a higher risk of being involved in a conflict because of the difficulty in monitoring. A parcel that has been acquired a long time ago is expected to have fewer pending conflicts but have more resolved conflicts in the past. The mode of acquisition may influence the probability of being involved in a conflict. Thus, we use a dummy variable for purchased parcels. The base group is the parcels that are acquired via inheritance. It is not clear which acquisition mode has a high probability of being involved in a conflict since each mode has potential causes of land conflicts. For instance, inherited parcels may suffer from land conflicts among siblings and relatives, while on purchased parcels someone other than the land seller may claim ownership of the purchased land. Finally, we include the rent of the parcel for one season as discussed earlier.

Household characteristic variables, $\boldsymbol{X}_i$, require some explanation. We stratify female-headed households in two groups: (female) widow-headed households and nonwidow-headed households (due to migration of the husband). We only include a dummy variable for nonwidow-headed households in the estimation models because many of the widow-headed households have parcels whose titles bear the name of their deceased husbands, and they are represented by the dummy variable for deceased husbands as titleholders. If we include both the dummy for deceased husbands as titleholders and female-headed households that include widows, we may have a multicollinearity problem.

Other household characteristics include the household head's age, its squared term, the maximum education levels of men and women in the household, the household size, the value of assets, the number of cattle per hectare, and the distance to the nearest urban center from the household. Finally, to represent access to informal institutions for conflict resolution, we include the number of elder groups per 1,000 households in a sub-location. In order to save space, we do not show the estimated coefficients of household characteristics except for female-headed household variable, as well as district dummies. Full results, however, are available upon request.

Estimation Results

Determinants of Land Conflicts

In Table 13-5, we present the results of the determinants of land conflicts that are from the Multinomial Logit model of equation (13.1). We find that when the land titleholder is the living parent, there are no significant differences in land conflicts compared with parcels whose titles belong to the household head/spouse. However, if the land title belongs to deceased parents, then the landowners are more worried about future conflicts than when the titles belong to the household head/spouse. In contrast, when the land titles belong to deceased husbands, widows are 12% more likely to have pending conflicts than when they have their own titles. Turning to nonwidow-female-headed households, we find that they also are 22% more likely to have pending conflicts than do male-headed households. These findings are similar to what Deininger and Castagnini (2006) find in Uganda and other anecdotal evidence found elsewhere (Wanyeki 2003).

Even when the households do not have a title on a plot, the results in Table 13-5 indicate that "no title" does not raise concerns about future land conflicts or cause pending conflicts. Compared with the cases when titles are under the names of the deceased husbands or parents, "no title" seems to be a better situation. It is possible that households do not feel strong needs to own titles on their plots when there are no threats from the siblings of the deceased parents or family members from the deceased husbands. For these cases, there is little need for issuing land titles to the owners.

The results in Table 13-5 suggest that when the number of elder groups per 1,000 households increases by one, the probability of pending conflicts decreases by 0.2%. Although the size of the impact is small, the result suggests that local informal institutions can play important roles in preventing land conflicts. We discuss the importance of including the local institutions in conflict resolutions in the Conclusions section of this chapter.

Impacts of Land Conflicts on Input Application

The results of the determinants of input application are presented in Table 13-6. The data indicate that the organic fertilizer application is significantly lower when there are pending conflicts or the households are concerned about future conflicts, whereas resolved conflicts do not have any impacts on organic fertilizer application. Based on the results in column 2, a simple simulation indicates that the amount of organic fertilizer application decreases by 31% when there are pending conflicts versus no conflicts. When farmers are concerned about future conflicts, the organic fertilizer application also decreases by 19%, according to a simple simulation. Organic fertilizer is considered to be

TABLE 13-5 Determinants of land conflict status at the parcel level (multinomial logit) "No experience of past conflicts and no concern about future conflicts" is the base group

	Land Conflict Status		
	Concerned about Future Conflicts (A)	*Pending* (B)	*Resolved* (C)
Title holder (ref. head or spouse)			
Parents	–0.010	–0.009	0.026
	(0.63)	(0.44)	(0.61)
Deceased parents	0.162	0.012	0.028
	(3.93)**	(0.88)	(1.26)
Deceased husband	–0.048	0.124	0.017
	(0.57)	(2.60)**	(0.64)
Others	–0.021	0.078	0.071
	(0.83)	(1.24)	(1.88)
No title	0.015	0.058	0.013
	(1.68)	(1.27)	(0.97)
Parcel characteristics			
ln (parcel size)	–0.008	0.001	–0.001
	(0.85)	(0.09)	(0.12)
ln (walking time to home)	–0.008	0.002	–0.028
	(1.43)	(0.15)	(2.75)**
ln (years since acquisition)	–0.010	0.014	0.008
	(1.24)	(1.21)	(0.08)
Purchased parcel (= 1)	–0.006	–0.028	–0.011
	(0.95)	(0.78)	(1.01)
ln (rent for one season)	0.027	–0.008	0.004
	(2.49)*	(0.26)	(0.60)
Household characteristics			
Female-headed HH	0.005	0.218	0.006
Non-widow (= 1)	(1.16)	(3.58)**	(1.15)
Community characteristics			
Number of elder groups per 1,000 households	0.001	–0.002	–0.001
	(0.15)	(1.98)*	(0.61)
Constant	–0.434	–0.083	–0.277
	(3.27)**	(0.82)	(1.73)
Number of observations		1,110	

* and ** indicate 5 and 1% significance levels, respectively. Coefficients on continuous variables indicate marginal changes in the probability of each outcome category evaluated at the mean values; and coefficients on dummy variables indicate changes in the probability for each outcome category when the value of the dummy variables changes from zero to one. The estimated coefficients of household characteristics except for female-headed household dummy and district dummies are not presented in the table but included in the model. Full results are available upon request.

a long-lasting investment in soil structure, at least for two to three years. Thus, it is understandable that households reduce organic fertilizer application on parcels under conflicts.

By the same argument, it is understandable that we find a negative impact of concerned future conflicts on mulching. Yet, we do not find a similar negative impact of pending conflicts on mulching. We are not sure about the results on mulching. On the other hand, chemical fertilizer has an immediate impact on crop production for one season but does not have a strong residual impact after one cropping season. Thus, households apply more or less equal amount of fertilizer, even if they worry about losing the land in the future because of land conflicts.

TABLE 13-6 Impacts of land conflicts on soil management at the plot level

	Chemical Fertilizer in kgs/ha (*Tobit*)	*Organic Fertilizer in 100 kgs/ha* (*Tobit*)	*Pr(Mulching)* (*Probit*)
Land conflict status			
Pending	27.96	–52.05	0.044
	(0.94)	(3.18)**	(1.63)
Resolved	2.697	–17.41	0.011
	(0.13)	(1.55)	(0.60)
Concerned about future conflicts	11.76	–29.53	–0.049
	(0.57)	(2.58)**	(2.73)**
Parcel characteristics			
ln (farm size)	–27.24	–29.86	0.007
	(3.44)**	(7.18)**	(1.02)
ln (walking time)	12.68	–22.52	–0.030
	(2.20)*	(6.74)**	(5.36)**
ln (years since acquisition)	–15.12	5.472	–0.015
	(2.11)*	(1.42)	(2.36)*
Purchased parcel (= 1)	11.89	–4.503	0.029
	(0.80)	(0.57)	(2.15)*
Household characteristics			
Female-headed household (= 1)	–32.869	–9.281	0.002
	(2.08)*	(1.13)	(0.15)
Constant	–186.460	–214.519	
	(3.26)**	(6.96)**	
Number of observations	5,954	5,954	5,954

* and ** indicate 5 and 1% significance levels, respectively. The estimated coefficients of household characteristics except for female-headed household dummy and district dummies are not presented in the table but included in the model. Full results are available upon request.

Turning to parcel characteristics, we find parcels that are farther away from the homestead receive less organic fertilizer and mulching. This could be because organic fertilizer and mulching materials are heavy and bulky and thus difficult to carry. Thus, they tend to be applied on parcels that are closer to the homestead where animals are kept in stalls. Also, the larger the farm size is, the less likely it is for the parcel to receive chemical and organic fertilizer. Again, the weight and bulkiness make it cumbersome to apply chemical and organic fertilizer to large farms.

Finally, we estimate the determinants of tree stocks of fodder, coffee, tea, and timber (Table 13-7). On coffee, tea, and timber, we do not find any significant impacts of land conflict variables on the stocks of these trees. However, we find that when farmers are concerned about future land conflicts, the number of fodder trees per hectare is significantly lower than when farmers have no concerns about conflicts. This could be because, by planting fodder trees, farmers may not want to signal to their relatives that they are not using the land productively. Although fodder trees may contribute to improving soil fertility by fixing nitrogen from the air, some farmers tend to think that fodder trees have low economic returns. Thus, they further assume that farmers who plant fodder trees have more land than they can operate. This could invite land conflicts. Farmers who want to avoid such conflicts may replace fodder trees with commercial crops such as maize.

Conclusions

Because of the increasing importance of land conflicts in rural Kenya, we have examined the determinants of land conflicts and estimated the impacts of such conflicts on soil management and tree stocks by using recent survey data of 897 households in central and western regions of Kenya. The results indicate that the sampled households have pending land conflicts on 4% of their parcels and are concerned about future conflicts on more than 9% of their parcels. Because of the advanced land registration system in Kenya, we find that more than 80% of the owned parcels in our data have been registered. However, even if the parcels are registered, many of the registrations are outdated because landowners neglected to update them due to high registration fees or disputes among relatives over ownership. Outdated titles raise concerns among *de facto* landowners about future conflicts. For instance, when titles are registered under the names of deceased parents, we find that children who cultivate the parcels are concerned about future conflicts on about 20% of their parcels. Thus, simply having a title may not mean the land tenure is secure.

We also find that when parcels are registered under the names of deceased husbands, widows are 12% more likely to have pending conflicts than landowners who own titles. Because the HIV/AIDS epidemic has increased the number of widows in Kenya, as well as in some other African countries, this

TABLE 13-7 Impacts of land conflicts on soil management at the plot level

	Fodder Trees per ha Tobit	*Coffee Bushes per ha Tobit*	*Tea Bushes per ha Tobit*	*Timber Trees per ha Tobit*
Land conflict status				
Pending	–18.31	233.6	1,068	49.04
	(1.10)	(1.26)	(0.80)	(0.45)
Resolved	1.117	118.6	101.0	56.33
	(0.10)	(0.93)	(0.12)	(0.73)
Concerned about future conflicts	–24.93	40.53	41.06	–11.68
	(2.07)*	(0.31)	(0.05)	(0.16)
Parcel characteristics				
ln (farm size)	–17.15	–201.2	609.9	–16.44
	(4.34)**	(4.23)**	(1.91)	(0.62)
ln (walking time)	–5.778	–52.61	–368.8	–106.0
	(2.10)*	(1.66)	(1.79)	(5.88)**
ln (years since acquisition)	12.33	206.8	372.4	40.55
	(3.55)**	(4.88)**	(1.45)	(1.80)
Purchased parcel (= 1)	–3.560	–31.67	400.0	116.3
	(0.47)	(0.33)	(0.65)	(2.23)*
Household characteristics				
Female-headed household (= 1)	–11.10	–134.0	365.8	–126.3
	(1.36)	(1.37)	(0.60)	(2.24)*
Maximum education of men	–1.032	10.97	101.8	6.815
	(1.02)	(0.88)	(1.37)	(1.03)
Maximum education of women	0.896	–0.264	80.25	–0.365
	(0.88)	(0.02)	(1.09)	(0.05)
Number of men	5.105	52.16	–338.0	–39.12
	(1.77)	(1.45)	(1.41)	(1.98)*
Number of women	0.795	40.88	198.3	39.08
	(0.28)	(1.20)	(0.90)	(1.96)
Number of oxen	3.225	78.37	–286.7	–8.497
	(0.82)	(1.45)	(0.69)	(0.35)
ln (asset value)	–1.320	6.723	–117.3	12.83
	(0.43)	(0.18)	(0.51)	(0.62)
Constant	–103.4	–1,526	–4,619	–212.1
	(3.36)**	(4.26)**	(2.06)*	(1.06)
Number of observations	1431	1431	1431	1431

* and ** indicate 5 and 1% significance levels, respectively. Fourteen district dummies are also included.

finding raises concerns about the welfare of widows and their children. Although there have been proposals to strengthen the land rights of widows and women in general, the implementation of such policies in practice must be sensitive to local customs. It may not be wise to impose a law that simply guarantees the land ownership of widows because the law may increase conflicts against women, not decrease, from husbands' relatives who fear of losing their ancestral land. This is exactly what Deininger and Castagnini (2006) find in Uganda. Sometimes well-intentioned interventions to improve land tenure may unintentionally increase conflict and social polarization (Atwood 1991; Pinckney and Kimuyu 1994). In Tigray region in Ethiopia, Holden et al. (2007b) found that land registration and certification had reduced land-related disputes both during and after the reform based on a survey covering more than 400 local conflict mediators who are responsible for resolving the majority of such local disputes. In Oromiya and SNNP regions of Ethiopia, Holden and Tefera (2008) found that joint certification of husbands and wives have strengthened the position of women, and especially polygamous wives in relation to their land, but they face problems enforcing their rights through the court system. However, also here local conflict mediators perceived that land registration and certification has contributed to reducing the number of land-related disputes.

Regarding the impacts of the land conflicts on farm management, we find that pending conflicts and concerns about future conflicts reduce organic fertilizer application significantly. The results indicate that organic fertilizer application decreases more than 31% when there are pending land conflicts. These results suggest that pending land conflicts should be solved to sustain land quality. On conflict resolution, we find that people resort to informal institutions first to resolve the conflicts and then to formal institutions only when the informal institutions are unable to resolve the conflicts. The estimation results also suggest that there are fewer pending conflicts in communities with more elder groups. Thus, it is important to recognize the ability of the local informal institutions and clarify the institutional responsibilities of different institutions. Otherwise, the lack of clarity of institutional responsibilities could be exploited by powerful individuals and may have negative consequences for equity.

In sum, in order to improve tenure security in a sustainable manner, not only must official land titles be provided but they must also be updated. In addition, roles of the local informal institutions in protecting land rights must be fully recognized. In fact, delegating the authority to update the land titles to the local institutions could be one possible solution of enhancing land tenure security in rural Kenya. This policy implication is especially important after the postpresidential election violence in Kenya. To resolve the root cause of the violence, the Kenyan government must tackle the land issues with the local informal institutions.

NOTES

1. Support for the data collection used in this chapter is provided by the 21st Century Center of Excellency project at National Graduate Institute for Policy Studies and by the Japan Trust Fund of the World Bank. The authors thank Takeshi Sakurai, Keijiro Otsuka, and the participants of the FASID Monthly Seminar and the Land Markets and Poverty Meeting in Addis Ababa for valuable comments, and Paul Kandasamy for his editorial assistance.
2. Jayne et al. (2003) show significant declines in land under crop cultivation relative to rural population in some African countries since 1960. In Kenya, it has shrunk by half, from 0.5 hectare per person to 0.25 hectare per person. Land is also found to be highly correlated with income among small-scale rural households who have limited nonfarm-income-producing opportunities.
3. We define a parcel as a piece of land that was acquired as a continuous piece of land. A parcel can be divided into small plots that are cultivated separately under different crops.
4. The RePEAT Project is a collaborative research project of National Graduate Institute for Policy Studies, the World Agroforestry Centre, and Tegemeo Institute in Kenya. It was financed by the 21st Century Center of Excellency Grant, which was provided by the Ministry of Education and Science of Japan, through National Graduate Institute for Policy Studies. See Yamano et al. (2005) for details about the data collection and preliminary results.
5. This practice is called "widow inheritance." Although this practice has traditionally functioned as a safety net mechanism for widows and their children, it has become dysfunctional as a safety net in the presence of the HIV/AIDS epidemic since it is considered to have contributed to the spread of HIV.
6. We have tried to use the titleholder variables as instrumental variables. However, because the titleholder variables are not strongly correlated with the land conflict variables, we have failed to identify the land conflicts. Ideally, the amount of land owned by parents and the number of siblings who competed for land inheritance would be good candidates for instrumental variables. Unfortunately, we did not collect such data.

CHAPTER 14

Land Markets and Poverty in Perspective

STEIN T. HOLDEN, KEIJIRO OTSUKA, AND FRANK M. PLACE

We introduced this book by giving an overview of the existing literature on land markets in Africa, emphasizing the historical context as well as political, economic, and demographic changes that have contributed to intensifying the demand for land. The increased pressure on land has brought land rights and land distribution high up on the policy agendas because of land's continued crucial role as a basis for social welfare of the majority of African populations. Increasing international and national focus on land rights and land policy reforms as illustrated by the establishment of the Commission for Legal Empowerment of the Poor and an expansion of the number of World Bank–funded land reform projects (World Bank 2006) demonstrate an increased emphasis on security of property rights and market development as crucial to promote economic development and poverty reduction. Likewise, the Global Land Tools Network under UNHabitat illustrates the increased emphasis on developing more pro-poor land reform approaches within a "continuum of rights" perspective. Still, there are very few land reform successes to build on in Africa (an exception may be some of the recent reform attempts in Ethiopia), and poverty reduction strategies are vague on land tenure issues, with many skeptical to the formalization agenda and many fear that the poor will lose out.

Much of the earlier empirical literature on land markets has tended to focus primarily on their efficiency implications. Our approach in this book is more in line with the approach of Sadoulet et al. (2001), who broadened the perspectives on land markets by linking them more explicitly to issues of poverty, equity, and policies for poverty reduction. Particularly land rental markets turn out to be important for the poor when there are significant imperfections in markets for credit, insurance, traction power, and labor, but these imperfections may also hinder the poor's access to land.

This book has studied the evolution of informal as well as formal markets for land in a number of densely populated countries to assess how these markets may affect the poor and contribute to economic development. Below we summarize our findings drawing on our conceptual framework and hypotheses developed in Chapter 2 and the various country case studies presented in Chapters 3 to 13. We have divided the summary into five parts. The first focuses on land market access, poverty, and equity; the second looks at transaction costs and the efficiency of land markets; the third examines land contract choice, production efficiency, and equity; and the fourth focuses on the implications of land markets for land management. In the fifth part, we draw some overall conclusions based on our cross-country comparisons, draw relevant policy recommendations, and suggest areas for future research.

Land Market Access, Poverty, and Equity

We present an overview of our hypotheses HA1 through HA8 and the findings when testing them in the various case studies in the book in Table 14-1. A main rationale of land rental markets is that they contribute to enhancing allocative efficiency by reallocating land from households poor in nonland resources to households rich in nonland resources relative to their land endowments. This resource reallocation enhances the productivity of land as well as of nonland resources. We found supportive evidence of this in six of our case studies from Ethiopia, Kenya, and Malawi, providing solid evidence in favor of our hypothesis HA1.

Our hypothesis HA2 stated that land rental markets reallocate land from land-rich to land-poor households. As we can see from Table 14-1, we found mixed evidence on this in the case studies. Findings in Kenya, Malawi and Uganda were in line with this hypothesis, but the findings from four studies in Ethiopia were to the contrary. A major distinction between Ethiopia and the three other countries is that Ethiopia has had an egalitarian land distribution after the land reform in 1975. That may explain why land primarily moves from households poor in nonland resources to households relatively richer in nonland resources, and this causes a more skewed distribution of operational holdings than of own holdings in Ethiopia. An important explanation is that land typically moves from poor households without oxen to wealthier households with oxen (draught power), manpower, and managerial ability. Unlike many parts of Asia, where rental markets of draught animals function effectively (Hayami and Otsuka 1993b), the ownership of draught animals is a critical predictor of the direction of land rental transactions in oxen-based rainfed cropping systems in sub-Saharan Africa. Land distribution is in-egalitarian in the other three countries where land rental markets serve an important role by providing access to land for landless and near landless households.

Our HA3 hypothesis stated that land rental markets serve as a safety net for poor tenant households. Also for this hypothesis we found mixed evidence with the same geographical distinction as for the previous hypothesis. Land

TABLE 14-1 Land market access, participation and equity: Testing of hypotheses HA1–HA8.

Hypotheses	*Chapter/Country*								
	3 Ethiopia	*4 Ethiopia*	*5 Kenya*	*6 Malawi*	*7 Uganda*	*8 Ethiopia*	*9 Ethiopia*	*10 Ethiopia*	*12 Uganda*
HA1: Land rental markets contribute to enhance allocative efficiency such that households rich in non-land resource endowments rent in land from households that are poor in non-land resource endowments relative to their land endowments	++	++	++	++			++	++	
HA2: Land rental markets reallocate land from land-rich to land-poor households.	––	––	++	++	++	––		––	
HA3: Land rental markets serve as a safety net for poor tenant households	––	––		++	++			––	
HA4: Land rental markets serve as a safety net for poor landlord households	++	++		+?				++	
HA5: Poor households may be rationed out of land rental markets.	++	++		–?	–?			++	+?
HA6: Land markets can serve as a ladder out of poverty.					++				
HA7: Land sales markets cause a concentration of land in the hands of the land-rich and wealthy due to distress sales by the poor and lack of credit markets to assist the poor in buying land			––		––				
HA8: Increasing population pressure cause emergence and expansion of both land purchase and rental markets				++	+?				

++ indicates positive and strongly significant; + positive and significant; ? insignificant or mixed evidence; – negative and significant; and –– negative and strongly significant.

? insignificant or mixed evidence

rental markets may provide a safety net for poor landless and near landless households in Malawi, Uganda, and Kenya, but the prevalence of fixed-rent contracts in places like Kenya may inhibit the very poor from accessing the amount of land desired. A different pattern emerged in Ethiopia where tenants face larger entry barriers in the land rental market because of the need for oxen, plows, and managerial skills in order to enter the rental market. Kin relationships can sometimes help households to access land and such necessary resources to farm rented in land. We may therefore conclude that the land rental market may serve as a safety net for poor potential tenants in countries with skewed land distributions.

Similarly, hypothesis HA4 stated that land rental markets can serve as a safety net for poor landlord households. This turned out to be an important role of land rental markets in Ethiopia, while we found some indications of this being the case also in Malawi (Chapter 6 by Lunduka, Holden and Øygard), where labor-poor households were more likely to rent out land. Particularly poor female-headed households in Ethiopia depend on sharecropping out their land and the output share was an important source of food and income. Restrictions on land renting may undermine their tenure security and this may also undermine the land rental market's function as a safety net.

The next hypothesis HA5 stated that poor households are rationed out of land rental markets. We found that poor and near landless households in Ethiopia, such as young married couples, faced severe difficulties accessing land unless they could get such land from their relatives. The usual solution to this was comanagement of land with their parents while very few other landless and very poor households succeeded in getting access to land as tenants. With growing population pressure this is an increasingly important problem and contributing to a more skewed land distribution also in Ethiopia. There was less evidence of such rationing out of poor households in Malawi, Kenya, and Uganda. The fixed rent for the land would typically have to be paid at the beginning of the season such that cash-poor households would face difficulties renting in land. Borrowing of land was also common in Uganda and Malawi and may serve to reduce the entry barrier. We see that rationing is having direct implications for the safety net function of the land rental market. Households that are rationed out cannot use this market as a safety net. Still, the market may serve as a safety net for those on the other side of the market and for those that are not rationed if rationing is selective.

Can land markets serve as a ladder out of poverty for the poor (HA6)? Such a ladder effect may materialize in several ways. The classical pattern is that landless households may start out as landless workers and advance from there into tenancy, first as sharecroppers and then as fixed-rent tenants before they may be able to buy their own land (Spillman 1919; Wehrwein 1958). A more modern perspective sees land markets as creating more flexible opportunities to rent in or out and buy or sell land and thus allowing more dynamic adjustments through migration and exploitation of off-farm opportunities.

Regarding the classical version of the hypothesis, Place et al. (Chapter 5) found evidence of landless households buying land in Kenya and Uganda. However, we are not sure whether those buying land initially were poor except in land. Rented plot sizes were very small, which may suggest that they may serve as a safety net, but not as a big step in a ladder. Better data with life histories of people and household panel data should be collected to test this hypothesis more rigorously.

We may draw an important distinction between Uganda and Ethiopia. In Uganda, Deininger and Mpuga found (Chapter 7) evidence of land markets providing a ladder out of poverty for the poor. Land sales markets there offer an opportunity for households to get a decent income from their land if they sell it. This can then provide them with a starting capital such that they can afford to migrate and establish a business in town or a farm somewhere else where population density is lower. This may help them escape poverty but it will crucially depend on the conditions in the area they migrate to. We may contrast this with the situation in Ethiopia, where land sales are prohibited and where households are required to farm a share of the farm themselves, making migration more difficult and risky, as people may lose their land rights without compensation if they stay away for more than two years (Tigray region). Only households that are able to invest in their land for future sales can escape from poverty through land market transactions.

We found little evidence to support the fear that land sales markets cause a concentration of the land in the hands of the land-rich and wealthy due to distress sales by the poor and lack of credit markets to assist the poor in buying land (HA7). In smallholder communities in Kenya and Uganda, sales markets have been utilized by households with limited access to inherited land. Consequently, the land purchases have improved equality of land holdings as compared to the distribution of inherited land only. Looking more at a national level, the skewed distribution of land appears to be a primary result of colonial and subsequent land policies that have favored the wealthy or elite, the land-rich, and government officials. These large landowners typically did not even have to pay for the land they were allocated by the previous regimes. Market imperfections tend to cause low utilization levels for the land of land-rich owners in some of our case study countries, such as Malawi. A World Bank project is trying to reallocate such poorly utilized estate lands to poor landless households through buy-back and distribution. It remains to be seen how cost effective this approach can be.

We proposed that increasing population pressure causes emergence and expansion of both land purchase and land rental markets (HA8). From Malawi (Chapter 6) Lunduka, Holden, and Øygard provide new evidence of emerging land rental markets in central and southern parts of the country, and increasing population pressure appears to be an important reason for this. In Chapter 7, Deininger and Mpuga provide evidence of increased participation in land rental markets in Uganda from 1992 to 1999, and this may partly be due to increasing population pressure but may also have been facilitated by conducive policies and market development in general. Population pressure alone may not be sufficient to induce land markets to operate. Policy restrictions and traditional rules may also prevent land sales markets from developing even in very densely populated areas. There exists anecdotal evidence of illegal land sales also in peri-urban areas in Ethiopia. The functioning of informal rental (sharecropping) markets depends on past policies, their degree of implementation, local leadership, tenure security, and trust among people and in the government, and all these affect the economic incentives for land market participation. Better policies that enhance tenure security may also enhance the functioning of land markets, and this may be particularly useful in areas with high population pressure.

Transaction Costs and the Efficiency of Land Markets

In Chapter 2, we developed six hypotheses regarding transaction costs and their implications for the efficiency of land markets. These are presented in Table 14-2, with the findings from the tests in the case study chapters in the book.

Our first hypothesis (HB1) was that there are significant transaction costs in the land rental markets causing allocative inefficiency. We assessed this by looking at the coefficient for own land in land renting models on either side of the market. Earlier studies in Africa (Kevane 1997; Tikabo et al. 2008) provide mixed evidence on the allocative efficiency of African land rental markets. The new case studies presented in this book (Chapters 3, 4, 5, 6, 7, 9, and 10) provide overwhelming evidence that land rental markets are active, but they are characterized by significant transaction costs. The coefficients on own land are summarized in Table 14-3. Efficient allocation (assuming linear response to nonland factors of production) would imply that these coefficients should be equal to +1 in the rent-out models and –1 for the rent in models. We see that most of the coefficients are closer to zero than to 1 or –1 and most of them are significantly different from 0 as well as 1 and –1.

This may not be a surprise, given that land is an immobile and geographically dispersed resource causing the need to transport all other resources to be used in production to the land and outputs to be transported from the land and to the locations for consumption. However, with high population densities, such distances should be shorter and facilitate higher allocative efficiency of local rental markets for land when policies are conducive. Tenure insecurity and lack of trust among contracting parties in the market may explain high transaction costs and could indicate considerable potential for improvement in allocative efficiency. Policies that enhance tenure security and removal of restrictions on land rental markets should therefore contribute to reduce the related transaction costs and enhance allocative efficiency. Given our earlier findings, this should also be good for the land-poor, who should get better

TABLE 14-2 Transaction costs and the efficiency of land markets: Tests of hypotheses HB1–HB6

	Chapter/Country					
	3	*4*	*5*	*6*	*9*	*10*
Hypotheses	*Ethiopia*	*Ethiopia*	*Kenya*	*Malawi*	*Ethiopia*	*Ethiopia*
HB1: There are significant transaction costs in the land rental markets causing allocative inefficiency.	++	++	++	++	++	++
HB2: High fixed costs related to farming due to indivisibility of factors of production (like oxen for plowing) in combination with small farm sizes increase the importance of the land rental market to achieve efficiency in production.	++	++				
HB3: High transaction costs in the land rental market (and labor market) explain the inverse farm size-land productivity relationship because land-rich large farmers do not rent out land to land-poor small farmers up to the point in which the ratios of land to nonland factors are equalized.			+			
HB4: Land fragmentation and nonconvex transaction costs in relation to land cause land transfers to be lumpy (primarily transactions of whole plots).		++				
HB5: Kinship contracts enhance allocative efficiency in the land rental market.		++				
HB6: There is asymmetric adjustment in the rental market related to higher transaction costs on the tenant side than on the landlord side of the market.	++	++				

++ indicates positive and strongly significant; + positive and significant. Blank spaces indicate that these have not been tested.

TABLE 14-3 Estimated coefficients of owned area in land rent in and rent out functions

Location	*Rent out*	*Rent in*
Amhara–Ethiopia	0.27***	–0.18***
Tigray–Ethiopia	0.63–0.70**	0.29–0.33***
Kenya	.06***	–0.11– –0.24**
Uganda	0.07***	–0.29***
Malawi	0.35–0.38***	–0.22– –0.28***

***, ** and * indicate that the coefficients on own land are different from 1 for area rented out and from -1 for area rented in at 1%, 5% or 10% significance levels. n.s. indicate a non-significant result

access to land and for poor landowners who may get a better income by renting out their land.

The coefficients on own land in the rent-out models in the Amhara region in Ethiopia were substantially lower than those in the Tigray region; this may be due to more unfavorable past policies in the Amhara region than in the Tigray region. Land registration and certification programs like those introduced in the Tigray region in 1998 and in the Amhara region in 2003 may contribute to improvement of allocative efficiency in these markets and reduce market inefficiency differences between the regions. Land certification may enhance the tenure security of landlords and make them more willing to rent out their land, which should also improve the access to land by tenants (Holden et al. 2007a).

Would it be beneficial to formalize the land rental market? We have largely studied the emerging informal land rental markets and their rationale and implications in this book. It may be argued that there are benefits from formalizing these transactions to ensure that contracts are not violating the law and to provide legal support to the contract parties in cases of contract violations. This is a highly relevant issue since de Soto (2000) and the Commission for Legal Empowerment of the Poor have put such a strong emphasis on formalization of land rights and land markets. It is also highly relevant because Ethiopia has created new land laws at federal and regional levels in 2005 and 2007 that require formalization of land rental contracts. The questions are whether the benefits of such formalization are larger than the costs, and whether the local institutions have the capacity to handle the additional administrative work in a good way. Such a local land rental registry would need annual updating and would be outdated even much more quickly than a land ownership rights registry. It is not obvious that written and registered contracts always are better than oral and unregistered contracts. It is possible that a system with voluntary registration of such contracts is more cost effective, involves less bureaucracy, and decreases transaction costs for those with trustworthy contract partners than a system with compulsory registration. However, this topic needs further research.

Our second hypothesis regarding transaction costs and efficiency of land rental markets stated that high fixed costs related to farming due to indivisibility of factors of production (such as oxen for plowing) in combination with small farm sizes increase the importance of the land rental market to achieve efficiency in production (HB2). This appears to be an issue primarily in the oxen-based (cereal) cropping systems in Ethiopia and less so in the hoe-based cropping systems in the other countries of study. The poor functioning or nonexistence of oxen rental markets cause most households without oxen to fail to access oxen on time to cultivate their land. Households without oxen therefore rent out their land. Households with one ox may pair their ox with another household also owning one ox, since a pair of oxen is required for the heavy plowing activity. This may also explain why it is so difficult for landless households without oxen to enter the land rental market as tenants. It may also explain the surprising finding of a positive sign for own land for area rented in in Table 14-3 for the Tigray region (Chapter 4 by Ghebru and Holden). The new land proclamations in Ethiopia also impose minimum farm size restrictions, which make sense when further fragmentation is likely to cause more inefficient and costly land use. The new policy promotes comanagement in cases of inheritance and divorce when farm size is as low as 0.5 hectare. Similar development into very small farms is also taking place in southern Malawi, southwestern Uganda, and western Kenya but the efficiency implications of this fragmentation may not be serious in these hoe-based systems, as the hoe may be used efficiently on very small pieces of land. The same may also be the case in the perennial crop zone in southern Ethiopia. The policy implication of this is that overemphasis on land regulations without due attention to other factor markets can actually lead to the outcome that the policies were intended to prevent—increasing inequality in land use.

High transaction costs in the land rental market (and labor market) explain the inverse farm size–land productivity relationship because land-rich large farmers do not rent out land to land-poor small farmers up to the point in which the ratios of land to nonland factors are equalized. There is support for this hypothesis (HB3) from the Kenya analysis, although the explanation is complex. Maize productivity is found to be higher on smaller farms after controlling for a number of plot and household characteristics. The effect is subdued after accounting for soil fertility differences because smaller farms use fertilizer and manure more intensively than larger farms (Chapter 13). This implies that the possible poorer access to credit and fertilizer markets for smaller farms has not negatively affected their input use. On the contrary, imperfections in land and labor markets cause more intensive input use on small farms and the persistence of the inverse relationship.

We developed a theoretical model in Chapter 2 that demonstrated some of the implications of nonconvex transaction costs. Our hypothesis HB4 is derived from this, which states that land fragmentation and nonconvex transaction costs in relation to land cause land transfers to be lumpy (primarily transactions of whole plots). This theory is most likely to be relevant in the oxen-based cropping systems in Ethiopia and in parts of Uganda where land fragmentation is high and where the use of oxen cause economies of scale at small plots. Land distributions in Ethiopia were typically organized at village level, where the land was first divided by land quality. Afterward, each household received a share of each land quality class. This also typically resulted in households having dispersed plots and facing the difficulty of transporting their oxen and the plow between these dispersed plots. Further subdivision of these plots for renting purposes would not reduce the need for transportation of animals and equipment unless the whole plot is rented out. The study in Tigray (Chapter 4) revealed that it is rare for only shares of plots to be rented out. Rental decisions, therefore, tend to be whether a plot should be rented out or not; rarely is it about fine-tuning the area rented out by dividing a plot into sub-plots for renting purposes. This may also explain some of the fairly low coefficients on own land that we summarized in Table 14-3. Also in Uganda, in some areas, plots are very far from homes—perhaps an hour walk. Splitting the plot for renting would make no sense, as the costs of managing a small plot or monitoring a tenant would be almost prohibitive. The policy implications of this are whether it would reduce overall transaction costs to reduce the degree of land fragmentation. Policies in Ethiopia currently encourage voluntary land consolidation among individual households. The rationale of providing all households with a share of each land quality class in the village was to spread their risk and to enhance their ability to be self-sufficient with a range of crops that each may grow only on some types of land. Risks due to pests, droughts, and hailstorms may also affect the various plots differently. However, we are not aware of any good studies of the actual insurance premiums that households have to pay through this fragmented farm holding structure and whether other forms of insurance could be more cost effective. Relying less on self-sufficiency, and more on markets, on other safety net mechanisms, and possibly on rainfall insurance are alternatives that should be further explored. The pure subsistence strategy is also increasingly infeasible with the increasing population pressure in rain-fed agricultural systems.

Our hypothesis HB5 stated that kinship contracts enhance allocative efficiency in the land market. High trust among kin may reduce tenure insecurity and the need for monitoring and enforcement of land contracts and therefore reduce the transaction costs in the land rental market. In Chapter 4 (by Ghebru and Holden), it was found that land rental markets functioned better in communities with a high share of kinship contracts among all land rental contracts in line with our hypothesis. However, a high share of kinship contracts could also be interpreted as a sign of tenure insecurity, poorly developed land rental markets, and less efficient land use, as was found in Chapter 9 (by Holden and Bezabih) and in Chapter 10 (by Kassie and Holden). In Chapter 9, it was found that female landlords may be forced to rent out their land to their in-laws and these in-laws appear to be less efficient land managers. The female landlords appeared to be less able to evict their in-law tenants in order to find better tenants. Better policies may contribute to reduce this tenure insecurity and strengthen the bargaining position of women in the land rental market. The new land laws put in place in Ethiopia aim to enhance tenure security and to strengthen women's land rights. New evidences appear to demonstrate positive effects on women (Dokken 2007; Holden et al. 2007a; Holden and Tefera 2008). The option to formalize the land rental market by reporting contracts and requiring the consent of the family in relation to renting out family land may also contribute to reduce within-family moral hazard problems, but this requires further investigation. A preference for kinship contracts may in many cases be a sign of high trust and lower transaction costs rather than coercion and higher transaction costs.

Our final hypothesis in this section was that there is asymmetric adjustment in the land rental market related to higher transaction costs on the tenant side than on the landlord side. In Chapter 4, Ghebru and Holden found in Ethiopia that tenants are more likely to be rationed in the land rental market than landlords, indicating that the land rental market does not clear with a single rental price, which is due to the dominance of sharecropping, resulting in an asymmetry and rationing on the tenant side of the market. This asymmetry appeared to be less pronounced in the other countries where fixed-rent contracts dominated, but further research is needed to assess this.

Land Contract Choice, Production Efficiency, and Equity

We have listed hypotheses HC1 through HC10 from Chapter 2 in Table 14-4 with the summary of findings in the different case studies in the book.

Our first hypothesis on contract choice stated that contract choice is determined by the bargaining power of the contract parties such that the more powerful receives a larger share of the benefits. We were able to test this hypothesis in only one of the case studies (Chapter 8 by Tadesse, Holden, and Øygard) in southern Ethiopia, where we found no support for it. To the contrary, we found support for the second hypothesis that contract choice is associated with resource pooling such that it balances the factor ratios across contract partners for mutual benefit (HC2). This may indicate that sharecropping contracts in southern Ethiopia are not as exploitative as they used to be before the land reform in 1975, and this may be because most landlords now are poorer than their tenants and therefore in a weaker bargaining position.

Our third hypothesis was that sharecropping and cost-sharing contracts are a response to missing or imperfect credit markets (HC3). Tadesse, Holden, and Øygard found supportive evidence of this in Chapter 8, where we learned that tenants with access to credit were more likely to go for fixed-rent contracts by paying the rent at the beginning of the season. The advantage for tenants to choose sharecropping contracts is that they can delay payment of the rent until after harvest as a share of the output. However, sharecropping contracts are not common in our other case study countries, and imperfections in credit

TABLE 14-4 Contract choice, production efficiency and poverty: Tests of hypotheses HC1–HC10.

	Chapter/Country					
Hypotheses	*3* *Ethiopia*	*4* *Ethiopia*	*6* *Malawi*	*8* *Ethiopia*	*9* *Ethiopia*	*10* *Ethiopia*
HC1: Contract choice is determined by the bargaining power of the contract partners such that the more powerful gets a larger share of the benefits.				−−		
HC2: Contract choice is associated with resource-pooling such that it balances factor ratios for mutual benefit.	+	+		++	+	+
HC3: Sharecropping and cost-sharing contracts are a response to missing or imperfect credit markets.			−−	++		
HC3a: Fixed-rent contracts are chosen by landlords who need urgent cash (distress rentals) when credit markets do not work (Note that fixed rents are paid in the beginning of crop season).				−		
HC3b: Fixed-rent contracts are preferred by tenants with access to credit.				++		
HC3c: Sharecropping contracts are chosen by tenants who want to delay payment of land rent till after harvest.			−−	++		
HC4: Fixed-rent contracts are more common in areas with better market access and where cash-cropping is common because of better access to cash and credit in these economies.	+?	+				
HC5: Sharecropping dominates in food insecure and drought-prone areas (response to risk).		+	−−	+	+	+
HC6: Sharecropping is associated with Marshallian inefficiency.					+	−
HC7: Sharecropping tends to be chosen when the landlords are able to enforce efficient land management by tenants.						++
HC8: Opportunities for contract renewal enhance input use on sharecropped land and this counteracts Marshallian inefficiency.						++
HC9: Sharecropping among kin is inefficient because such contracts may be driven by other concerns than production efficiency.					++	+
HC10: Poor female landlord households face higher transaction costs in the land rental market and are associated with inefficient land use.					++	

++ indicates positive and strongly significant; + positive and significant; ? insignificant or mixed evidence; – negative and significant; and –– negative and strongly significant.

? insignificant or mixed evidence

markets appear not to induce formation of sharecropping contracts in these communities, even though this could have made it easier for poor tenants to rent in land. It is somehow contradictory that sharecropping is used in Ethiopia where tenants tend to be wealthier than the landlords; however, having asset wealth (mainly in the form of oxen) does not necessarily mean the household will have disposable cash at the onset of the rainy season.

We also have a number of related hypotheses (HC3a, HC3b, and HC3c) shown in Table 14-4. The first of these hypotheses (HC3a) states that fixed-rent contracts are preferred by landlords who need urgent cash. The case study in southern Ethiopia (Chapter 8) did not find evidence to support this but many of the studies revealed that landlords tended to be much poorer than tenants in Ethiopia, so the hypothesis may still be valid although we have not been able to test it rigorously in other case studies. We found evidence of tenants with access to credit preferring fixed-rent contracts in Chapter 8 (HC3b) and, likewise, tenants without access to credit preferring sharecropping contracts (HC3c). These findings are also in line with those by Laffont and Matoussi (1995) in Tunisia and by Tikabo and Holden (2003) in Eritrea.

Our next hypothesis stated that fixed-rent contracts are more common in areas with better market access and where cash cropping is common because of better access to cash and credit in these economies (HC4). We were unable to test this hypothesis rigorously in our case studies. Fixed-rent contracts dominated in all land rental markets in Kenya, Malawi, and Uganda, whereas sharecropping dominated in all our study areas in Ethiopia, including areas in the south, which are more cash crop oriented. The difference may therefore be more driven by culture and tradition that affect the transaction costs than the degree of market integration of the economies.

Hypothesis HC5 stated that sharecropping dominates in food-insecure and drought-prone areas (response to risk). The dominance of sharecropping in Ethiopia could also be because large parts of Ethiopia may be characterized as food insecure and drought prone. However, sharecropping also dominated in areas that are less drought prone while sharecropping was not common in food-insecure and drought-prone areas in Malawi. This is not because sharecropping contracts are unknown in Malawi, since such contracts have been widely used by tobacco landlords. We were therefore unable to impose a rigorous test of the risk explanation of sharecropping, as it can be confounded with other explanations.

Hypothesis HC6 is the classical Marshallian inefficiency hypothesis stating that sharecropping is associated with lower land productivity than on owner-operated land. The findings by Kassie and Holden in Chapter 10 clearly contradict this hypothesis, whereas another study in northern Ethiopia by Holden and Bezabih (Chapter 9) found significant evidence of Marshallian inefficiency. It should be added, however, that the Kassie and Holden chapter was able to impose a stricter test of this using household fixed effects, but this was not possible given the objectives in Chapter 9 by Holden and Bezabih.

Other studies from Ethiopia have also provided mixed results on this (Pender and Fafchamps 2006), as have studies in other parts of the world, but the majority of studies have not found significant Marshallian inefficiency (Otsuka and Hayami 1988; Otsuka et al. 1992; Otsuka 2007). It appears that Marshallian inefficiency is not as large as the Marshallian theory suggests. However, most of these studies did not control for endogeneity of contract choice, and so the findings may therefore be flawed. The policy implication is that the fear of Marsahllian inefficiency should not be used as a basis to prohibit sharecropping by law. Rather, sharecropping may compensate for multiple market imperfections and enhance social welfare in a second-best world.

Hypothesis HC7 proposes that sharecropping tends to be chosen when the landlords are able to enforce efficient land management by tenants. Kassie and Holden provide strong evidence of this in Chapter 10 in a case study area in northern Ethiopia where land productivity and use of fertilizer was found to be significantly higher on sharecropped plots than on tenants' owner-operated plots, and particularly so for non-kin tenants' plots. The explanation for this apparently surprising finding is that tenants want to renew their contracts and the best way for them to achieve this is by using more resources on their rented in land to increase the output to make their landlord happy. Landlords who have non-kin tenants are more likely to emphasize the managerial skills and productivity of their tenants when they consider renewal of contracts, but for kin partners other concerns may also matter, as the findings of Holden and Bezabih indicate (Chapter 9).

Our hypothesis HC8 states that opportunities for contract renewal enhance input use on sharecropped land and this counteracts Marshallian inefficiency. As we just discussed in relation to hypothesis HC7, the evidence of Kassie and Holden in Chapter 10 appears to be in line with this hypothesis. The primary mechanism of contract enforcement appeared to be open-ended sharecropping contracts with contract renewal based on performance. Kassie and Holden found that tenants used significantly more fertilizer on sharecropped plots than on their owner-operated plots in line with this hypothesis. Such selection by landlords is considered to be the main reason for general absence of Marshallian inefficiency in Asia (Hayami and Otsuka 1993).

Our next hypothesis states that sharecropping among kin is inefficient because such contracts may be driven by other concerns than production efficiency (HC9). Chapter 9 by Holden and Bezabih provides evidence that land productivity is lower on land rented out by female landlords to in-law tenants than on land rented out to other tenants. At the same time, they found that these kinship contracts were of longer duration than their contracts with other tenants. Therefore, female landlords appeared less interested or less able to evict their in-law tenants, even though they were less productive. This implies that kinship contracts do not necessarily enhance land use efficiency, although Ghebru and Holden found in Chapter 4 that kinship contracts may enhance allocative efficiency in the land rental market. We recommend further research to investigate the gender-kinship-coersion-moral hazard issues within families and their policy implications.

The final hypothesis relates to the previous one and states that poor female landlord households face higher transaction costs in the land rental market and are therefore associated with inefficient land use (HC10). Holden and Bezabih (Chapter 9) found evidence in support of this hypothesis. They show that tenants of female landlords were significantly older and had fewer oxen while duration of partnership also was significantly longer than that of male landlord households. This may indicate that women have a weaker bargaining position vis-a-vis their tenants and face higher transaction costs when attempting to evict their less productive tenants. Female landlords were also more likely to have contracts with in-law and blood-related tenants than male landlords were. Female landlords who rent out their land to in-law tenants are likely to live in the village of their dead or divorced husbands, and perhaps that explains their relatively weak position in land market negotiations. Strengthening of women's land rights through land registration and certification, like what has recently been introduced in Ethiopia, may reduce this problem and may thus enhance both equity and efficiency of land use.

Implications of Land Markets for Land Management

How do land markets affect land management and investment in land improvement? The effects may be different for land rental markets versus land sales markets. We present our key hypotheses from Chapter 2 in Table 14-5 together with a summary of the findings in Chapters 10 to 13.

Hypothesis HD1 stated that longer-term investments are lower on rented land than on land acquired through inheritance or purchase. The findings by Benin and Pender in two regions in Ethiopia (Chapter 11) were somewhat surprising and contradicted this hypothesis. Tenants were not found to invest less on rented land than on their own land. This could possibly be because tenants in this way may qualify for renewal of their contracts with their landlords, similar to what Kassie and Holden found in Chapter 10.

In Uganda, Nkonya, Pender, Benin, and Kato (Chapter 12) found significantly more investments in trees on purchased and owner-operated land than on rented plots in their descriptive analysis, whereas in their regression analysis they found a positive correlation between agroforestry investments and other NRM investments on rented plots. So, also in Uganda, it is not clear that land rental markets cause less investment on rented land.

In Kenya, Yamano and Deininger (Chapter 13) found evidence of less investment on rented land, as numbers of all types of trees were lower on rented than inherited or purchased plots. However, this may be due to the preference of landlords to rent out land with fewer investments, introducing a selection bias that could not be controlled for with the existing data. Mulching

TABLE 14-5 Land markets and land management: Tests of hypotheses HD1–HD7

	Chapter/Country			
	10	*11*	*12*	*13*
Hypotheses	*Ethiopia*	*Ethiopia*	*Uganda*	*Kenya*
HD1: Longer-term land investments are lower on rented land than land acquired through inheritance or purchase.		–	+?	++
HD2: Short-term investments are preferred on rented land.	++	–	+	+
HD3: Rented land is associated with less sustainable land management practices, including conservation structures.		–?	+	+
HD4: Tenants invest more in rented land in order to strengthen their property rights and probability of contract renewal.	+?	++		
HD5: HD1 will be offset by the allocative efficiency effect of land markets so that production efficiency on rented land is not lower than on permanently held land.	+			–
HD6: Land with land disputes is associated with lower use of farm inputs.				++
HD7: Poor female-headed households are more likely to suffer from land disputes.				++

++ indicates positive and strongly significant; + positive and significant; ? insignificant or mixed evidence; – negative and significant; and — negative and strongly significant.

? insignificant or mixed evidence

and manure were also found to be lower on maize and sometimes other crops. However, not all types of inputs followed this pattern and more hired labor was used on rented plots than on inherited plots.

Our second hypothesis (HD2) stated that short-term investments are preferred on rented land. Benin and Pender (Chapter 11) found in the Amhara region in Ethiopia that use of contour plowing, crop residues, and crop rotation were significantly more common on owner-cultivated plots than on rented plots, but no such significant differences were found in the Tigray region. Therefore, there is very limited evidence of land management being less active or worse on rented land than on owner-operated land in Ethiopia. The enhanced security of tenure of tenants and the possibility of contract renewal seems to create incentives for good land management on rented land. The exception may be the Amhara region, where a politically motivated land redistribution in 1997 may have discouraged short-term investments on rented land.

In Uganda, Nkonya, Pender, Benin, and Kato (Chapter 12) found more common adoption of slash and burn practices and less use of animal manure on rented land, in line with hypothesis HD2, while they also found more use of crop residues on rented plots, contradicting HD2.

Is rented land associated with less sustainable land management practices like we propose in hypothesis HD3? After controlling for reverse causality, Benin and Pender found little evidence of this in Ethiopia, although they found evidence of more intensive land use on owner-operated than sharecropped land in the Amhara region. The findings by Nkonya et al. in Chapter 12 in Uganda that rented land is associated with more slash and burn practices and less use of manure support hypothesis HD3 and may give reasons to be concerned.

Do tenants invest more in rented land in order to strengthen their property rights and the probability of contract renewal (HD4)? The findings by Kassie and Holden in Chapter 10 indicate that tenants use more fertilizer on rented-in plots, while Benin and Pender in Chapter 11 provide evidence of tenants investing on rented land even though their tenure is insecure due to land rental contracts being relatively short-term. As noted earlier, tenants in Kenya do not invest much in tree planting, which might be thought to help increase their land rights. However, the specific issue was not rigorously tested in Kenya.

More rigorous tests of hypotheses HD3 and HD4 could not be implemented with the existing data, making this an interesting area for future research building on household panel data. Given the strong emphasis on sustainable land management in the new land proclamations in Ethiopia, which also state that the tenant has the responsibility for sustainable management of rented in land, it is interesting to see whether such command-and-control policies may have any impact. It will depend crucially on the local land administrations and their ability to train and motivate the village level Land Administration Committees that so far have received no compensation for their work related to land registration and certification.

Hypothesis HD5 stated that HD1 is offset by the allocative efficiency effect of land markets so that production efficiency on rented and permanently held land can be similar. Since we found little evidence supporting hypothesis HD1 in Uganda and Ethiopia, it may be difficult to test this hypothesis. The findings of Kassie and Holden in Chapter 10 indicate that there are substantial allocative efficiency effects on land productivity. In Kenya, there is some evidence of lower productivity on rented land in Chapter 5 (recall that fixed rentals predominate) and there is further evidence that maize yields remain lower on rented plots as compared to inherited plots (but not compared to purchased plots) after controlling for soil fertility. This may be because of lower initial investment levels on rented land and other unobservable plot characteristics but could also be due to lower incentives for tenants to undertake short-term investments.

Finally, we have two hypotheses assessing the importance of land disputes. HD6 proposes that land with land disputes is associated with lower use of farm inputs. Yamano and Deininger found in Chapter 13 that pending conflicts were associated with significantly lower fertilizer application in their study of the importance of land conflicts in central and western Kenya. They also found that eviction and inheritance related conflicts have particularly strong negative impacts on organic fertilizer application. These findings are also in line with findings of Deininger and Castagnini (2004) in Uganda.

Hypothesis HD7 states that poor female-headed households are more likely to suffer from land disputes than are other households. Yamano and Deininger also found support for this hypothesis. They discovered that widows with land registered under the names of their deceased husbands were significantly more likely to have pending conflicts. The HIV/AIDS epidemic has contributed to a higher number of widows in Kenya as well as in other African countries. Lack of updating of land registries and land titles may therefore threaten the land rights of these vulnerable households. The possibility of strengthening their position would depend on local administrative capacity and the legal support they could access as well as traditional customs and local power structures. Even though Kenya has one of the most advanced land titling systems in Africa, it does not seem to be capable of protecting the rights of poor female-headed households. With the expansion of more low-cost land registration and certification programs in several regions in Ethiopia, it becomes important to assess how these affect local land conflicts, and particularly so for poor women who also have weak traditional land rights there. Holden et al. (2007b) find evidence that the reform has contributed to reduce the number of land-related disputes in Tigray, both during and after the reform. Holden and Tefera (2007) find similar effects in southern Ethiopia. Although a proposed land reform has not passed the parliament in Malawi, it may have important implications for poor female-headed households there, and the effects of the reform if or when it is implemented may be different in areas with matrilineal than in areas with patrilineal inheritance systems.

Overall Conclusions and Policy Implicatons

This book represents the first attempt to undertake a systematic assessment of the role of land rental and sales markets across African countries by drawing on household surveys that allow statistical tests of many issues that previously had been studied only in isolated case studies. Our effort brings out new evidence on a number of important and highly policy-relevant issues.

Our first broad question was whether land markets have an important role to redistribute land from land-rich to land-poor households in countries with an in-egalitarian land distribution while this is not the case in countries with more egalitarian land distribution. Our research findings seem to strongly support this in the case of land rental markets in Kenya, Malawi, and Uganda. Even though the land rental markets are not very active in these countries, we have strong statistical evidence that they reallocate land from land-rich to land-poor households. Future research should focus on how transaction costs could be reduced in land rental markets in these countries such that their land redistribution effect could be further enhanced.

We also find that land markets have an important role as a safety net for poor (potential) landlords in Ethiopia where land distribution is more egalitarian. Food and income from rented out land are very important, particularly for poor female-headed households that lack the capacity to farm the land themselves. Recent restrictions on how much land households are allowed to rent out may have negative effects on such poor households, increase their tenure insecurity and harm land use efficiency as well.

Our second broad question was whether land rental markets play a more important role in countries where land sales markets are prohibited than in countries where they are not. It appears that land rental markets play a more important role in Ethiopia where land sales markets are prohibited than in the other countries where land sales markets have been legalized. This could be because the land rental market substitutes for land sales and contributes to long-term reallocations of land, even though land rental contracts formally are only of a short-term nature. Evidence shows that it is very common to renew short-term rental contracts that tend to be open-ended. The evidence from Malawi and Uganda also shows that land rental markets are expanding and are more frequent than land purchases and sales, but the shares of households that participate in land rental markets in these countries were lower than in the case studies in Ethiopia. In Uganda, where land sales are legal, it is quite common for households to sell their land and migrate to town or to another rural area with lower population pressure to obtain more land there at a lower price.

Our third broad question was whether formal legalization enhances; and traditional (customary) land rights inhibit; development of land markets, and can land markets undermine the rights of people with only customary rights? Although we could not carry out a comprehensive assessment of this, our findings in Malawi indicate that traditional norms and power structures can constrain land market development. The findings by Holden et al. (2006) in Malawi also confirm this. However, it is not obvious that formalization of land rights and of land markets always enhances market development, as formalization also may add transaction costs to land transactions and thus have the opposite effect. Therefore, it becomes an empirical issue whether benefits of formalization for market development outweigh the costs. An example is the recent attempt to formalize land rental markets in Ethiopia where informal land rental markets work quite well. Holden and Tefera (2008) find that the requirement of joint certification of husbands and wives in Oromiya and SNNP regions of Ethiopia and the requirement of consent of the family before land is rented out may have reduced the amount of land renting.

Our fourth broad question was whether sharecropping is more common in food-insecure areas and in high-trust societies. There are likely to be cultural reasons for the strong dominance of sharecropping contracts in Ethiopia, whereas such contracts are rare in Kenya, Malawi, and Uganda. Occurrence of sharecropping may be due to risk-sharing motives of risk-averse tenants and landlords but could also originate from long traditions in feudal Ethiopia and as ways to get around the legal bans on land renting and hiring of labor during the Derg regime (1974–1991). Sharecropping has also been commonly used by tobacco landlords in Malawi and is common also in Eritrea, in cash crop areas in Ghana, and in intensive rice-growing areas in Madagascar. Our findings in this book add to the literature and shows that credit market imperfections and resource-pooling motives are important reasons for variations in sharecropping and cost-sharing contracts in southern Ethiopia. It is therefore possible that sharecropping is a sign of high trust, but it could also be a sign of strong enforcement power. These interpretations are consistent with the review of sharecropping literature in Asia (Hayami and Otsuka 1993b; Otsuka 2007). We could speculate and suggest that high trust and higher capital requirements explain why sharecropping is used to relax the credit constraint of tenants in some societies where tenants are wealthier than landlords, like in Ethiopia, but it is not used in other societies where tenants may be poorer than their landlords, like in Uganda and Kenya.

Our fifth broad question was whether land rental markets are more active in areas with oxen-based cropping systems than in hand hoe-based cropping systems. It is possible that the oxen-traction based farming systems that dominate in much of Ethiopia have a higher initial fixed cost and lower factor substitution possibilities, which increase the benefits from land rental markets when oxen rental markets do not work, as compared to hoe-based cropping systems that dominate in Kenya, Malawi, and Kenya. However, land rental markets are also very active in the perennial zone in southern Ethiopia where oxen are less important as traction power.

Labor markets function better than oxen rental markets but we also found evidence of household labor endowments, gender and age of household heads, and the managerial skills of household heads to affect both access to and participation in land rental markets, demonstrating significant market imperfections in labor markets as well. These imperfections again increase the importance of land rental markets to enhance flexibility of adjustment.

Our sixth broad question was whether land markets are a threat to poor women and female-headed households. Our findings with respect to the role of land rental markets, on the contrary, indicate that land rental markets are an important safety net for poor female-headed households in Ethiopia. Although we found signs in Chapter 9 that female landlords were less able to choose the best tenants and maximize the benefits from their sharecropping contracts, they tend to be poor in nonland resources and are less able to farm the land efficiently themselves. We found little evidence that land sales markets cause poor female households to sell their land (distress sales). Although anecdotal evidence on this exists, it appears to be so rare that random household surveys did not pick up much of this, at least in the countries where data were collected for the analyses in this book.

In Uganda, it was found that female-headed households were not participating as much in land rental markets as were male-headed households, but more research is needed here to identify why and whether they would benefit from more active participation. In Malawi, it was found that land rental markets were active in matrilineal as well as patrilineal areas, although the land rental markets appear to be somewhat less active in matrilineal areas. Households in matrilineal areas were less likely to rent out land, and matrilocal households were less likely to rent in land than patrilocal households. Patrilocal households rented in significantly more land than did matrilocal households. It is uncertain how the planned land policy reform (that has been put on hold due to resistance among local chiefs) will affect women in matrilineal areas in Malawi, while it may contribute to strengthen women's rights in patrilineal areas.

Our last broad question was related to what policies are most relevant to think of to enhance the role of land markets related to improving production efficiency, investment, and poverty reduction under the observed variations in the countries of study. Most of the past land policy reforms in Africa have not been pro-poor, even though they may have been well intended. Past land titling programs have tended to benefit the wealthy and powerful at the expense of the poor and marginalized, due to poor implementation, the high cost of obtaining titles in complex and often corrupt and inefficient bureaucracies, and due to limited or no formal recognition of customary land rights (Besley 1995; Cotula et al. 2004; Deininger 2003). An important question is: Has enough has been learned to avoid redoing the same mistakes in new land reforms? We will return to this when we discuss new land reforms below. But first we want to draw some policy lessons from our findings in this book.

The most important policy lessons that may be drawn from this book are

1. Land markets, and especially land rental markets, have important roles to play to enhance efficiency of land use, improve investments, and provide a safety net for the poor in rural areas where population densities are high and growing. Many land reforms have tended to ignore, misunderstand, or underestimate the potential importance of land rental markets.
2. The role of land markets may be substantially enhanced as devices for poverty reduction by conducive policies that strengthen tenure security and the land rights of the poor.
3. Land rental markets may be good for the poor both in countries with egalitarian and in-egalitarian land distributions. Removal of restrictions on such markets may lower the transaction costs, which may be good for efficiency as well as equity in environments with pro-poor land policies.
4. Sharecropping does not necessarily lead to less efficient land use and should be stimulated as such contracts provide credit to poor tenants, share risk among the parties, and provide higher food security to poor landlords. Fixed-rent contracts are not necessarily preferable to sharecropping contracts and may sometimes be a sign of distress rentals.
5. Restrictions on land rental markets often have unintended negative effects and should not be implemented unless it can be demonstrated that the interventions are necessary and provide the intended benefits. We advocate a cautious pilot project approach to test new interventions.

Some of the important new land reform initiatives, including pro-poor approaches, include

1. Formalization of private property rights to land to "wake up" the poor's dead capital (de Soto 2000), which is being implemented in Tanzania.
2. Formalizing customary land rights as a way to build on traditional land rights systems like what is being tried in Tanzania and proposed in Malawi.
3. Low-cost, broad-based, and high-speed land registration and certification like what is being implemented in Ethiopia.
4. Strengthening of women's land rights such as joint certification of husbands and wives, like what has been implemented in some regions in Ethiopia.
5. Changes in restrictions on land sales and land rental markets, such as allowing land sales in Uganda, allowing more long-term rental contracts in Ethiopia, and limiting the share of the farm that can be rented out (Ethiopia).
6. Formalizing the land rental markets by requiring land rental contracts to be written, reported, accepted, and registered in the local land administration or community administration (planned implemented in Ethiopia).
7. Imposing stronger land management restrictions on land users to enhance more sustainable land management, such as making secure tenure rights conditional on proper land use (recently introduced in Ethiopia).

This book should provide a good starting point for assessing whether these new land reform initiatives enhance the development of land markets and whether land markets undermine the rights of people with only customary rights. So far, there has been limited research on these new reforms in Africa. Recent exceptions include Holden et al. (2007a, 2007b, 2008), who provide evidence that the third type of reform above—low-cost, broad-based, and high-speed land registration and certification—has increased land rental market activity, stimulated investments on land and land productivity, and reduced land-related disputes in the Tigray region in Ethiopia. Holden et al. (2006) also provide evidence that customary rules and local power structures (local chiefs) tend to constrain land markets development in southern and central Malawi. Lack of transparency and power concentration may make it difficult to introduce pro-poor reforms. Important keys to successful reforms are broad-based local participation, commitment by land administrations, support by local governments, transparent processes, and local demand for the reform (Holden and Tefera 2008).

The recent low-cost registration and certification programs that have been implemented in four regions in Ethiopia deserve close scrutiny because they may not exclude the poor in the same way as some other titling programs did due to their low-cost, broad-based, and rapid implementation. They may provide interesting lessons that have relevance in other African countries. It remains an open question whether such a low-cost approach can succeed in controlling corruption and provide a qualitative improvement that can yield sustainable benefits for the poor. A key question is the ease with which the poor can update subdivisions and transfers and the costs involved. Kenya's registration and titling program was initially surely pro-poor, though not pro-women, but land registers are currently less likely to be updated in the poorer regions. Similar research on the land reform in Tanzania can provide additional insights. There is some anecdotal evidence from Tanzania that the type 1 and type 2 forms of land reform stated above may have marginalized and displaced pastoralists and that land grabbing and sales of customary lands may have had negative effects on other vulnerable groups (Benjaminsen and Sjaastad 2007). Attempts at formalizing customary land rights may lead to more conflicts when applied in settings where there are conflicts of interest and imbalances in powers among the different stakeholders (Chimhowu and Woodhouse 2006).

Land policy reforms may be important to strengthen women's land rights that traditionally are weak in many African countries. Women are in many places perceived as the property of men, and it may be a long step from there to becoming equal owners. Our research has demonstrated that strengthening women's land rights may strengthen their rights to rent out their land, but it may also affect the amount of land they retain in cases of divorce and death of their husbands (Dokken 2007; Holden and Tefera 2008). However, the inheritance rights may be equally important in the longer run and for how equitable the future distribution of land and land rights will be. Secure tenure rights and the rights to rent out land are also very important for old, sick, and orphan households that lack nonland resources to farm the land efficiently themselves. Such policies would be pro-poor and enhance both efficiency and equity. One should, however, be realistic about how difficult it is to change deep-rooted cultural gender roles. Changing the laws may not make any difference unless legal support is provided at the local level.

The large-scale social experiments that the new land reforms represent are also an excellent opportunity for implementing high-quality impact assessments that can provide valuable new insights. A pilot program approach allowing learning by doing, starting out simple (low-cost approach), while ensuring strong local participation, rather than using a top-down high-tech approach beyond the comprehension of local people may also have many advantages. There can be high risks of strategic responses by the informed and powerful both before and during such reform programs, and this may cause displacement of the poor and vulnerable unless a very cautious, inclusive, and broad participatory process is chosen. Such reforms may also cause land markets to "go wild" and cause highly distorted land prices (see, e.g., Benjaminsen and Sjaastad 2002 for an example from Mali).

So far there are very few good impact assessments of how such new land reforms and their impacts on land markets, investments, and land disputes affect the welfare of the poor. Systematic research combining analysis of household panel data with experimental designs should be able to tease out valuable policy lessons on what works and what does not work so well in these reforms. And we must hope that African policymakers are willing to listen to the lessons from research that would be a crucial step toward better land policymaking in the years to come.

Overall, this book has demonstrated that land markets play an increasingly important role in many African countries and that land markets have a potential to benefit the poor more than representing a threat. Land markets serve to overcome market failures in markets for credit, insurance, traction power, and managerial skills, and can thus enhance resource use efficiency and welfare of all parties involved as long as such transactions are voluntary. Policies that strengthen the land rights of the poor and vulnerable are, however, crucial to ensure that the poor get a larger share of these benefits and to prevent exploitation and violation of their basic human rights. Formalization of the land rights of the poor may be an important step in the right direction if formalization does not undermine their informal rights. Development assistance will be crucial to mobilize sufficient resources for even low-cost land administration reforms, since poor African countries may be unable to carry the burden of such large-scale implementation projects alone. International organizations also have an important role to play to facilitate exchange of experiences and to provide expertise to the reform countries.

References

Abate, T. 1995. Land Redistribution and Intra-Household Relations: The Case of Two Communities in Northern Ethiopia. *Ethiopian Journal of Development Research* 17 (1): 23–42.

Ackerberg, D. A., and M. Botticini. 2000. The Choice of Agrarian Contracts in Early Renaissance Tuscany: Risk Sharing, Moral Hazard, or Capital Market Imperfection? *Explorations in Economic History* 37 (3): 241–257.

Adams, M., S. Sipho, and S. Turner. 1999. Land Tenure Reform and Rural Livelihoods in Southern Africa. Natural Resource Perspectives Number 39. London: Overseas Development Institute.

Adesina, A., and J. Chianu. 2002. Determinants of Farmers' Adoption and Adaptation of Alley Farming Agroforestry Technology in Nigeria. *Agroforestry Systems* 55 (2): 99–112.

Ahmed, M. M., B. Gebremedhin, S. Benin, and S. Ehui. 2002. Measurement and Sources of Technical Efficiency of Land Tenure Contracts in Ethiopia. *Environment and Development Economics* 7 (3): 507–527.

Alemu, T. 1999. *Land Tenure and Soil Conservation: Evidence from Ethiopia.* Göteborg: Department of Economics, Göteborg University.

Allen, F. 1982. On Share Contracts and Screening. *Bell Journal of Economics* 13 (2): 541–547.

Amare, Y. 1995. *Household Economic Status, Seasonality and Food Strategies in Ethiopia: The Case of Wogda.* Northern Shewa: Thesis (Ph.D.), Boston University.

Amsalu, T. A., and J. de Graaff. 2006. Farmers' Views of Soil Erosion Problems and Their Conservation Knowledge in the Beressa River Catchment, central Highlands of Ethiopia. *Agriculture and Human Values* 23 (1): 99–108.

Andre, C., and J. P. Platteau. 1998. Land Tenure under Unbearable Stress: Rwanda Caught in the Malthusian Trap. *Journal of Economic Behavior and Organization* 34 (1): 1–47.

ANRS (Amhara National Regional State). 2000. Proclamation Issued to Determine the Administration and Use of the Rural Land in the Amhara National Region. Proclamation No. 46/2000. Bahir Dar, Ethiopia: Zikre Hig.

ANRS (Amhara National Regional State). 2006. Revised Rural Land Administration and Use Proclamation. Proclamation No.133/2006. Bahir Dar, Ethiopia: Zikre Hig.

Atwood, D. A. 1991. Land Registration in Africa: The Impact on Agricultural Production. *World Development* 18 (5): 659–671.

Baland, J. M., F. Gaspart, F. Place, and J. P. Platteau. 2003. The Distributive Impact of Land Markets in Central Uganda. Working Paper. 252-2003/13. University of Namur.

Baland, J., F. Gaspart, F. Place, and J. Platteau. 2007. The Distributive Impact of Land Markets in Uganda. *Economic Development and Cultural Change* 55 (2): 283–312.

Bamwerinde, W., B. Bashaasha, W. Ssembajjwe, and F. Place. 2006. The Puzzle of Idle Land in the Densely Populated Kigezi Highlands of Southwestern Uganda. *International Journal for Environment and Development* 3 (1): 1–13.

Banerjee, A. V., J. P. Gertler, and M. Ghatak. 2002. Empowerment and Efficiency: Tenancy Reform in West Bengal. *Journal of Political Economy* 110 (2): 239–280.

Bardhan, P. K. 1984. *Land, Labor and Rural Poverty: Essays in Development Economics.* New York: Columbia University Press.

Barrows, R., and M. Roth. 1990. Land Tenure and Investment in African Agriculture: Theory and Evidence. *Journal of Modern African Studies* 28 (2): 265–297.

Basu, K. 1992. Limited Liability and the Existence of Share Tenancy. *Journal of Development Economics* 38 (1): 203–220.

Becker, S. A., and A. Ichino. 2002. Estimation of Average Treatment Effects Based on Propensity Scores. *Stata Journal* 2 (4): 358–377.

Bell, C. 1977. Alternative Theories of Sharecropping: Some Tests Using Evidence from Northeast India. *Journal of Development Studies* 13 (4): 317–346.

Bell, C., and C. Sussangkarn. 1988. Rationing and Adjustment in the Market for Tenancies: The Behavior of Landowning Households in Thanjavur District. *American Journal of Agricultural Economics* 70 (4): 779–789.

Bell, C., and P. Zusman. 1976. A Bargaining Theoretic Approach to Crop Sharing Contracts. *American Economic Review* 66 (4): 578–588.

Bellemare, M. 2006. Testing between Competing Theories of Reverse Share Tenancy. Terry Sanford Institute of Public Policy Working Paper. Durham: Duke University.

Benin, S. 2006. Policies and Programs Affecting Land Management Practices, Input Use and Productivity in the Highlands of Amhara Region, Ethiopia. In *Policies for Sustainable Land Management in the East African Highlands,* edited by J. Pender, F. Place, and S. Ehui (pp. 218–256). Washington, DC: IFPRI.

Benin, S., M. Ahmed, J. Pender, and S. Ehui. 2005. Development of Land Rental Markets and Agricultural Productivity Growth: The Case of Northern Ethiopia. *Journal of African Economies* 14 (1): 21–54.

Benjamin, D. 1995. Can Unobserved Land Quality Explain the Inverse Productivity Relationship? *Journal of Development Economics* 46 (1): 51–84.

Benjamin, D., and L. Brandt. 2002. Property Rights, Labor Markets, and Efficiency in a Transition Economy: The Case of Rural China. *Canadian Journal of Economics,* 35 (4): 689–716.

Benjaminsen, T. A., and E. Sjaastad. 2007. *Property Rights Formalisation in Africa.* Uppsala: Nordiska Afrikainstitutet.

Benneh, G. 1970. The Impact of Cocoa Cultivation on the Traditional Land Tenure of the Akan of Ghana. *Ghana Journal of Sociology* 6 (1): 43–61.

Berry, R. A. and W. R. Cline. 1979. *Agrarian Structure and Productivity in Developing Countries.* Baltimore: Johns Hopkins University Press.

Berry, S. 1975. *Cocoa, Custom and Socio-economic Change in Rural Western Nigeria.* Oxford: Clarendon Press.

Besley, T. 1995. Property Rights and Investment Incentives: Theory and Evidence from Ghana. *Journal of Political Economy* 103 (5): 903–937.

Beyene, A. 2004. Land Tenure Institutions and the Smallholder-Farming Systems: Challenges for Agricultural Development. Paper presented at the 12th Annual Conference on the Ethiopian Economy, Ethiopian Economic Association: Addis Ababa.

Bezabih, M., and S. T. Holden. 2006, August. Tenure Insecurity, Transaction Costs in the Land Lease Market and their Implications for Gendered Productivity Differentials. Paper presented at the 26th International Conference of the International Association of Agricultural Economists, Gold Coast, Australia.

Bidinger, P. D., T. S. Walker, B. Sarkar, A. R. Murty, and P. Babu. 1991. Consequences of Mid1980s Drought: Longitudinal Evidence from Mahbubnagar. *Economic and Political Weekly* 26 (4): 105–114.

Bikaako, W., and J. Ssenkumba. 2003. Gender, Land and Rights: Contemporary Contestation in Law, Policy and Practice in Uganda. In *Women and Land in Africa—Culture, Religion and Realizing Women's Rights*, edited by L. M. Wanyeki (pp. 232–277). London, New York: Zed Books.

Binswanger, H. P., K. Deininger, and G. Feder. 1993. Agricultural Land Relations in the Developing World. *American Journal of Agricultural Economics* 75 (5): 1242–1248.

Binswanger, H. P., K. Deininger, and G. Feder. 1995. Power, Distortions, Revolt and Reforms in Agricultural Land Relations. In *Handbook of Development Economics IIIB*, edited by J. Behrman and T. N. Srinivasan (pp. 2659–2772). Amsterdam: Elsevier.

Binswanger, H. P., and M. Rosenzweig. 1986. Behavioral and Material Determinants of Production Relations in Agriculture. *Journal of Development Studies* 22 (3): 503–539.

Blarel, B. 1994. Tenure Security and Agricultural Production under Land Scarcity: The Case of Rwanda. In *Searching for Land Tenure Security in Africa*, edited by J. W. Bruce and S. E. Migot-Adholla (pp. 71–96). Dubuque, IA: Kendall/Hunt.

Bliss, C. J., and N. H. Stern. 1982. *Palanpur: The Economy of an Indian Village*. Delhi and New York: Oxford University Press.

Boadu, F. 1992. The Efficiency of Share Contracts in Ghanaís Cocoa Industry. *Journal of Development Studies* 29 (1): 108–120.

Boeder, R. B. 1984. *Silent Majority: A History of the Lomwe People in Malawi*. Pretoria: African Institute of South Africa.

Brandao, A. S. P., and G. C. Rezende. 1992. Credit Subsidies, Inflation and the Land Market in Brazil: A Theoretical and Empirical Analysis. Working Paper. Washington DC: World Bank.

Braselle, A., F. Gaspar, and J. Platteau. 2002. Land Tenure Security and Investments Incentives: Puzzling Evidence from Burkina Faso. *Journal of Development Economics* 67 (2): 373–418.

Braverman, A., and J. E. Stiglitz. 1986. Cost-Sharing Arrangements under Sharecropping: Moral Hazard, Incentive Flexibility, and Risk. *American Journal of Agricultural Economics* 68 (3): 642–652.

Bruce, J. 1988. A Perspective on Indigenous Land Tenure Systems and Land Concentration. In *Land and Society in Contemporary Africa*, edited by R. E. Downs and S. P. Reyna (pp. 23–52). London: University Press of New England.

Bruce, J. W., A. Hoben, and D. Rahmato. 1994. *After the Derg: An Assessment of Rural Land Tenure Issues in Ethiopia*. Madison: Land Tenure Center, University of Wisconsin-Madison.

Bruce, J. W., and S. E. Migot-Adholla. 1994. *Searching for Land Tenure Security in Africa*. Dubuque, IA: Kendall/Hunt.

Burgess, R. 2001. Land and Welfare: Theory and Evidence from China. Working Paper. London: London School of Economics.

Cain, M. 1981. Risk and Insurance: Perspectives on Fertility and Agrarian Change in India and Bangladesh. *Population and Development Review* 7 (3): 435–474.

Carter, M. R. 1984. Identification of the Inverse Relationship between Farm Size and Productivity: An Empirical Analysis of Peasant Agricultural Production. *Oxford Economic Papers* 36 (1): 131–145.

Carter, M. R., K. Wiebe, and B. Blarel. 1994. Tenure Security for Whom? Differential Effects of Land Policy in Kenya. In *Searching for Land Tenure Security in Africa*, edited by J. W. Bruce and S. E. Migot-Adholla (pp. 141–168). Dubuque, IA: Kendall/Hunt.

Carter, M. R., and F. J. Zimmerman. 2000. The Dynamic Cost and Persistence of Asset Inequality in an Agrarian Economy. *Journal of Development Economics* 63 (2): 265–302.

Chauveau, J. P. 2000. Question Fonciere et Construction Nationale en Cote d'Ivoire. *Politique Africaine* 78: 94–125

Cheung, S. N. S. 1968. Private Property Rights and Sharecropping. *Journal of Political Economy* 76 (6): 1107–1122.

Cheung, S. N. S. 1969. *The Theory of Share Tenancy*. Chicago: University of Chicago Press.

Chimhowu, A., and P. Woodhouse. 2006. Customary vs. Private Property Rights? Dynamics and Trajectories of Vernacular Land Markets in Sub-Saharan Africa. *Journal of Agrarian Change* 6 (3): 346–371.

Chunrong Al, J. L. Arcand, and F. Ethier. 1996. *Moral Hazard and Marshallian Inefficiency: Evidence from Tunisia*. Cahier 0896. Quebec, Canada: University of Montreal, Center for Economic Development Research.

Clay, D. C., T. Reardon, and J. Kangasniemi. 1998. Sustainable Intensification in the Highland Tropics: Rwandan Farmers' Investments in Land Conservation and Soil Fertility. *Economic Development and Cultural Change* 46 (2): 351–378.

Collier, P., and D. Lal. 1986. *Labor and Poverty in Kenya 1900–1980*. Oxford: Clarendon Press.

Commission for Africa. 2005. Our Common Interest: Report of the Commission for Africa. http://www.commissionforafrica.org/english/report/introduction.html

Cotula, L., C. Toulmin, and C. Hesse. 2004. *Land Tenure and Administration in Africa: Lessons of Experience and Emerging Issues*. London: International Institute for Environment and Development.

Daudelin, J. 2002, April. *Land and Violence in Post-Conflict Situations* (conference draft). Ottawa: North-South Institute for the World Bank.

Deaton, A. 1991. Saving and Liquidity Constraints. *Econometrica* 59 (5): 1221–1248.

Deaton, A. 1997. *The Analysis of Household Surveys: A Microeconometric Approach to Development Policy*. Washington, DC: World Bank.

Dehejia, R. H., and S. Wahba. 2002. Propensity Score Matching Methods for Non-Experimental Causal Studies. *Review of Economics and Statistics* 84 (1): 151–161.

Deininger, K. 2003. *Land Policies for Growth and Poverty Reduction: A World Bank Policy Research Report*. Oxford and New York: World Bank and Oxford University Press.

Defoer, T., A. Budelman, C. Toulmin, and S. Carter. 2000. Building Common Knowledge: Participatory Learning and Action Research. *In Managing Soil Fertility in the Tropics. A Resource Guide for Participatory Learning and Action Research*, edited by T. Defoer and A. Budelman (pp. 39–46). Amsterdam: Royal Tropical Institute.

Deininger, K., D. A. Ali, S. T. Holden, and J. Zevenbergen. 2007. Rural Land Certification in Ethiopia: Process, Initial Impact, and Implications for Other African Countries. World Bank Policy Research Working Paper 4218. Washington, DC: World Bank.

Deininger, K., and R. Castagnini. 2004. Incidence and Impact of Land Conflict in Uganda. World Bank Policy Research Working Paper 3248. Washington, DC: World Bank.

Deininger, K., and R. Castagnini. 2006. Incidence and Impact of Land Conflict in Uganda. *Journal of Economic Behavior and Organization* 60 (3): 321–345.

Deininger, K., and S. Jin. 2002. Land Rental Markets as an Alternative to Government Reallocation? Equity and Efficiency Considerations in the Chinese Land Tenure System. World Bank Policy Research Working Paper 2930. Washington, DC: World Bank.

Deininger, K., and S. Jin. 2005. The Potential of Land Markets in the Process of Economic Development: Evidence from China. *Journal of Development Economics* 78 (1): 241–270.

Deininger, K., and S. Jin. 2006. Tenure Security and Land-Related Investment: Evidence from Ethiopia. *European Economic Review* 50 (5):1245–1277.

Deininger, K., S. Jin, and H. K. Nagarajan. 2007. Determinants and Consequences of Land Sales Market Participation: Panel Evidence from India. Policy Research Working paper 4323. Washington, DC: World Bank.

Deininger, K., S. Jin, and M. K. Nagarajan. 2008. Efficiency and Equity Impacts of Rural Land Rental Restrictions: Evidence from India. European Economic Review 52(5):892–918

Deininger, K., and P. Mpuga. 2002. Land Markets in Uganda: Incidence, Impact and Evolution over Time. World Bank Discussion Paper. Washington DC: World Bank.

Deininger, K., and L. Squire. 1998. New Ways of Looking at Old Issues: Inequality and Growth. *Journal of Development Economics* 57 (2): 259–287.

de Janvry, A., G. Gordillo, J. P. Platteau, and E. Sadoulet. 2001. *Access to Land, Rural Poverty, and Public Action*. Oxford: Oxford University Press.

Delville, P. L., C. Toulmin, J. P. Chauveau, and P. P. Colin. 2002. *Negotiating Access to Land in West Africa: A Synthesis of Findings from Research on Derived Rights to Land.* London: International Institute for Environment and Development.

de Soto, H. 2000. *The Mystery of Capital: Why Capitalism Triumphs in the West and Fails Everywhere Else.* New York: Basic Books.

de Zeeuw, F. 1997. Borrowing of Land, Security of Tenure, and Sustainable Land Use in Burkina Faso. *Development and Change* 28 (3): 583–595.

Dijkstra, A., and J. Donge. 2001. What Does the Show Case Show? Evidence and Lessons from Adjustment in Uganda. *World Development* 29 (5): 841–863.

Dokken, T. 2007. *Women's Land Rights in Tigray, Ethiopia.* Norway: Norwegian University of Life Sciences, Ås.

Dong, X. Y. 1996. Two-Tier Land Tenure System and Sustained Economic Growth in Post-1978 Rural China. *World Development* 24 (5): 915–928.

Dorward, A. 1999. Farm Size and Productivity in Malawian Smallholder Agriculture. *Journal of Development Studies* 35 (5): 141-–161.

Drimie, S. 2002. *HIV/AIDS and Land: Case Studies from Kenya, Lesotho, and South Africa.* Pretoria, South Africa: Human Science Research Council.

Dubois, P. 2002. Moral Hazard, Land Fertility and Sharecropping in Rural Area of the Philippines. *Journal of Development Economics* 68 (1): 35–64.

Dubuisson, J. F. 1988. Evolution du Régime Foncier et Intensification de L'Agriculture en Afrique Sub-Saharienne: Le Cas d'un Village du Sud Bénin. M.A. Thesis, Faculté des Sciences économique, sociales et de gestion, University of Namur.

Duncan, A. 1990. *The Impact of Pricing Policy on Tobacco Tenancy in Malawi.* Oxford: Food Studies Group, Oxford University.

Ege, S. 1997. The Promised Land: The Amhara Land Redistribution of 1997. SMU-rapport 5/97. Dragvoll, Norway: Norwegian University of Science and Technology, Centre for Environment and Development.

Ervin, C. A., and D. E. Ervin. 1982. Factors Affecting the Use of Soil Conservation Practices: Hypotheses, Evidence, and Policy Implications. *Land Economics* 58 (3) 277–292.

Eswaran, M., and A. Kotwal. 1985. A Theory of Contractual Structure in Agriculture. *American Economic Review* 75 (3): 352–367.

Fafchamps, M. 2004. *Market Institutions in Sub-Saharan Africa. Comparative Institutional Analysis* (CIA) Series. Cambridge, MA & London: MIT Press.

FAO. 2005. *The State of Food Insecurity in the World 2005: Eradicating World Hunger, Key to Achieving the Millennium Development Goals.* Rome: Food Agriculture Organization of the United Nations.

FAOSTAT. 2006. www.faostat.fao.org

FDRE (Federal Democratic Republic of Ethiopia). 1997. Rural Land Administration Proclamation of the Federal Government of Ethiopia. Proclamation No. 89/1997. Addis Ababa, Ethiopia: Federal Negarit Gazeta.

FDRE (Federal Democratic Republic of Ethiopia). 2002. *Sustainable Development and Poverty Reduction Program.* Addis Ababa, Ethiopia: MOFED.

FDRE (Federal Democratic Republic of Ethiopia). 2005. Rural Land Administration and Use Proclamation. Proclamation No. 456/2005. Addis Ababa, Ethiopia: Federal Negarit Gazeta.

Feder, G. 1985. The Relation between Farm Size and Farm Productivity : The Role of Family Labor, Supervision and Credit Constraints. *Journal of Development Economics* 18 (2–3): 297–313.

Feder, G., and D. Feeny. 1991. Land Tenure and Property Rights: Theory and Implications for Development Policy. *World Bank Economic Review* 5 (1): 135–153.

Feder, G., L. J. Lau, J. Y. Lin, and X. Luo. 1992. The Determinants of Farm Investment and Residential Construction in Post-Reform China. *Economic Development and Cultural Change* 41 (1): 1–26.

Feder, G., and R. Noronha. 1987. Land Rights Systems and Agricultural Development in Sub-Saharan Africa. *World Bank Research Observer* 2 (2): 143–169.

Fred-Mensah, B. K. 1999. Capturing Ambiguities: Communal Conflict Management Alternative in Ghana. *World Development* 27 (6): 951–965.

Gavian, S., and S. Ehui. 1999. Measuring the Production Efficiency of Alternative Land Tenure Contracts in a Mixed Crop-Livestock System in Ethiopia. *Agricultural Economics* 20 (1): 37–49.

Gavian, S., and M. Fafchamps. 1996. Land Tenure and Allocative Efficiency in Niger. *American Journal of Agricultural Economics* 78 (2): 460–471.

Ghatak, M., and P. Pandey. 2000. Contract Choice in Agriculture with Joint Moral Hazard in Effort and Risk. *Journal of Development Economics* 63 (2): 303–326.

Ghebru, H., and S. T. Holden. 2006. Factor Market Imperfections and Rural Markets in Northern Ethiopian Highlands. Norway: Norwegian University of Life Science. Mimeo.

Giller, K. E., G. Cadisch, C. Ehaliotis, E. Adams, W. D. Sakala, and P. L. Matongoya. 1997. Building Soil Nitrogen Capital in Africa. In *Replenishing Soil Fertility in Africa,* edited by R. J. Buresh, A. Sanchez, and F. Calhoum (pp. 151–183). Madison: American Society of Agronomy and Soil Science Society of America.

Golan, E. 1994. Land Tenure Reform in the Peanut Basin of Senegal. In *Searching for Land Tenure Security in Africa,* edited by J. W. Bruce and S. E. Migot-Adholla (pp. 231–250). Dubuque, IA: Kendall/Hunt.

Goldstein, M., and C. Udry. 2005. Landrights and Agricultural Investment in Ghana. Working papers 929. Economic Growth Center, Yale University.

Green, J. 1987. Evaluating the Impact of Consolidation of Holdings, Individualization of Tenure, and Registration of Title: Lessons from Kenya. Land Tenure Center Paper No. 129. Madison: University of Wisconsin.

Greene, W. H. 2003. *Econometric Analysis.* New York: Macmillan.

Gunjal, K., S. Williams, and R. Romain. 1996. Agricultural Credit Subsidies and Farmland Values in Canada. *Canadian Journal of Agricultural Economics* 44 (1): 39–52.

Hagos, F. and S. Holden. 2003. Incentives for Conservation in Tigray, Ethiopia: Findings from a Household Survey. Department of Economics and Social Sciences, Agricultural University of Norway. Mimeo.

Hallagan, W. S. 1978. Self-Selection by Contractual Choice and the Theory of Sharecropping. *Bell Journal of Economics* 9 (2): 344–354.

Haugerud, A. 1989. Land Tenure and Agrarian Change in Kenya. *Africa* 59 (1): 61–90.

Hayami, Y., and K. Otsuka. 1993a. *Kasugpong* in the Philippine Rice Bowl: The Emergence of New Labor Institutions after the Land Reform, In *The Economics of Rural Organization: Theory, Practice, and Policy,* edited by A. Braverman, K. Hoff, and J. E. Stiglitz (pp. 292–304). Oxford: Oxford University Press.

Hayami, Y., and K. Otsuka. 1993b. *The Economics of Contract Choice: An Agrarian Perspective.* Oxford: Oxford University Press.

Hill, P. 1963. *The Migrant Cocoa-Farmers of Southern Ghana.* Cambridge: Cambridge University Press.

Holden, S. T., and M. Bezabih. 2006. Why Is Land Productivity Lower on Land Rented Out by Female Landlords? Theory and Evidence from Ethiopia. Norway: Norwegian University of Life Sciences. Mimeo.

Holden, S. T., K. Deininger, and H. Ghebru. 2007a. Land Certification and Land Market Participation in Tigray: A Household Panel Model with Unobservable Heterogeneity and State Dependence Nordic. Paper presented at Development Economics Conference, Copenhagen.

Holden, S. T., K. Deininger, and H. Ghebru. 2007b. *Impacts of Land Certification on Land-Related Conflicts in Tigray Region.* Ås, Norway: Norwegian University of Life Sciences. Mimeo.

Holden, S. T., K. Deininger, and H. Ghebru. 2008. Land Certification, Land-Related Investment and Land Productivity Impacts. Samfunnsøkonomenes Årskonferanse, Oslo: University of Oslo.

Holden, S. T., and H. Ghebru. 2006. Kinship, Transaction Costs and Land Rental Market Participation. Working Paper. Ås, Norway: Norwegian University of Life Sciences.

Holden, S. T., R. Kaarhus, and R. Lunduka. 2006. *Land Policy Reform: The Role of Land Markets and Women's Land Rights in Malawi.* As, Norway: Norwegian University of Life Sciences.

Holden, S. T., B. Shiferaw and J. Pender. 2001. Market Imperfections and Land Productivity in the Ethiopian Highlands. *Journal of Agricultural Economics* 52 (3): 62–79.

Holden, S. T., B. Shiferaw, and M. Wik. 1998. Poverty, Market Imperfections and Time Preferences: Of Relevance for Environmental Policy? *Environment and Development Economics* 3 (1): 105–130.

Holden, S. T., and Tefera, T. 2008. From Being Property of Men to Becoming Equal Owners? Early Impacts of Land Certification on Women in Southern Ethiopia. Report written for UNHABITAT. Nairobi: UNHABITAT.

Holden, S. T., and H. Yohannes. 2002. Land Redistribution, Tenure Insecurity, and Intensity of Production: A Study of Farm Households in Southern Ethiopia. *Land Economics* 78 (4): 573–590.

Jansen, H. G. P., J. Pender, A. Damon, and R. Schipper. 2006. Rural Development Policies and Sustainable Land Use in the Hillside Areas of Honduras: A Quantitative Livelihoods Approach. Research Report No. 147. Washington, DC: International Food Policy Research Institute.

Jayne, T. S., T. Yamano, M. T. Weber, D. Tschirley, R. Benfica, A. Chapoto, and B. Zulu. 2003. Smallholder Income and Land Distribution in Africa: Implications for Poverty Reduction Strategies. *Food Policy* 28 (3): 253–275.

Johnson, D. G. 1950. Resource Allocation under Share Contracts. *Journal of Political Economy* 58 (2): 111–123.

Kallon, K. 1996. Land Tenure, Entrepreneurship and Agricultural Development: A Study of the Mende Land Tenure System in Sierra Leone. Colorado: Department of Economics, University of Northern Colorado. Mimeo.

Kassie, M., and S. T. Holden. 2006, August. Sharecropping Efficiency in Ethiopia: Threats of Eviction and Kinship. Paper presented at 26th International Conference of the International Association of Agricultural Economists. Gold Coast, Australia.

Kassie, M., and S. T. Holden. 2007. Sharecropping Efficiency in Ethiopia: Threats of Eviction and Kinship. *Agricultural Economics* 37 (2–3): 179–188.

Kebede, B. 2002. Land Tenure and Common Pool Resources in Rural Ethiopia: A Study Based on Fifteen Sites. *African Development Review/Revue Africaine de Developpement* 14 (1): 113–149.

Kevane, M. 1996. Agrarian Structure and Agricultural Practice: Typology and Application to Western Sudan. *American Journal of Agricultural Economics* 78 (1): 234–245.

Kevane, M. 1997. Land Tenure and Rental in Western Sudan. *Land Use Policy* 14 (4): 295–310.

Kimura, S., K. Otsuka, and S. Rozelle. 2007. Efficiency of Land Allocation through Tenancy Markets: Evidence from China. Tokyo: National Graduate Research Institute for Policy Studies. Mimeo.

Kisamba-Mugerwa, W., and R. Barrows. 1989. *Land Tenure and Agricultural Development in Kampala and Uganda.* Madison: Makerere Institute for Social Research and the Land Tenure Center.

Kranton, R. E., and A. V. Swamy. 1999. The Hazards of Piecemeal Reform: British Civil Courts and the Credit Market in Colonial India. *Journal of Development Economics* 58 (1): 1–24.

Kutcher, G. P., and P. L. Scandizzo. 1981. *The Agricultural Economy of Northeast Brazil.* Washington, DC: World Bank.

Laffont, J. J., and M. S. Matoussi. 1995. Moral Hazard, Financial Constraints and Sharecropping in El Oulja. *Review of Economic Studies* 62 (3): 381–399.

Lanjouw, J. O. 1999. Information and the Operation of Markets: Tests Based on a General Equilibrium Model of Land Leasing in India. *Journal of Development Economics* 60 (2): 497–527.

Lastarria-Cornhiel, S. 1997. Impact of Privatization on Gender and Property Rights in Africa. *World Development* 25 (8): 1317–1333.

Lastarria-Cornhiel, S. 2001. Privatization of Land Rights and Access to Factor Markets: A Path to Gender Equity? Madison: Land Tenure Center, University of Wisconsin-Madison.

Lavigne Delville, P., C. Toulmin, J. P. Colin. and J. P. Chauveau. 2002. *Negotiating Access to Land in West Africa: A Synthesis of Findings from Research on Derived Rights to Land.* London: IIED/GRET.

Lawry, W. S., and M. D. Stienbarger. 1991. Tenure and Alley Farming in the Humid Zone of West Africa: Final Report of Research in Cameroon, Nigeria, and Togo. Land Tenure Center Research Paper 105. Madison: University of Wisconsin.

Lunning, H. A. 1965. The Impact of Socio-Economic Factors on the Land Tenure Pattern in Northern Nigeria. *Journal of Administration Overseas* 4 (3): 173–182.

Macours, K., A. de Janvry, and E. Sadoulet. 2004. *Insecurity of Property Rights and Matching in Tenancy Market.* Baltimore MD and Berkeley CA: John Hopkins University and University of California, Berkeley.

Maddala, G. S. 1983. *Limited-Dependent and Qualitative Variables in Econometrics.* Cambridge, UK: Cambridge University Press.

Malcolm, D. M. 1953. *Sukumaland: An African People and Their Country.* London: Oxford University Press.

Manji, A. 2003. Capital, Labor and Land Relations in Africa: A Gender Analysis of the World Bank's "Policy Research Report on Land Institutions and Land Policy." *Third World Quarterly* 24 (1): 97–114.

Manyong, V., and V. Houndekon. 2000. Land Tenurial Systems and the Adoption of Mucuna Planted Fallow in the Derived Savannas of West Africa. CAPRi Working Paper No. 4, Washington, DC: International Food Policy Research Institute.

Marshall, A. 1890. *The Principles of Economics. Vol. I.* London: Macmillan.

Mathenge, G. 2008, February 10. Answer to Land Problems Depend on Political Will. *The East African Standard (Nairobi).*

Matlon, P. 1994. Indigenous Land Use Systems and Investments in Soil Fertility in Burkina Faso. In *Searching for Land Tenure Security in Africa*, edited by J. W. Bruce and S. E. Migot-Adholla (pp. 41–70). Dubuque, IA: Kendall/Hunt.

Migot-Adholla, S. E., G. Benneh, F. Place, and S. Atsu. 1994b. Land, Security of Tenure, and Productivity in Ghana. In *Searching for Land Tenure Security in Africa*, edited by J. W. Bruce and S. E. Migot-Adholla (pp. 169–198). Dubuque, IA: Kendall/Hunt.

Migot-Adholla, S. E., P. Hazell, B. Blarel, and F. Place. 1991. Indigenous Land Rights Systems in Sub-Saharan Africa: A Constraint on Productivity? *The World Bank Economic Review* 5 (1): 155–175.

Migot-Adholla, S., and F. Place. 1998. The Economic Effects of Land Registration on Smallholder Farms in Kenya: Evidence from Nyeri and Kakamega Districts. *Land Economics* 74 (1): 360–373.

Migot-Adholla, S. E., F. Place, and W. Oluoch-Kosura. 1990. Tenure Security and Land Productivity in Kenya. Washington, DC: World Bank Mimeo.

Migot-Adholla, S. E., F. Place, and W. Oluoch-Kosura. 1994a. Security of Tenure and Land Productivity in Kenya. In *Searching for Land Tenure Security in Africa*, edited by J. W. Bruce and S. E. Migot-Adholla (pp. 119–140). Dubuque, IA: Kendall/Hunt.

Ministry of Lands and Settlement. 2004. National Land Policy Formulation Process Paper. Nairobi (mimeo).

Mkandawire, R. M., R. Jaffee, and S. Bertoli. 1990. *Beyond "Dualism": The Changing Face of the Leasehold Estate Sub-sector of Malawi*. Lilongwe and Binghampton, NY: University of Malawi and Institute of Development Anthropology.

Mtawali, K. M. 1989. *An Analysis of Characteristics of Households Facing Food Insecurity in Malawi*. East Lansing, MI: Kellogg International Programme in Food Systems

Mukherjee, C., H. White and M. Wuyts, 1998. *Econometrics and Data Analysis for Developing Countries*. London: Routledge.

Mukwaya, A. B. 1953. *Land Tenure in Buganda: Present Day Tenancies*. Kampala: Eagle Press.

Nankumba, J. S. 1990. A Case Study of Tenancy Arrangements on Private Burley Tobacco Estates in Malawi. African Rural Social Science Series. Morrilton, AR: Winrock International Institute for Agricultural Development.

Nankumba, J. S. 1994. The Potential of Agroforestry Technologies under Customary Land Tenure in Malawi: A Case Study of Lilongwe District. Lilongwe: Bunda College. Mimeo.

Nega, B., B. Adenew, and S. Gebre-Sellasie. 2003. Current Land Policy Issues in Ethiopia. *Land Reform, Land Settlement, and Cooperatives* 11 (3): 103–124.

NEPAD. 2003. *Comprehensive African Agriculture Development Programme*. Pretoria, South Africa: New Partnership for African Development.

Newell, A., K. Pandya, and J. Symons. 1997. Farm Size and the Intensity of Land Use in Gujarat. *Oxford Economic Papers* 49 (2): 307–315.

Nkonya, E., C. Kaizzi, and J. Pender. 2005a. Determinants of Nutrient Balances in a Maize Farming System in Eastern Uganda. *Agricultural Systems* 85 (2): 155–182.

Nkonya, E., J. Pender, P. Jagger, D. Sserunkuuma, C. Kaizzi, and H. Ssali. 2004. Strategies for Sustainable Land Management and Poverty Reduction in Uganda. Research Report No. 133. Washington, DC: International Food Policy Research Institute.

Nkonya, E., J. Pender, C. Kaizzi, K. Edward, and S. Mugarura. 2005b. Policy Options for Increasing Crop Productivity and Reducing Soil Nutrient Depletion and Poverty in Uganda. EPT Discussion Paper 134. Washington, DC: International Food Policy Research Institute.

Nyanda, M. E., and G. Shively. 1989. Sources of and Factors Influencing the Incomes of Tenants on Agricultural Estates: The Case for Malawi. Ithaca, NY: Cornell University Food and Nutrition Policy Program.

Obunde, P. O. 2004. Effects of Land Tenure on Agricultural Productivity, and the Environment: A Case Study of Suba and Laikipia Districts. Report No. 7. Kampala, Uganda: IFPRI Eastern Africa Food Policy Network.

Okoth-Ogendo, H. 1999. Land Policy Development in East Africa: A Survey of Recent Trends. http://www.oxfam.org.uk/issues/livelihoods/landrights/africa_east.htm

Olinto, P. V. 1995. *Land Quality and the Inverse Relationship between Farm Size and Productivity: A Panel Data Analysis of Paraguayan Farm Households.* Madison: Department of Agricultural Economics, University of Wisconsin-Madison.

ONRS (Oromia National Regional State). 2002. Oromia Rural Land Use and Administration Proclamation. Proclamation No. 56/2002. Finfinnee, Ethiopia: ONRS.

Otsuka, K. 2002. Enhancing Land Access and Land Rights for the Marginalized: Regional Overview in International Context. Paper presented at World Bank Regional Land Workshop in Phnom Penh, Cambodia.

Otsuka, K. 2007. Efficiency and Equity Effects of Land Markets. In *Handbook of Agricultural Economics,* Vol. III, edited by R. E. Evenson and P. Pingali (pp. 2672–2703). Amsterdam: Elsevier.

Otsuka, K., H. Chuma, and Y. Hayami. 1992. Land and Labor Contracts in Agrarian Economies: Theories and Facts. *Journal of Economic Literature* 30 (4): 1965–2018.

Otsuka, K., and Y. Hayami. 1988. Theories of Share Tenancy: A Critical Survey. *Economic Development and Cultural Change* 37 (1): 32–68.

Otsuka, K., and K. Kalirajan. 2005. An Exploration of a Green Revolution in Sub-Sahara Africa. *Journal of Agricultural and Development Economics* 2 (1): 1–7.

Otsuka, K., and K. Kalirajan. 2006. Rice Green Revolution in Asia and Its Transferability to Africa: An Introduction. *Developing Economies* 44 (2): 107–122.

Otsuka, K., and F. Place. 2001. *Land Tenure and Natural Resource Management: A Comparative Study of Agrarian Communities in Asia and Africa.* Baltimore: Johns Hopkins University Press.

Pedro Sanchez. 2005. *Halving Hunger: It Can Be Done: UN Millennium Project Task Force on Hunger 2005.* London: Earthscan.

Pender, J. 1996. Discount Rates and Credit Markets: Theory and Evidence from Rural India. *Journal of Development Economics* 50 (2): 257–296.

Pender, J., and M. Fafchamps. 2001. Land Lease Markets and Agricultural Efficiency: Theory and Evidence from Ethiopia. Environmental and Production Technology Division Discussion Paper No. 81. Washington, DC: International Food Policy Research Institute.

Pender, J., and M. Fafchamps. 2006. Land Lease Markets and Agricultural Efficiency in Ethiopia. *Journal of African Economies* 15 (2): 251–284.

Pender, J., and B. Gebremedhin. 2004. Impacts of Policies and Technologies in Dryland Agriculture: Evidence from Northern Ethiopia. In *Challenges and Strategies for Dryland Agriculture,* edited by S. C. Rao (pp. 389–416). CSSA Special Publication

32, Madison, WI: American Society of Agronomy and Crop Science Society of America.

Pender, J., and B. Gebremedhin. 2006. Land Management, Crop Production and Household Income in the Highlands of Tigray, Northern Ethiopia. In *Strategies for Sustainable Land Management in the East African Highlands*, edited by J. Pender, F. Place, and S. Ehui (pp. 107–139). Washington, DC: IFPRI.

Pender, J., and B. Gebremedhin. 2007. Determinants of Agricultural and Land Management Practices and Impacts on Crop Production and Household Income in the Highlands of Tigray, Ethiopia. *Journal of African Economies* 17 (3): 395–450.

Pender, J., P. Jagger, E. Nkonya, and E. Kato. 2004. Livelihood and Land Use Options in Northwestern Uganda. Washington DC: International Food Policy Research Institute (IFPRI). Mimeo.

Pender, J., and J. Kerr. 1998. Determinants of Farmers' Indigenous Soil and Water Conservation Investments in India's Semi-Arid Tropics. *Agricultural Economics* 19 (1–2): 113–125.

Pender, J., and J. Kerr. 1999. The Effects of Land Sales Restrictions: Evidence from South India. *Agricultural Economics* 21 (3): 279–294.

Peters, P. 2002. Bewitching Land: The Role of Land Disputes in Converting Kin to Strangers and in Class Formation in Malawi. *Journal of Southern African Studies* 28 (1): 155–178.

Pinckney, T., and P. Kimuyu. 1994. Land Tenure Reform in East Africa: Good, Bad or Unimportant? *Journal of African Economies* 3 (1): 1–28.

Place, F. 1995. *The Role of Land and Tree Tenure on the Adoption of Agroforestry Technologies in Zambia, Burundi, Uganda, and Malawi: A Summary and Synthesis.* Madison: Land Tenure Center, University of Wisconsin.

Place, F., and P. Hazell. 1993. Productivity Effects of Indigenous Land Tenure Systems in Subsaharan Africa. *American Journal of Agricultural Economics* 75 (1): 9–10.

Place, F., Masupayi, R., and K. Otsuka. 2001. Tree and Cropland Management in Malawi. In *Land Tenure and Natural Resource Management: A Comparative Study of Agrarian Communities in Asia and Africa*, edited by K. Otsuka, and F. Place (pp. 144–191). Baltimore: Johns Hopkins University Press.

Place, F., and S. E. Migot-Adholla. 1998. Land Registration and Smallholder Farms in Kenya. *Land Economics* 74 (3): 360–373.

Place, F., and K. Otsuka. 2000. Population Pressure, Land Tenure, and Tree Resource Management in Uganda. *Land Economics* 76 (2): 233–251.

Place, F., and K. Otsuka, 2001a. Population, Land Tenure, and Natural Resource Management: The Case of Customary Land Areas in Malawi, *Journal of Environmental Economics and Management* 41 (1): 13–32.

Place, F., and K. Otsuka. 2001b. Tenure, Agricultural Investment, and Productivity in the Customary Tenure Sector of Malawi. *Economic Development and Cultural Change* 50 (1): 77–100.

Place, F., and K. Otsuka. 2002. Land Tenure Systems and Their Impacts on Agricultural Investments and Productivity in Uganda. *Journal of Development Studies* 38 (6): 105–128.

Place, F., J. Ssenteza, and K. Otsuka. 2001. Customary and Private Land Management in Uganda. In *Land Tenure and Natural Resource Management: A Comparative Study of Agrarian Communities in Asia and Africa*, edited by K. Otsuka and F. Place (pp. 195–233). Baltimore: Johns Hopkins University Press.

Platteau, J. P. 1995. The Evolutionary Theory of Land Rights as Applied to Sub-Saharan Africa: A Critical Assessment. Cahiers de recherche, No. 145, CRED. Belgium: Department of Economics, University of Namur.

Platteau, J. P. 1996. The Evolutionary Theory of Land Rights as Applied to Sub-Saharan Africa: A Critical Assessment. *Development and Change* 27 (1): 29–86.

Platteau, J. P., and J. M. Baland. 2001. Impartial Inheritance versus Equal Division: A Comparative Perspective Centered on Europe and Sub-Saharan Africa. In *Access to Land Rural Poverty and Public Action,* edited by A. de Janvry, G. Gordillo, J. P. Platteau, and E. Sadoulet (pp. 27–67). Oxford: Oxford University Press.

Quibria, M. G., and S. Rashid. 1986. Sharecropping in Dual Agrarian Economies: A Synthesis. *Oxford Economic Papers* 38 (1): 94–111.

Quisumbing, A. R., J. B. Aidoo, E. Payongayong, and K. Otsuka. 2001a. Women's Land Rights in the Transition to Individualized Ownership: Implications for Tree-Resource Management in Western Ghana. *Economic Development and Cultural Change* 50 (1): 157–181.

Quisumbing, A. R., J. B. Aidoo, E. Payongayong, and K. Otsuka. 2001b. Agroforestry Management in Ghana. In *Land Tenure and Natural Resource Management: A Comparative Study of Agrarian Communities in Asia and Africa*, edited by K. Otsuka and F. Place (pp. 53–96). Baltimore: Johns Hopkins University Press.

Quisumbing, A. R., and J. A. Maluccio. 2003. Resources at Marriage and Intrahousehold Allocation: Evidence from Bangladesh, Ethiopia, Indonesia, and South Africa. *Oxford Bulletin of Economics and Statistics* 65 (3): 283–327.

Quisumbing, A. R., E. Payongayong, J. B. Aidoo, and K. Otsuka. 2003. Land Tenure and the Management of Land and Trees: The Case of Customary Land Areas of Ghana. *Environment and Development Economics* 8 (1): 77–104.

Rabe-Hesketh, S., A. Skrondal, and A. Pickles. 2004. GLLAMMs (Generalized Linear Latent and Mixed Models) Manual. Working paper series No. 160. Berkeley: University of California, Berkeley, Division of Biostatistics.

Radner, R. 1981. Monitoring Cooperative Agreements in a Repeated Principal Agent Relationship. *Econometrica* 49 (5): 1127–1148.

Radner, R. 1985. Repeated Principal-Agent Games with Discounting. *Econometrica* 53 (5): 1173–1198.

Rahmato, D. 1984. *Agrarian Reform in Ethiopia.* Uppsala, Sweden: Scandinavian Institute of African Studies.

Rahmato, D. 2003. *Access to Resources and Livelihood Insecurity.* Addis Abbaba: Forum for Social Studies.

Ray, T., and N. Singh. 2001. Limited Liability, Contractual Choice, and the Tenancy Ladder. *Journal of Development Economics* 66 (1): 289–303.

Republic of Kenya. 2004. National Land Policy Formulation Process Paper. Nairobi: Ministry of Lands and Settlement. Mimeo.

Republic of Uganda. 1998. *Land Act, 1998.* Kampala, Uganda: Government Printer.

Robison, L. J., D. A. Lins, and R. VenKataraman. 1985. Cash Rents and Land Values in U.S. Agriculture. *American Journal of Agricultural Economics* 67 (4): 794–805.

Roden, D. 1971. Changing Patterns of Land Tenure among the Nuba of Central Sudan. *Journal of Administration Overseas* 10 (4): 294–309.

Rosenzweig, M. R., and K. I. Wolpin. 1993. Credit Market Constraints, Consumption Smoothing, and the Accumulation of Durable Production Assets in Low-Income Countries: Investment in Bullocks in India. *Journal of Political Economy* 101 (2): 223–244.

Rosett, R. N. 1959. A Statistical Model of Friction in Economics. *Econometrica* 27 (2): 263–267.

Roth, M., J. Unruh, and R. Barrows. 1994. Land Registration, Tenure Security, Credit Use, and Investment in the Shebelle Region of Somalia. In *Searching for Land Tenure Security in Africa*, edited by J. W. Bruce and S. E. Migot-Adholla (pp. 199–230). Dubuque, IA: Kendall/Hunt.

Rozelle, S., K. Otsuka, and J. Huang. 2005. The Engines of a Viable Agriculture: Advances in Biotechnology, Market Accessibility and Land Rentals in Rural China. *China Journal* 53: 81–111.

Rugadya, M. 2003. Current Status and Challenges in the Land Reform Process in Uganda: An NGO Perspective. Kampala, Uganda: Uganda Land Alliance. Mimeo.

Sadoulet, E., and A. de Janvry. 1995. *Quantitative Development Policy Analysis*. Baltimore: Johns Hopkins University Press.

Sadoulet, E., A. de Janvry, and C. Benjamin. 1998. Household Behavior with Imperfect Labor Markets. *Journal of Industrial Relations* 37 (l): 85–108.

Sadoulet, E., A. de Janvry, and S. Fukui. 1997. The Meaning of Kinship in Sharecropping Contracts. *American Journal of Agricultural Economics* 79 (2): 394–407.

Sadoulet, E., S. Fukui, and A. de Janvry. 1994. Efficient Share Tenancy Contracts under Risk: The Case of Two Rice-Growing Villages in Thailand. *Journal of Development Economics* 45 (2): 225–243.

Sadoulet, E., R. Murgai, and A. de Janvry. 2001. Access to Land via Land Rental Markets. In *Access to Land Rural Poverty and Public Action,* edited by A. de Janvry, G. Gordillo, J. P. Platteau, and E. Sadoulet (pp. 196–229). Oxford: Oxford University Press.

Shaban, R. A. 1987. Testing between Competing Models of Sharecropping. *Journal of Political Economy* 95 (5): 893–920.

Shaban, R. A. 1991. Does the Land Tenancy Market Equalize Holdings? Working Paper. University of Pennsylvania.

Sharma, N., and J. Dreze. 1996. Sharecropping in a North Indian Village. *Journal of Development Studies* 33 (1): 1–39.

Shetty, S. 1988. Limited Liability, Wealth Differences and Tenancy Contracts in Agrarian Economies. *Journal of Development Economics* 29 (1): 1–22.

Shipton, P. 1988. The Kenya Land Tenure Reform: Misunderstandings in the Public Creation of Private Property. In *Land and Society in Contemporary Africa*, edited by R. E. Downs and S. P. Reyna (pp. 91–113. Hanover and London: University Press of New England.

Singh, N. 1989. Theories of Sharecropping. In *The Theory of Agrarian Institutions*, edited by P. K. Bardham (pp. 37–72. Oxford: Clarendon Press.

Sjaastad, E., and D. W. Bromley. 1997. Indigenous Land Rights in Sub-Saharan Africa: Appropriation Security and Investment Demand. *World Development* 25 (4): 549–562.

Skoufias, E. 1995. Household Resources, Transactions Costs, and Adjustment through Land Tenancy. *Land Economics* 71 (1): 42–56.

SNNPR (Southern Nation, Nationalities and Peoples Region). 2003. Rural Land Adminstration and Utilization Proclamation. Proclamation No. 53/2003. Ethiopia.

Solomon, A. H. 2004, November. A Reality Check on the Quest for Tenure Security and Land Administration in Ethiopia. Paper Presented at the Expert Group Meeting on Secure Land Tenure: New Legal Frameworks and Tools. Nairobi.

Spillman, W. J. 1919. The Agricultural Ladder. *American Economic Review* 9 (S1): 170–179.

Stata Corp. 2006. *Stata Statistical Software: Version 9.0.* Texas: Stata Corporation.

Stiglitz, J. E. 1974. Incentives and Risk Sharing in Sharecropping. *Review of Economic Studies* 41 (2): 219–255.

Tefera, B., G. Ayele, Y. Atnafe, P. Dubale, and M. Jabbar. 2000, May. Nature and Causes of Land Degradation in the Oromiya Region: A Review of Literature. Paper Prepared for the National Seminar on Policies for Sustainable Land Management in the Highlands of Ethiopia. Addis Ababa, Ethiopia: International Livestock Research Institute.

Teklu, T., and A. Lemi. 2004. Factors Affecting Entry and Intensity in Informal Rental Land Markets in Southern Ethiopian Highlands. *Agricultural Economics* 30 (2): 117–128.

Tikabo, M. O., and S. T. Holden. 2003. *Land Contract Choice: Poor Landlords and Rich Tenants—Evidence from the Highlands of Eritrea.* Norway: Norwegian University of Life Sciences.

Tikabo, M. O., and S. T. Holden 2004. *Land Tenancy and Efficiency of Land Use in the Highlands of Eritrea.* Norway: Ås, Department of Economics and Resource Management, Norwegian University of Life Sciences.

Tikabo, M. O., S. T. Holden, and O. Bergland. 2008. Factor Market Imperfections and the Land Rental Market in the Highlands of Eritrea: Theory and Evidence. Mimeo.

TNRS (Tigray National Regional State). 1997. Proclamation to Define Land Use in Tigray National Regional State. Proclamation No. 23/1989(EC). Ethiopia: Negarit Gazeta, Mekelle.

Udry, C. 1996. Gender, Agricultural Production, and the Theory of the Household. *Journal of Political Economy* 104 (6): 1010–1046.

Udry, C., J. Hoddinott, H. Alderman, and L. Haddad 1995. Gender Differentials in Farm Productivity: Implications for Household Efficiency and Agricultural Policy. *Food Policy* 20 (5): 407–423.

UN Millenium Project. 2005. Halving Hunger: It Can Be Done. Task Force on Hunger.

Unruh, J. 2001. Land Dispute Resolution in Mozambique: Institutions and Evidence of Agroforestry Technology Adoption. CAPRi Working Paper No. 12. Washington, DC: International Food Policy Research Institute.

van Donge, J. K. 1999. Law and Order as a Development Issue: Land Conflicts and the Creation of Social Order in Southern Malawi. *Journal of Development Studies* 36 (2): 373–399.

Vaughan, M., and G. H. R. Chipande. 1986. Women in Estate Sector of Malawi: The Tea and Tobacco Industries. World Employment Programme Research Working Paper. Geneva, International Labor Organization.

Vranken, L., and J. Swinnen. 2006. Land Rental Markets in Transition: Theory and Evidence from Hungary. World Development 34 (3): 481–500.

Walker, C. 2002. Land Reforms in Southern and Eastern Africa: Key Issues for Strengthening Women's Access to and Rights in land. Harare Zimbabwe: United Nations Food and Agriculture Organization. Mimeo.

Wan, G. H., and E. Cheng. 2001. Effects of Land Fragmentation and Returns to Scale in the Chinese Farming Sector. *Applied Economics* 33 (2): 183–194.

Wangila, J. 1999. Socio-Economic Characteristics of Households in the Soil Fertility Replenishment Pilot Project Area in Vihiga and Siaya Districts. Nairobi: ICRAF.

Wanyeki, L. M. 2003. *Women and Land in Africa: Culture, Religion and Realizing Women's Rights*. London: Zed Books.

Wehrwein, C. F. 1958. An Analysis of Agricultural Ladder Research. *Land Economics* 34 (4): 329–337.

Whitehead, A., and D. Tsikata. 2003. Policy Discourses on Women's Land Rights in Sub-Saharan Africa: The Implications of the Return to the Customary. *Journal of Agrarian Change* 3 (1): 67–112.

Wilde, J. 2000. Identification of Multiple Equation Probit Models with Endogenous Regressors. *Economics Letters* 69 (3): 309–312.

Wilson, R. 1972. Land Tenure and Economic Development: A Study of the Economic Consequences of Land Registration in Kenya's Smallholder Areas. *Journal of Statistical and Social Inquiry* 22 (3): 124–151.

Wood, S., K. Sebastian, F. Nachtergaele, D. Nielsen, and A. Dai. 1999. Spatial Aspects of the Design and Targeting of Agricultural Development Strategies. Environment and Production Technology Division. Discussion Paper 44. Washington, DC: International Food Policy Research Institute.

Wooldridge, J. 2002. *Econometric Analysis of Cross Sectional and Panel Data*. London: MIT Press.

World Bank. 2006. *Land Policy and Administration*. Washington, DC: World Bank.

Yamano, T., and T. S. Jayne. 2004. Measuring the Impact of Working-age Adult Mortality on Small-Scale Farm Households in Kenya. *World Development* 32 (1): 91–119.

Yamano, T., K. Otsuka, F. Place, Y. Kijima, and J. Nyoro. 2005. The 2004 REPEAT Survey in Kenya (First Wave): Results. GRIPS Development Database No.1. Tokyo: National Graduate Institute for Policy Studies.

Zeldes, S. P. 1989. Consumption and Liquidity Constraints: An Empirical Investigation. *Journal of Political Economy* 97 (2): 305–346.

Zeller, M. A., A. Diagne, and V. Kisyombe. 1997. Adoption of Hybrid Maize and Tobacco in Malawi's Smallholder Farms: Effects on Household Food Security. In *Food Security and Innovations: Successes and Lessons Learned*, edited by F. Heidues, and A. Fadani (pp. 621–634). Frankfurt: Peter Lang Press.

Zimmerman, F. J., and M. R. Carter. 1999. A Dynamic Option Value for Institutional Change: Marketable Property Rights in the Sahel. *American Journal of Agricultural Economics* 81 (2): 467–478.

Zusman, P., and C. Bell. 1989. The Equilibrium Vector of Pair-Wise Bargained Agency Contracts with Diverse Actors and Principals Owning a Fixed Resource. *Journal of Economic Behavior and Organization* 11 (1): 91–114.

Index

M

N

O

P